Fireworks
Principles and Practice

CONTENTS

PREFACE TO THE THIRD AND FOURTH EDITIONS

The idea of this book took place more than 30 years ago, and it is gratifying that is still has a place in the world of fireworks.

The original intention was to produce a straightforward description of firework manufacture in the Western world. It was an attempt to describe what had happened in the past and to make suggestions for good practice at the present time.

It also was an attempt to be fairly basic and thereby not offend friends and competitors in the trade who had to make an attempt to make a living in fireworks. Amateurs have wonderful enthusiasms and, like true scientists, they need to know everything—for its own sake. They also want to share their findings with everyone else. However, this is in complete contrast to commerce, where survival may depend on the quality of the product or the price at which it might be produced. Needless to say, this advantage can be very costly in terms of hours of research and capital expenditure.

Recent years have seen the dramatic decline of the Western firework industry. The story is the same for almost every country where it has become uneconomical to make small fireworks compared to the price at which they can be bought from China. In the United Kingdom, for example, there were 10 manufacturers of small shop fireworks in 1960, but there are none left now. In the United Kingdom, only Kimbolton Fireworks makes a range of display fireworks, with two other firms making special effects for the stage. There also are a couple of individuals with licenses to manufacture from basic raw materials.

Much of the material from the Far East is inexpensive, and importers must be experienced to guarantee good quality. It also is a boon to the ever-increasing numbers of unspecialized (and often legally ignorant) importers who import explosives in much the same way as they might handle inert products. The Civil Service, in the United Kingdom at least, has been less than effective in the control of these illegalities in recent years. They frequently plead the problems of open EC borders, but it was not so in the past, and while the EC gets the blame for most things, the United Kingdom goes the extra mile in gold-plating EC rules. On the other hand, it is clear that some of our partners in the EC have been much better at "looking after their own" than is the case in the United Kingdom. Oddly enough, in 2004 attempts at new EC standards for fireworks with committees dominated by bureaucrats from government departments and testing houses are further destroying the industry and making it unworkable and uneconomical.

This industrial decline is a great tragedy, but it is just possible that there will be a place for those manufacturers able to make a good-quality product—but for how long? Those people willing to make this capital outlay need a Civil Service that creates a level playing field. In 2004 a high-profile manufacturer is constantly bombarded with rules, regulations, bureaucratic nonsense, and more and more costs at every stage. Every sizable company has to employ unproductive safety advisers, subscribe to suppliers' safety information—much of which is not applicable—and make space for records and risk assessments. Before a display on a roof in the east part of London, we had to spend time making risk assessments with no fewer than 10 interested parties. No one can deny that the *simple* desire for good health and safety management is laudable. In reality, it has become a burden, with those working in it leaving no stone unturned and sometimes reaching absurdum as they justify their existence. It is a regular joke that health and safety is the only growth industry that we have, and it is expensive too. Even the Bible indicates that things do not change very much:

> Luke XI v 46 "Yes you Lawyers and Pharisees, you load men with intolerable burdens and will not put a single finger to the load."

In the meantime, companies disappear or transfer their production to the Third World. All this time, the importers increase in number and have the financial gain.

It is difficult to predict what the next years will bring. Importations from the Far East will increase, but it is clear that there are too many Chinese exporters. Most of them are selling much the same products, and it is always more important to remain competitive than to produce superior products. A Chinese supplier may well sell the same products to several people in a limited market—a policy doomed to price-cutting—and in the United Kingdom the market is already saturated. The expertise of the EEC industry is already being lost. However, conflicts in the Middle or Far East could have dire consequences for imports.

Once again, I am grateful to the many friends who have helped to make this edition possible; in particular, to friends mentioned in Chapter 1 who have filled out the details about the firework scene in their own countries.

I would particularly mention Dr. Takeo Shimizu (b. 1912), whom I have known for more than 30 years and who has been absolutely prolific in his research for the firework trade. What would we have done without him?

Mention also must be made of the late Chris Philip, who died in January 1998. The importation of fireworks into the United Kingdom had always been a major problem because of the prohibition on the admixture of chlorates and sulfur. However, Chris Philip set out to challenge a somewhat negative attitude toward importation at that time. A total ban had been easy to control, but his success has done no favors to the home-based industry some 30 years later. The British Standard for Fireworks created a flurry of activity at the time, but the government has not been particularly proactive in controlling the quality of what can be sold from abroad in recent years. Some of the fines seen in other parts of the world might clarify a few minds.

Sadly, the firework world lost George Plimpton in 2003. A journalist and lifelong firework enthusiast, he was much better known, revered, and respected in the United States than elsewhere. Nevertheless, he created a place in the firework world for his book *Fireworks: A History and Celebration* (37).

Nearer to home, I am grateful to Mark Lancaster, elected a member of Parliament in 2005, Dr. Tom Smith, Tony Cardell, Roy Butler—my old teaching colleague and friend—and John Bennett, the editor of the excellent UK magazine "Fireworks." This magazine has done much to encourage an interest in fireworks and to keep the available history intact. But also to my wife, Kath, who tends to see herself less as a "firework widow" in these latter years but who has always maintained that I eat and sleep fireworks and talk about them—but less so—in bed these days.

I also am grateful to my son-in-law, Stuart Adlam, architect by profession but now our Commercial Director, who has been so helpful revising photographs and drawings, revising a disc, and telling me what to do when the computer asks awkward questions.

Ronald Lancaster
7, High Street
Kimbolton
Huntingdon
Cambs. PE28 0HB
UK

SIXTY YEARS' LOVE OF FIREWORKS—DIAMONDS IN THE SKY

Ron Lancaster

Yorkshire in the United Kingdom used to be the center of the firework industry in the north. We are not exactly sure when it seriously started in the middle of the nineteenth century, but it almost certainly was connected with the mining industry.

At the foot of the Pennines, Huddersfield is a town that grew with the Industrial Revolution. It grew around the River Colne but is surrounded by hills and magnificent scenery. Quite near to where I lived, at one of the high prominent points, there is a tower that was built to celebrate the Diamond Jubilee of Queen Victoria. From that point it is just possible to see the Standard Firework Factory on a disused quarry site, as the hills slope upward toward the Lancashire border. One or two miles to the east from my house I could see the Lion Fireworks Factory, which had grown from the first beginnings with Alan Jessop in 1847.

As a small boy, just eight years old in 1939, I found that fireworks were fascinating. On the way home from school around November time, we would go into the local bakery and general store and, after lovingly admiring all the wonderful shapes and colorful labels in the glass case, we would select one or two as our pocket allowed. Each night we would bring out the growing collection and play with them, and the fact that the colorful Witches Cauldron cone was leaking silver aluminum powder did not matter. There was little central heating in those days, just open fireplaces, and, needless to say, we did not take Dad seriously when he said that if we did not put them away, he would throw them on the fire back.

The wife of the man who delivered the coal worked at the firework factory, so we also had the benefit of buying a nice selection box at a good price. There is just one other memory of this early period when I was taken up the Castle Hill to see the big display that was put on in 1937 for the coronation of our much loved George VI. Sitting close to the biggest firework display I had ever seen, the impression seems to have been far-reaching.

November 5th was a lovely family day, when family and perhaps some friends gathered around the bonfire that had been carefully built over the previous weeks. Sometimes it was built more than once if some little monsters nearby had raided yours when you were not looking, and on some more nasty occasions it got set alight prematurely. Our local vicarage was surrounded by many mature trees, with the result that the vicar trimmed them at this time so that the local lads would tow the branches away.

The party always had to be accompanied by food, of course—sausages, treacle toffee, and Parkin, which was a kind of oatmeal sweetcake made with molasses. We even played little games like "apple bobbing"—extracting an apple floating in a bucket of water only with your mouth. It was an enjoyable family occasion that was not too noisy—a far cry from the year 2000, when the noise from Chinese material can make some people feel as if they are under siege.

The arrival of World War II in 1939 on September 3 had little impact on a boy of eight, but it had serious consequences for the firework manufacturers, whose magazines were full to capacity. The fireworks were all ready for distribution for November 5th, but they were destined to stay there until the end of the war in Europe.

The financial consequences for the various family-owned businesses, unable to sell their stock, the problems of insurance for storage, as well as the need for storage for work to come, can only be imagined. No provision had been made for war work, and other work had to be found for those who had not been taken for the forces or other war work. For a time Standard Fireworks made blinds and lampshades out of black paper for the mandatory "blackout" for protection against aerial attack. They also continued to make indoor fireworks and sell them for at least a season, but after a period, contracts were set up with the Ministry of Defense to manufacture such items as practice ammunition, Very pistol stars, Thunderflash sticks, and parachute flares.

For those of us who were at school, it was the parachute flares that created the most fun. Huddersfield had an excellent trolley bus system that covered the whole borough, and children could get a one-penny return ticket (a sum of money a mere fraction of a cent and virtually worthless today) with which you could travel anywhere in the large borough.

In waterproofs and rubber boots, we used to go up to Crosland Hill and about mid-morning run through the heather to catch the silk parachutes that gently floated down after the sodium nitrate/magnesium flare had burned out. If you had a long stick, you could hook a parachute before some other "little monster" who did not have a stick. On the other hand, as you start to ascend the Pennines at this point, the weather and the wind in particular can be quite a problem, so we could be traveling any kind of distance through the moor. On rare days the parachutes would land in the town itself. A school contemporary became a distinguished runner, but it was the pyro that inspired me, and we were delighted on the rare occasions when we found the odd unignited flare that the searchers had missed.

Naturally, our interest was sporadic, for the variety was not great and when we had seen a goodly number of Very stars and Thunderflashes at the Lion Factory, it became somewhat boring. Goodness knows what the reaction would be today when people even complain about church bells.

May 8, 1945 saw the end of the war in Europe, and the country burst into jubilation with music, dancing, and, of course, fireworks. It often amuses me when people ask if fireworks have a shelf life, but we can assure them that material made in 1939 was still fine in 1945. The labels still had the prewar prices, but a percentage was added to cover storage costs and insurance over five years. However, these sales also created a scarcity, and there was not enough time to manufacture for November. In fact, it took some time to organize firework production as some military work continued and materials were in short supply. Food was still scarce and as some wheat flour was used for making some fireworks, I seem to remember one company being prosecuted for using it. I remember the rationing of sweets as late as 1952 at university.

A characteristic of many amateurs is the assumption that the sight of a formulation and the purchase of some chemicals will enable them to make a firework; they can also be quite disappointed with the results. When my cousin and I began in the 1940s, we had the added disadvantage that chemicals were not so easy to procure. We could get potassium nitrate and sulfur, but we had to make our own charcoal and potassium chlorate. The pharmacy charged six old pence for 28 g of potassium chlorate, but for that money you could buy two meals from the local fish and chip shop. On the other hand, sodium chlorate sold as weedkiller was much cheaper, and in those days it was not adulterated with phlegmatizing chemicals. We calculated the quantities required from the equation to mix sodium chlorate and potassium chloride (used as a fertilizer) and mixed the hot saturated solutions to make potassium chlorate; the yield was quite good and we became quite efficient at recrystallization.

We also became quite skilled at distilling charcoal in sealed treacle tins, very quickly realizing the different kinds of sparks and burning speeds that can be obtained from different kinds of wood. It was also possible to get good willow charcoal powder from the pharmacy since they still used it occasionally in medicine. Aluminum was obtained from the larger paint dealers, but magnesium was much more of a problem. Occasionally we got some magnesium block to file down.

Oddly enough, we were not interested in making bangs; it was rockets that we spent most of the time investigating. One or two of the old books had furnished tube and spindle sizes, but we concentrated mostly on the English 4-ounce size with some degree of proficiency. Sadly, as all amateurs discover, it is the consistency of the burning speed of the powder that is all-important, and this cannot easily be achieved with small batches. Nevertheless, we learned a lot from the experience and, as is so often the case, the end of the war, the availability of fireworks, and the need to work more seriously in preparation for university contributed to a slowing down process—but not entirely.

Huddersfield Railway Station has a wonderful classical facade that has made it quite famous in railway architecture. As you got off the train, there was a huge hoarding that said "Huddersfield, The Home of Standard Fireworks." Steam was coming to an end, but I can still smell the steam and the soot from those huge locomotives as they entered the station and their high-pitched scream as they let off the excess steam. On many occasions we were going on to Manchester through the tunnel under the Pennines to a place called Belle Vue.

Brocks had a permanent site in this old world pleasure garden with regular displays in that postwar period. It was old-fashioned U.K. manufacture at its best. At each end of the raised "stage" there were sheds where huge lancework devices on wheels could be made and kept dry. When required, they were simply pushed out on the rails and ignited. There were battle enactments, trains, volcanic eruptions, saloons, and walking cowboys with guns. The best piece was the Weird White Waterfall, which stretched across the whole site. The waterfall sticks each had an electric igniter, and they were placed along a wire that could be kept quite taut as it was controlled with a winch. Two loud maroons and two large-bore military whistles created the stunning effect of instant ignition and silver drop. You were scared out of your wits with the bangs and tortured by the beats from those large whistles.

A body of water separated the audience from the firework arena, and it was this that gave a suitable and convenient distance away from the main fireworks, but it did not stop the shell hemispheres from dropping into the crowd. In those days papier-mâché shell halves were only lightly papered and glued together, with the result that it was not uncommon for 200-mm hemispheres to fall among us. We were never aware of anyone being injured, but it makes you think now when people are ready to complain over the smallest of matters. One feels that if they saw fireworks in some parts of the world, they would almost have a heart attack.

Looking back, it seems almost unbelievable that by the time I arrived at the University of Durham in 1950, it was going to be the end of my contact with fireworks for seven years. Apart from the University Regatta, I hardly ever saw any firework displays, and I certainly did not realize that one day I would be returning to that site to actually put on a display.

After three years at Durham, two on military service where I was expected to have an interest in the 25-1b guns in the Royal Artillery, I went on to study theology for two years at Oxford. Cuddesdon College was founded after the Oxford movement in the nineteenth century and was associated with quite a number of well-known figures in the Anglican Church.

It is apposite at this stage to mention this later display in Durham, which was for the 150th anniversary of the founding of Durham University. Although it had been a seat of learning for some hundreds of years, it did not actually become a university until this later time. The display was on the traditional site of Bede College, but at the end of the show there was a little meeting between the principal of the college, a Cuddesdon friend of mine, the Archbishop of York, and the Archbishop of Canterbury, both of whom had also been at Cuddesdon—and myself. The main difference was that they were suitably dressed while I was somewhat covered with carbon and perspiration!

This was not the only time that I had found myself in this position with the well-loved Archbishop Runcie, for I have also presented displays on two occasions for the ten-yearly Lambeth Conferences for the Anglican Bishops from all over the world. Mark Lancaster and I well remember the ending of the first of these two displays when Archbishop Runcie took us to a meeting of all the world's archbishops where Mark had to explain to the famous Desmond Tutu what we had been up to.

Ordination in 1957 meant a return to Yorkshire to work in an industrial parish. Morley was a town that had specialized in making an inexpensive kind of blazer cloth and also happened to be quite well-known for growing rhubarb. There was a very large youth club attached to the church and, as we had little money to spare for fireworks, I decided to make some of my own once again. To the relief of my landlady, the vicar's garage provided a suitable venue. More importantly, it was not too far to Huddersfield and it provided the opportunity to make contact with the firework trade there. The Lion Firework Company and Standard Fireworks were extremely friendly, and it did help that Richard and Edward Greenhalgh at Standard Fireworks were both dedicated churchmen.

Although I did not quite realize it at the time, these two men, along with Joseph Seymour at Lion, were to give that extra helping hand to bridge the gap that exists between amateur and professional practice.

Although the Lion Company was more than a century old, Standard had grown out of Lion at the beginning of that century to become, through remarkable effort, one of the largest firework makers in Europe, employing well over 500 people.

Standard did not really expand from smaller beginnings until 1922. James, the father, along with sons Edward and Richard and daughter Kate, built up the business with complete dedication, rivaling Brocks and Pains, who had most of the market. By 1939 they had as much as half of the U.K. trade. James was 79 when the war began, the magazines were full to the brim, and there was a huge overdraft with the bank. For a while only indoor fireworks were made, but the filling of 1" and 1 1/2" Very cartridges started very quickly and as materials became more scarce the indoor fireworks came to an end—never to recover properly again. The value of the stock and the purchase tax was frozen during the war, but a percentage increase, got rid of the bank debt.

By the standards of today, the Greenhalgh family would be regarded as excessively paternalistic, but their devotion to the Christian work ethic, their support of those associated with them, and their friendliness to me will never be forgotten. Just before "Mr. Edward" died, he sent for me and, as I talked to a very sick man in the garden, I was moved by his compliments about the work I had been doing.

A move in 1960 to a very different parish in Harrogate enabled me to build a small laboratory in the parish house. No one seemed to be much concerned except one of our choirmen, who happened to be an insurance assessor! He did not make too much of a fuss, but I suppose that if we had caused a fire, it was fairly obvious that the insurance would have been invalid. Looking back, it was somewhat naive, and even the police were happy to supply me with licenses for gunpowder.

After six years as a curate and the refusal of the offer of two university chaplaincies, I applied for the post of chaplain at Kimbolton

School, which is about 100 kilometers north of London. After the usual discussions about teaching divinity and chemistry and where I was to live, the headmaster asked if there was anything else. I asked him if there was anywhere that I could build a firework laboratory and waited for him to fall off his seat.

While I knew that the main part of the school was housed in a Tudor house largely rebuilt by Vanbrugh and Adam, and even that Queen Katherine of Aragon had died there after Henry VIII had got rid of her, I did not know that Sir John Popham also had lived there and that he was the judge at the trial of Guy Fawkes!

Amazingly, the headmaster told me that there was a suitable high-walled vegetable garden that had once belonged to the Duke who had formerly lived in the castle and that I could use that.

Applications to the old Home Office Explosives Inspectorate drew a polite refusal, but when I made the point that, provided the provisions of the law were met, a refusal might be unlawful, Dr. Jim Jeacock came to see me. We got on well and he was extremely helpful in ways that I had not considered. Now long in retirement, we are good friends to this day. One very helpful suggestion was the information that there was a firework factory in liquidation in Suffolk and that we might negotiate a couple of workshops from there. We did, in fact, get two brand new ones and with some of the pre-university boys and a flatback truck, we went over to Suffolk and brought them to Kimbolton. There was just one other thing bugging me, and that was the proximity over the high wall of a small Roman Catholic chapel used on Sundays. Clearly, it was too near for the U.K. Acts, and I well remember Dr. Jeacock's words, "You are not going to make fireworks on Sunday, are you? No, well then forget it."

There also was a tinge of sadness. I was not going to find it easy to get back to Huddersfield, but a strange coincidence came about. As I was occasionally in south London and quite near the old Pains factory in Mitcham, I wrote to ask them if I might visit them and see their plant.

John Deeker wrote back to say that he would be glad to show me around, and shortly afterward I went there to meet them. They had just been taken over by British Match, whose chief chemist happened to be there, which led to some interesting discussions about mutual problems. Soon afterward I was asked to work with them as a consultant, a job that was to last from 1963 to 1977. We have often laughed afterward about that first visit. It appears that the senior director (former owner) Philip Milholland had thought that it was a waste of time to show a parson around the plant, but John Deeker had disagreed, adding that he probably was a nice old man and that he would take him around. However, they were somewhat surprised to find that the old person was the same age as John Deeker and Tony Little—32.

The sad part was that it had to mean a break with Huddersfield, and it took a number of years for the rift to heal. I also was just beginning to realize that commerce and the amateur world are very different places. Looking back, it was at this time that I became involved in another political situation when I was staying with my old friend Lars Barfod of Tivoli Fireworks in Copenhagen. We went over to see a factory in Sweden for fun and interest only to find that I was unknowingly treading on dangerous ground involving competition with signals. Now it is ancient history and once again my Swedish friend Björn Söderberg and I can look back and laugh about it.

It took just about a year to get the explosives license, which arrived in September 1964. The new "factory" consisted of two buildings, one for chlorates and one for sulfur, and a concrete gunpowder store. It allowed me to work with about 7 kilos of composition and 35 kilos of gunpowder. In the course of time, we added a third building for assembling fireworks and two storage magazines further away from the community.

It is inevitable that I did not hear all that might have been said about placing such workshops quite near the center of the village, but the media produced misleading headlines such as "Firework Factory in School Grounds" or "Firework Factory in Back Garden." The school owns much of the former Duke's estate, which was considerable.

All was well for a while until the school bursar began to be alerted to potential insurance problems. We thought that we had fixed this, for I managed to extend my household policy with the Ecclesiastical Insurance until they sent an assessor one day. They asked me to find another insurer, adding that stained glass windows and organs were more in their line, and this was when we turned to Pains for help. In time, even this was a problem (and insurance was no problem in those days), and the issue was precipitated by the Queen's visit to Nottingham at the time of the Silver Jubilee. I had paced out the distance of 100 meters from the fireworks to the official party and allowed for the prevailing wind, but it changed and we dropped a spark on the mink stole of a prominent lady who had borrowed it for the night. The cost to repair it is insignificant today, but there was an argument not only with insurers but also about a little old lady much farther away who had got some hot dross on her cardigan. She wanted a new cardigan and was not concerned about the small blister on her arm! How things have changed in the insurance world, but at least we found a new way to insure with Lloyds through an old boy of the school and we were settled for many years to come.

The early 1960s were a period of research and consolidation, but big changes were afoot. The first signs of terrorism meant that site security was becoming tighter, and marriage in 1966 meant that the firework activities would have to fund themselves. It was decided that with the help of my geography teaching colleague, Roy Butler, we would do a few small displays in order to break even. Roy had arrived at the school a year after me, knowing nothing about fireworks, but he soon found himself making Roman candles and has done so ever since.

But there was another more sinister factor in the late 1960s. There had been a number of accidents with domestic fireworks as the industry grew after the war, and these were highlighted by a BBC program called "Man Alive." Presented with a very definite anti-firework slant, for a number of years it managed to polarize people—for and against—and tried to give the impression that organized displays were essential in the future. While this was good for us, we were not in agreement with the view and there was the fear that if they won this battle, they would turn their attention next to those organized displays. To some extent that did happen, but by the time the borders of the Common Market were opened, there was a blossoming of imports from China. Some old U.K. firms disappeared, while new purveyors of Chinese material started to create some of the old political problems again.

The anti-firework campaign led to a sharp decline in retail turnover, and from that time some smaller companies went out of business. James Arthur Kilner, the director of Lion Fireworks, decided to close the business and realize the assets, so 120 years of trading came to an end. By 1970, the Wells family also decided to call it a day. Sad as this was at the time, it was not all bad since Kimbolton was able to purchase considerable quantities of materials and tools and acquired quite a number of large displays at a time when we wished to expand.

Kimbolton exercised very little commercial activity in the 1960s, mainly because we were all working schoolmasters, but we had a distinct advantage in that, teaching in an independent boarding school, once we were off duty, we were free. We might well teach until 4 p.m., work in the workshops until 6 p.m., and then go back on duty until, say, 10 p.m. or sometimes until the next morning in a boarding house. Most of the displays were in the summer vacations and we did not have to commute to work. It was ideal.

Only once did we ask for a day off from school, and that was to put on the display for the 25th anniversary of the Coronation of the Queen in 1978, a year after her accession. If we had a show on a weekend in term time, we loaded the van after school on Friday, taught Saturday until lunchtime, left to put on a display after lunch, returning at 12 midnight and often later. Sunday morning was chapel, so the van was unloaded after lunch in time for tea and evening chapel. Needless to say, this routine only took place in the latter part of the summer term or around November. From the late 1960s this was a regular pattern and the business escalated.

Oddly enough, the first public display was put on by my colleagues as Kath and I were on our honeymoon. Domestic issues were already changing, and I was no longer allowed to keep a small amount of gunpowder in the house in a bread bin. As Kath would say today, she did not realize that from time to time she would become a firework widow. The nearby town thought that they could no longer afford Brock, and this was the reason we took on the show. However, we had our eye on Brock's show at Northampton, which, like many of his shows, was very ground "set piece" based, and we wanted to replace it with a more visible aerial display. They were a bit unsure then of a (possibly) eccentric Reverend, but it all changed in a strange way. The corporation has a fine suite of public baths, and one day, along with one or two colleagues, I realized that the man who organized the show was sharing the steam room with us. A whispered comment to persuade him while he had his pants down worked, so we took on the Northampton Show as our first commercial display in 1978.

It has already been mentioned that the late 1960s was a period of learning, and I have recorded my gratitude to Willy Stott, the display manager of Standard Fireworks, a man who loved fireworks and who taught me much of what I know about the old ways. He helped me even after he retired, because he wanted to do so, and I can still see him ill with a colostomy battling with a 5-meter lancework frame in a force 8 or 9 gale on the sea front at Blackpool. It was being televised by the BBC for a Miners' Gala. In the early days Willy and I had spent many happy hours at displays, which were then done by Standard for the Edinburgh Tattoo, in Ulster, and elsewhere.

This early period was also notable for the influence that German firework makers brought about. Again, I am grateful to Richard Greenhalgh, who pointed me in that direction as he had worked for some time at Silberhütte between the wars. On my first visit to Germany this was now part of the eastern zone, so I went to Eitorf instead to meet Hermann Weber and his chemist, Mr. Freiinger, both of whom had originally been with the firm of Nicolaus at Meiningen, also in the East. They called the firm Pyro-Chemie (now Weco) and were very kind indeed, but I shall never forget the first visit in 1959. On my way to Siegen, I waited at Köln for a train but could not believe my eyes as a battered old steam engine pulled alongside the platform. We meandered down the River Sieg and at Eitorf, found that the station was not rebuilt and that I had to jump down on to the line. It is somewhat different today.

Gratitude also is due to Fritz Lünig, who took me around with him to some of his displays and probably thought that I was a bit of a nuisance wanting to know about everything. Amateurs can never fully understand that their genuine interest may be suspect, and I have learned my lesson over the years. However, I am sure that the test of time has proved that a thirst for knowledge has not seriously damaged their businesses. I have much to thank Germany for: many fine contacts that have been made, particularly in recent years with Walter Zink of Cleebronn.

I joined the old firm of James Pain as a consultant in 1963 just in time. The old factory was getting hemmed in by a school and other buildings and part of the testing site was a graveyard. The closure of the plant to join with the rural Wessex factory made much sense, but, sadly, few skilled workers were willing to make the move from suburban London to rural Salisbury. However, the experience was invaluable and welcome. Being a consultant is a kind of balancing act because you are not really one of them. Appearing out of the blue every few weeks makes it difficult to be a part of the team, and when things do not work out, it is obvious that the blame falls on the person who is not present. Sometimes it was necessary to get changes made by either convincing the management that it is necessary or doing it by stealth. An example would be a blue star. At this time the formulation was not very good and suddenly the shells started detonating. It quickly became clear that the stars were crumbling to dust because whoever had rewritten the mix card had reduced the amount of dextrine. A suggestion that my formulation should be used prompted the reply that they had tried it and it would not dry properly. However, the factory made all my stars for my private use at that time, and I pointed out that, unknown to some, they had been making this formulation for me for some years, but not using them in normal production. Such is the human factor in a very conservative industry.

Serious explosions do not often happen, and when they do it is not always easy to know what the cause might be. One such explosion happened as people were arriving at work one day. A brick drier exploded and it contained magnesium flares in addition to whatever had caused the exothermic reaction. Burning flares along with masonry were thrown about, creating considerable damage, but happily before the factory had started production. It was a year or two after the death of an employee that our suspicions led to some clarification.

A key figure at Pains-Wessex in this period was a delightful man, Gordon Curtis. Gordon had worked since the war on smokes in particular and had always been referred to as the "Smoke King." From smoke on airplane wings in the early days to screening smokes and latterly for distress signals and insecticides, he was the expert and it was a delight to know him. The introduction of fireworks into the area was a fascination to him, as were all things pyrotechnic, even making the torch for one of the Olympic Games. Quiet in manner, Gordon wanted to hear other opinions about the subject and was always willing to share his experience. We have always said that fireworks are a mixture of chemistry and cooking, and as it happens he was an excellent cake maker.

Toward the end of the 1960s Pains-Wessex decided to take part in some of the European firework competitions, which were just beginning to gain in popularity. In 1968 it was decided to go to Cannes, but apart from Cowes, our experience was limited to firing off barges by traditional methods. Cannes presented a real challenge as it involved the coordination of three barges in the sea. By the evening we all felt ill as we had been gently rocked on the sea in the blazing sun and the primitive electrical firing system was useless. It was quite embarrassing, but at least we did not pound a barge so much as one company was said to have done, causing it to sink.

We got an award for the best color and, with my heavily pregnant wife on my arm, I cannot think now why I attended the award ceremony wearing a clerical collar. This was somewhat emphasized by a "lady of the streets," who tried to take my arm as we entered the building. Not again until 2001 did I go to Cannes when Kimbolton took the first prize and the Winner of Winners in 2002.

The Pains-Wessex display at San Sebastian in 1968 won the first prize. At that time many military noise units were being manufactured, and I recall large old-fashioned round shells being stuffed with these magnesium flash compositions to produce an absolutely stunning finale. It also appears that in excess of 100 marine parachute flares in red, green, and yellow was something special for the judges.

A year later we had a problem with smaller magnesium flash charges. The composition is so brisant that the Bickford fuse and the closing cork in small Roman candle flash units require no glue. The extreme pressure in large mine bags is another matter, and it was at San Sebastian that wooden crates of mines blew up into a mass of paper and wooden pieces. I was lucky to get caught only in the back.

It was a great learning curve, and there is insufficient space to include all of the fun. There was the drunken, swearing boatman who rowed us back over the river at Salcombe at the end of the show, wondering if we would need to swim some of it. There was the park at Brixham, where I tried to take the park entrance with me when I drove through the gates; and a drive to Monaco when at the last minute for some reason we could not take the van full of tackle and fireworks across the ferry to France. Hurriedly we had to remove the fireworks, which were all packed in tea chests, as was the custom in those days. The fireworks were flown across to Le Touquet, where we had to drive the van and collect them. Many of the shells were round in those days with the lift in cones, which were not too secure. The airport men said that they had found "tea" leaking from the chests on the floor of the plane. Now the chests were on the back of the van and were bounced all the way to Monaco. What a mess! We repaired as best we could, and the cones were replaced with gunpowder and tea mixed together. Old tea chests always had loose tea in them, but we also found some Greek shells in a store and used the lifting charges in them and hoped for the best.

Nearer home in Kimbolton, Pains had been asked to put on a four-yearly display at a nearby town that had been founded by Quakers as a garden town. As it is quite near to us, we still do this display, which has just had its centenary. On this occasion it was a dank, still night when the smoke and humidity hung everywhere. In those days the old manufacturers used huge posts 100 cm square and 6 meters high. On the top of one such post was a large horizontal wheel, but when the match was ignited, nothing happened. The operator went back, but there was no ignition. As we watched, we could see the fire creeping so slowly up the match pipe. I was amazed to see the operator, portfire in his teeth, climb up the post like a monkey and light the piece at the top. He was not young either and had originally worked for C. T. Brock.

San Sebastian–1968 was the time when I was to meet a lifelong firework friend, Dr. Felix Gõni, at that time a medical student and now professor of biochemistry at Bilbao University. Long associated with the Valencia firework trade, Felix has been associated with fireworks most of his life. Recently, discussing the problems that EEC bureaucracy is causing for the industry, he characteristically said, "All those years ago we thought that pyrotechnics was all about fireworks."

It also was about this time that we were starting to require some working income at Kimbolton Fireworks, and this led to our acquiring a big display in the center of Oxford for the Round Table at November time; it has continued to this day without a break.

A problem with November displays tends to be the requirement to have the traditional bonfire. At Oxford they used to have two, which were either side of the viewing area. After some time we persuaded them not to have the fires until after the show. It is not so much the light that spoils the fireworks, although this is bad enough, but the light shows just how much smoke is created by the display. It not only conceals the firework effects, but also destroys anything with weak light emissions such as charcoal gold effects. At one of the early displays, I have often wondered which caused the most fun because shortly after the show had started, the wind blew sparks into the fires, which were well primed with flammable oils and I got hotter and hotter. Gradually I stripped off many of my clothes while the crowd produced a tune we all tend to associate with this kind of entertainment.

Around this time we shall never forget a display in Brighton, where we found to our surprise that we had only about 100 centimeters of soil on top of solid chalk. The stakes would not go in the ground, and it was the first time that we had taken a Crown Wheel (Girandola) as we thought that we had perfected this. Most people know that you need only one erratic motor for the device to go off course, and ours did. It lifted about 4 meters, turned 90 degrees, and moved toward the crowd. It is an unnerving experience as you realize what might happen and you cannot run after it and get it back. As it reached the crowd, but becoming lighter, it just lifted over their heads and headed for a huge marquee full of people and drink. People were running in all directions and the mood changed again. Happily, it burned out in time and fell to the ground, but we were besieged by people who said "What a wonderful firework! How did you make it do that?"

I had met the Wells brothers (Joseph, Albert, and Wilfred) long before 1972. They were well respected in the industry, kindly, honest, old-fashioned as you might say in the nicest possible way. Things were not going well and they were elderly by then. The factory had moved from Honor Oak Park to Dartford marshes about 1940, but it was the dramatic flooding of the Thames in 1953 that crippled them. With magazines crammed full for the coronation, they lost almost everything and one magazine created much devastation when it blew up as it dried out. We have often speculated that it would be ammonium perchlorate, barium chlorate, potassium chlorate, and seawater. By 1965 they had 120 employees, but it was never easy and they sold out to Schermuly in 1968. Ian Craig from Schermuly became the manager until 1971, when it was decided to stop making fireworks. Ian carried on the business from then onwards as a display organization called Phoenix Fireworks. The time was also significant to Kimbolton, who took on a small number of their shows—Henley Regatta, Aldeburgh, and Thorpeness, which we also still do in 2005.

Pains-Wessex took over Schermuly in 1971, and then Unwins became involved and the factory was closed in 1984.

The three displays among the others that the Wells brothers had kindly passed on to us have their own distinctions to this day. Henley is, of course, one of the great social occasions in the United Kingdom. The events culminate on the first Saturday in July with great pomp and a splendid display, which we have always been proud to execute. Aldeburgh Carnival is a lovely event at the seaside in a small town, but the display takes place on the shingle beach. It is easy to drive the posts in the shingle, but walking to and fro for a night to set up just about cripples the calves of your legs. Thorpeness has its display on an island in a mere, which is just a flooded river and not very deep. The children create a water pageant in the summer, which culminates in this display. When we first took it on in 1972, we found that Wells had sunk great posts in the mere using barges, but we persuaded the client to go aerial. As the island is small and a haven for ducks, we used to park a barge at the edge and fire the large rockets from frames on the barge. On one occasion when my wife was the assistant, she had a bit of an accident stepping back from the barge to the island, the ducks had characteristically greased the land, and Kath's foot went into the water between the barge and the island. It was a double drama because the trees and bushes on the island were useful for natural functions but, to my wife's horror, they had all been cut down that year.

Clacton Carnival was another Wells display, and when we first arrived the groundsman proudly showed us his collection of 6-m by 100-cm posts and two rugby posts for the waterfall. As we needed little of it for an aerial show, he was clearly worried, but they were happy with the change and simpler set piece arrangements—we had adopted aluminum scaffolding to save weight. However, the organizers had a temporary enclosure at that time to collect revenue, and one cheeky youth who had climbed up it refused to get off. When we fired the first maroon, he was scared, fell off, and broke his arm.

Another interesting annual display from Wells was at a naval training establishment for boys called HMS Ganges. The display took place at the end of a day's celebration but included the famous gun moving races and the Sunset Ceremony, complete with band and uniformed figures manning the huge mast. The event ended with the fireworks.

HMS Ganges was closed in 1972, with the final salute being taken by Earl Mountbatten. In those days we imported some very powerful gerbs in steel tubes with a heavy steel nozzle, and we had realized that one of them had exploded and blown this nozzle out. It was unique and caused some concern—in fact, we stopped using them. But in 1977, at the Silver Jubilee, Earl Mountbatten reminded me jokingly of HMS Ganges and told me that I nearly got him with a piece of metal that landed at his feet and presumably came from the fireworks!

They often say rags to riches to rags in three generations and, although not strictly applicable here, the Wells tradition came to an end. The first Wells lived from 1816 to 1871. Wilfred continued working with the new Pains Fireworks until he retired in 1996 at the age of 82. He died in 1998, greatly loved by many people.

It was in the early 1970s that some of the firework trade started to turn its attention to importation. The Home Office Explosives Inspectorate had something of a problem in that the law prohibited admixtures of chlorate and sulfur, so they were under some obligation to check what came in. Early experience taught them that these mixtures were common abroad, particularly in blue stars and Chinese crackers, and they were often proved to be right at that time. They got around the problem by virtually forbidding importation, even if it was not quite lawful.

At a party I was introduced to a man called Chris Philip who liked fireworks, had a property in Malta, and was looking for an extension of his business. I warned him of the problems, but he was determined to overcome them and over many years we became firm friends. To begin with, he brought in cylinder shells from Malta, but gradually he built up quite a wide range of fireworks, particularly from Spain, using many of my own contacts.

The idea was not popular with the established U.K. firms that had long had a monopoly, but it was extremely useful to the growing display fraternity since the U.K. trade was not especially good at shell making, perhaps with the exception of Pains and Kimbolton, who made shells based on the Italian system. As they say, if you cannot beat them, join them, and in the course of time Kimbolton started importing from other places, including China. In those days it was the income from importation that allowed Kimbolton to extend its magazines at first and later on to build a new factory. Could we reverse the importation business now if we had the chance? Not an easy answer, but there is no doubt that inexpensive imports from the Pacific Rim have all but destroyed the European firework industry in 2000+, with a great push from EEC bureaucracy.

Importation was hard at first because the U.K. government insisted on testing. Each importation license was handled by one of a number of inspectors; their behavior was idiosyncratic and inefficient by business standards. They were out on the road when you needed to contact them, it took weeks or even months to get results, and it was needlessly expensive. To make matters worse, there were continual arguments about labeling, an issue that was not settled until 1988 with the arrival of the British Standard for Fireworks. In time the government recognized its own problems and started to issue period import licenses. These ceased to exist with the opening of Common Market borders, but little did anyone know what a mistake this was to become.

The argument was that open borders would mean that fireworks could come across from the EEC without custom checks, so there was no check on transhipments either. It took some years, some time after Chris Philip's death in 1998, to realize that this was a boon to importers with no knowledge of the dangers of the industry. New importers have been handling explosives much in the way they might handle any inert object, with many known cases of illegal storage and distribution. Many people willing to learn have started in a small way but quickly increased in size and scope, but they too, without the specialized knowledge, have created nasty incidents in unsuitable places. Another problem was created in the United Kingdom by a gap in the British Standard, which did not specify the weight of composition in rockets. One importer brought in salute rockets with 65 grams of flashpowder, which has a TNT equivalent of about 20 grams. Youths used these to blow up cars and telephone kiosks. It is hoped that the new regulations (MSER) of 2005 will address many of these problems, but only with good enforcement, and the outlook is not good.

Little did Chris Philip know what a "can of worms" he was opening 30 or so years ago, but I know what his answer would be today. He was a fighter, no lover of bureaucracy, and he would have done the same again, but I have the feeling that he would now be pressing the bureaucracy to create more careful measures to control what is coming into the country and to make sure that it is adequately stored. This is right because there are many people who do not like fireworks and there are lots of people among the "chattering classes" who try to ban anything they do not like.

From 1971 onward we started a build-up of public displays, the beginning of a relationship with customers who are still with us in

2005. The venues are too many to mention, but many of them stand out for interesting things that happened.

We started the Kings Lynn Festival in 1971 and, being on the opposite side of the river from the audience, there is a curious echo that crosses the river every time we make a sound. This does not please the many ducks, who tend to depart noisily at the first maroon and then return later. There was one occasion when a bursting shell illuminated one slow leaver, and unfortunately we saw it plummet into the river.

At the Coventry Carnival we had to finish with a lancework portrait of the retiring mayor. A three-frame (4.6 m x 3 m) portrait took up a large amount of a limited budget, and we desperately tried to persuade the secretary to stop having it. He answered that it was as much as his job was worth, so we cut the frame to 10' x 10' (3 m x 3 m). The bottom half we made into the mayor's robes for permanency, and we changed the top annually by adding spectacles, hair, or whatever from year to year. In time, the whole event came to an end, but years later I spoke to an ex-employee of C. T. Brock. Recalling Coventry, he asked if we still did the mayor and said, "Well, when Brocks did it, they were all so boozed up at the end of the show that we used to use an old frame of the Queen and take the tiara off." At New Year's Eve, as we entered the EEC in 1973, we were asked to do a small display consisting of welcome lancework. The budget today would be derisory, and they even wanted to make their own woodwork to save money. At subzero temperatures we could not even drive the stakes properly. Truly miserable, they even objected to giving us a cup of coffee—and the estate was worth millions.

A lovely contrast was a display for the Duke and Duchess of Buccleuch for a special birthday for their daughter. Over tea made by the Duchess in the Orangery, we discussed the display details, but I was somewhat surprised to see several fine specimens of an Australian plant called *Humea elegans*. I had grown this as a youth in the vicarage conservatory, but in later years I had not been able to get the seed as the suppliers and even Kew had told me that it caused respiratory problems in some people. The display went well and when I got home the next afternoon after teaching in school, my daughter showed me a large *Humea* that the Duke himself had just brought for me—a lovely touch.

During this period we extended our display business into the Birmingham area. It really began for the municipality with a Tulip Festival (despite the problems that the local squirrels eat the bulbs), but later in the season with a number of November time Bonfire Carnivals. However, Cadbury's factory in the center of another garden estate has been a regular display event since that time. The factory never stopped working, and we were forbidden to take rockets in case they went through an open window and upset the production. Even today the earlier part of the day continues a tradition of dancing around a maypole, and I can hear the local brass band playing the "Teddy Bears' Picnic." On one occasion they gave us a music tape, asking us to set it to fireworks and end up with a set piece of a "Wispa Bar." We listened to the tape, which started with the Blue Danube, then moved to a brass band, and then to the staccato violins of a well-known TV program, and decided that it was hopeless. We started with the tape and finished with it, and everyone was delighted. We would not get away so easily today, but it also goes to prove that the average member of the public does not consciously coordinate the fireworks and the music. In fact, where precision is not required, it is sufficient to clearly coordinate certain pieces among a more general background.

The military vary enormously. There was the occasion when a bandmaster wanted one flash rocket at a certain bar in the *1812 Overture*, and we made the point that if he sent a man over to tell us when we had got to that point in the score, we would do it. Another occasion at a Tattoo, I went to have a chat with the bandmaster and said that I hoped that he would regard it as a bit of fun and that we would start and finish together as we had no preparation for this event. He replied that people would be more interested in our fireworks than in his music and that we could do what the hell we liked!

Music and firework shows have moved on in the past 20 years, but I will never forget a display for Kodak in 1984 at Lacock Abbey to celebrate the 150th anniversary of Fox Talbot's first photograph. Part of the display for the Royal Firework Music was to be accompanied by a famous orchestra and the famous conductor Sir Charles Groves. We arrived in time to hear their rehearsal, but naively did not realize that they only rehearsed certain bits of it. I asked Sir Charles for a time and he said that he had no idea. He told me to sit down and check my watch and, opening the score, he started to hum the music and at the end I had a time!

Another memorable musical event was a concert to mark the occasion when the firm J.C. Bamford received an export award from the Lord Lieutenant of the county. The orchestra in thin dinner jackets were seated on a floating platform on the lake, but it suddenly turned very cold and when their turn came to play, they proved to be in a hurry and rushed through it. They finished before us and we felt embarrassed by it, but the audience did not mind a bit. One man's description of a cellist sawing away, but at the same time wildly looking in the air in all directions wondering if anything would come down and hit him, certainly relieved our mild concerns.

There are a number of regattas on the Thames in the summer in addition to the famous Henley Regatta, and they have all been eventful in their time. At Sunbury on one occasion they had cut long grass, which turned out to be in a field that the cows had just vacated for the day. Cows do create problems on occasions, particularly in the dark, and another such occasion was a lovely surprise birthday party for the author Roald Dahl. The field looked fine when we saw it in daylight when he was not there, but we had to arrive in the dark and surprise him at the appropriate time; he loved fireworks. Just to complicate things a little, it also was raining, but he insisted afterward that we join his party, which we did in stockinged feet.

At Wargrave all the people are on the other side of the river for the regatta party and there was a local boatman to ferry people across. The disco was so loud that I decided that it might be good for business to have the announcer say who was doing the display. I should have sent a business card across for him to get the Kimbolton Fireworks correct, but it was just a verbal comment and when it was time to start he said, "The fireworks are starting now and they are being done by a guy called Cliff Bolton."

Displays on islands also have their snags. The biggest usually is that the boats provided are not tough enough to carry the tackle, and it can be very wearisome at the end of a show. It is inevitable for people and tackle at some stage to get in the water and there is usually a goodly number of insects.

The year 1977 was very important not only because it was the 25th anniversary of the Queen's accession, but also because it was the year that Kimbolton Fireworks became independent. The preparations for the Silver Jubilee started about 18 months before Jubilee Day, when the Silver Jubilee Committee was formed. I was asked to join the committee as the firework adviser along with a number of City gentlemen, chaired by the Earl of Drogheda and hosted by Lord Sterling. We planned five displays. The three main displays were planned for the Shell Building on the Thames and at the Alexandra Palace and the two displays at Richmond and Greenwich were to be fired by Brock. In April of that year, in discussions with Pains-Wessex about the many changes and takeovers that had taken place since 1963, it was thought that my work was complete there and it seemed that a young Australian of German descent named Peter Buddarick would fulfill their requirements. Peter had arrived some time before as an amateur and had devoured everything he could find about fireworks with complete enthusiasm.

Jubilee Day was quite an event. We were to attend a party on the penultimate top floor of the Shell Building, and we knew that we would be presented to Prince Charles. Early that day I was asked if I would go to talk about the display on the John Dunne music program in the early evening. I argued that it was not easy as we had to be in the Shell Building quite early for security reasons. In the end we drove down to Chris Philip's house in Highgate, changed into our formal clothes, and the BBC taxied us to the studio. On the way down I realized that I had left all my money in my other suit, so when the music played and the red light was off, I asked John if he could lend me a fiver! He was a bit surprised, but he got it back immediately on Monday, of course.

However, the shows went splendidly, much to the credit of Peter Buddarick, but after the display Kath and I, along with John and Pam Deeker, were taken upstairs, where to our surprise we found all the Royal Family. Other guests were the Prime Minister (James Callaghan) along with President Carter's son and his wife. Amazingly, the Duke of Kent commented on our shows, which he had seen, and, of course, Earl Mountbatten reminded me of HMS Ganges.

In time, the Duke of Edinburgh encouraged people to go down in the lift, where they were transferred in the state coaches to the balcony at the Palace. We went down later in the lift and caught the underground back to Highgate, wondering whether it had really happened.

In 1978 the Jubilee Committee reconvened to discuss a firework display in June for the anniversary of the coronation, which had taken place a year after accession. It was agreed that the newly independent Kimbolton would put on this display, and I was keen to do it in the Victoria Gardens outside the Palace. At first the Royal Bailiff was very much against this because they would have just planted all the geraniums in the beds and the next day was the Trooping of the Color Ceremony.

As it happened, we were now well into using felt wads in candles, and we assured him that we would not upset the flowers and that if he had 20 men on the next day to sweep up all the debris, it would be no problem. The aerial part of the display was a little further back on parkland. In those days the company was still small and, as Chris Philip had all his mortar stock much nearer to hand, we asked him to look after this with his men.

It was a memorable event and the BBC made a very fine film of it. Friends wrote or phoned to say that they had seen it in other parts of the world, but we shall never forget the huge crowds packing the Mall and

Constitution Hill, where there were alternating pregnant silences followed by roars of delight.

A TV crew from CBS set themselves up by a battery of mines and refused to move. We warned them of the problems, but they were accustomed to fireworks and they were staying there! I can still see them running for it after we pressed the electric switch to light the batteries of mortars. It was all good fun at the time, but as you get older, you think of problems that might happen—or it is that health and safety syndrome that bugs everyone these days.

To this day one of my favorite displays takes place at the Shrewsbury Flower Show. The show has ended with fireworks each night for more than 100 years and 1979 was our first time. There is a lovely atmosphere, with beautiful flowers, military bands, and arena entertainments, culminating with the display on two nights in August. People attend to watch the fireworks where the effect is enhanced as the site slopes to the river and then rises steeply on the other side up to Shrewsbury School. At the end of the evening the military bands all march on together with a choir, and the British Legion, for the Sunset Ceremony, and the English and Welsh national anthems. There are some fireworks to the *1812 Overture* or other music, which is then followed by the main display after a short break. Roy Butler, and I hope that there will be more to come, we started it with just the two of us, but now we just give a bit of help and watch. Those outside the trade never quite realize that you cannot fire a show and watch it at the same time.

Jersey in 1979 was an occasion for a visit from the Queen, and we were delighted to receive a request from the Greffier of the States to discuss a display for the occasion. Senator Dupre told us that a French firm had done the last display on the harbor wall and that there had been some boat damage. I think he was a little concerned when I suggested that we did the show at Elizabeth Castle, which was some distance from the promenade. It is possible to walk across the sand at low tide, but at high tide they had a service of landing craft of the type used in the war. Senator Dupre was so concerned that the fireworks would not be visible enough that he asked us to take some fireworks over for a demonstration.

The display was a huge success, and we had been told that, as it would be high tide after the show, we would be taken off the island by the fire service. In order to get into the dinghy, we had to wade in the sea on the ramp, but, to the embarrassment of our helpers, the outboard motor refused to start and we had to paddle in the harbor mouth some little distance away. As we neared the harbor mouth, the various currents made movement difficult and then, to our horror, the ferry boat came out of the entrance and we moved up and down on the sea and could almost see in the portholes. Clambering on to the slippery steps, we had to walk the length of the harbor only to find that the porter at the Pomme D'Or Hotel was clearly doubtful whether these bedraggled people were residents.

Our next trip to Jersey was for the Royal Wedding on July 29, 1981. There were no problems getting off Elizabeth Castle after a splendid display, although we had a longer walk to the Grand Hotel on the promenade opposite the castle. This time the manager was waiting for us with a drink in the bar even before we had a chance to wash and brush up.

We had been in discussion with Michael Parker about the display for the Royal Wedding in Hyde Park on the night before the wedding, but I was uneasy about the amount of lancework that he wanted and the position of the huge façade, which he wanted to recreate, because of the prevailing wind. The new Pains were happy to meet his needs, and we traveled to Jersey that day to find that the Greffier had arranged to entertain us to dinner and provide TV coverage of the London display. This is not an appropriate place for my views on the London display.

Our later displays in Jersey were very enjoyable, but they too were somewhat eventful. In 1983, for the occasion of the visit of the Queen Mother, at a presentation with the island's senators, Her Majesty told me that she loved fireworks and was looking forward to seeing the show from the royal yacht outside the harbor. Although the weather was fine, the sea was very rough, which made for problems for officials traveling by tender to the dinner on the yacht. We only just managed to jump safely off a quay covered in wet slime onto a much more substantial boat, which had come to collect us after the show. We returned again to Jersey for the anniversary of the Liberation in 1985.

The year 1980 was rather special. In the first place, it was the 80th birthday of the Queen Mother, and we were called to put on a large display on the Thames in August. This was in the very center of London near the Festival Hall. It became an annual event for the next six years. Although there is plenty of space at that point in the river, much depends on the strength and direction of the wind. I recall one dear lady who complained that we had dropped some debris in her bosom and burned her suit. She said that she had been so inspired by the show that she would be there next year and would be happy if we would just replace her suit. Those were the days! On another occasion, after we had loaded the barge lower down the river, it started to rain at lunchtime and never stopped. We are better organized these days, but at that time the mortar crates were covered with tarpaulins, which are quite heavy. The rain was torrential, so we decided to uncover only sections of the mortar crates at a time as the show progressed. The great fear was that it would get alight under the tarp and what then? In fact, toward the end it did just that, with no problem—the shells just punched holes in the tarp. It is much safer now with polythene.

George Solti was a great fireworks enthusiast who used our services, but one Thamesday was to coincide with a concert he was conducting in the Royal Festival Hall with the Chicago Symphony Orchestra. We were not to start until the concert was over and he had arrived at the terrace. We were given a little military man to act as a coordinator with the hall. He was very nervous and asked where he could stand once the display had started. We advised him to find a spot near the beginning of the show as that would quickly become "vacant" so to speak. All through the show I could see him kneeling on the deck, anorak over his head, face down. When the show ended, I heard him say, "It's all over now and I reckon that it has taken a year off my life."

Those were the days when we could start a show with a long 200-mm cylindrical multiple break maroon shell. The steel mortar sometimes cut a neat disc out of the bottom of the steel drum filled with sand, the hollow steel barge rocked slightly and sounded like a bell. One old Italian friend used to say, "Everyone is nervous as you fire the first shell in these circumstances, but if you are still nervous when you fire the second, you are in the wrong business." Much the same applies when preaching: If you lose that tension, it is time to retire.

There can be no doubt that this period in the United Kingdom marked the beginning of the large modern type of display. It was the end of the old ways, and Kimbolton set the pace for some years to come. It also inspired a number of other people to do the same thing, but without the capacity to manufacture. As we often say, "There is room at the top, but no room to sit down."

The year 1980 was also special because we were asked to put on a firework display for the opening of the famous Humber Bridge, which the Queen had officially opened during the day. There was just one snag in that the entrance to the bridge on the Hull side did not lend itself easily for such a display, so the show was in the docks of the City itself—where most of the people could easily attend. The same year we put on a huge display for the Newcastle 900 celebrations and then two more later on for the openings of the Eldon Center and the new Metro.

The Eldon display was "tricky." In the first place, it was in the center of the complex with mostly hardstanding, but to complicate matters there had been a shower of rain on top of the ground, which was already frozen. Quite the most slippery occasion I remember. The Metro display was on the River Tyne, which is in a deep gorge, and it was on this occasion that the first maroons set off the burglar alarms all along the river.

The York Festival in 1981 is well remembered as we used the railway bridge across the Ouse toward the east coast. The site required the use of the riverbank as well, and this was manned by my father-in-law, Cyril Smith. On one or two occasions I needed to shout to him, not realizing that the crowd was picking this up, with the result that we kept receiving great shouts of "Nice one, Cyril," which was a song currently being used at football matches.

Two royal openings took place in 1983. One was the opening of the rebuilt Temperate House in Kew Gardens. As a gardener from childhood, I have always had a great attraction to Kew Gardens, particularly in the four-year period when I lived in North London. We have done many displays there since that time in addition to the display when Princess Diana opened the new glasshouse in 1987.

A display at Worcester College, Oxford, in 1983 certainly was different. The professor of music asked us to set off fireworks to accompany Handel's Royal Firework Music, which was to be different because all the student musicians were to play an original version mainly with wind instruments. We naturally were anxious about this and had planned a show based on one agreed-upon time period. By the time we got set up, the professor was a little worried about how it should be done and at that late stage asked us to separate the fireworks into five sections, with each section to be fired after they had played their section! The display was to be very late indeed, and some locals apparently complained, but at least being entirely on our own on the site for much of the evening, we were entertained by a baby fox that spent its time walking around us.

Another regular site for our display even today is at the center of the Barbican Centre in the City. Her Majesty opened the center in 1982, with the fireworks as part of the proceedings. We were offered more money than we could use since the site consists of a small terrace at the side of the large water feature. We had the Barbician Centre on one side, a school and a block of flats on adjacent sides, and the church of St. Giles Cripplegate at our backs. In such confined circumstances the use of big gerbs and candles can be very dramatic along with a water feature for floating fireworks. The Vicar of St. Giles naturally looked after our needs, and when we went the following year to repeat the display, he told me that he had been speaking with the Queen at a local function after the display and that she had made the following remark: "I can see that you did not trust the priest friend of yours at the display. I could see that you had a fireman on the church roof."

Our first visit to Japan took place in 1983. The plan was to take the whole family on a package tour and to meet Dr. Shimizu, who had been in correspondence with me since 1970. The trip was a lovely experience, the family making the most of the new things to see, while Dr. Shimizu and I visited Hosoya and traveled on the "Bullet Train" to Shizuoka to meet Ikebun and Mr. Hirata of Goyo Limited. The photograph on page 46 was taken at that time, but another photograph of myself standing against a dummy 900-mm shell shows me holding my hand on the right side of my abdomen. The next day, having bid farewell to Dr. Shimizu, I did not feel very well on the train to Kyoto and in the night I was very ill. The following day they removed my appendix in the American Baptist Hospital. I was very well looked after, but that is another story and, because I did not want to burden Dr. Shimizu and Mr. Hirata, they did not hear of this until we got home.

Returning to Japan in 1993, we again met up with Dr. Shimizu, and this time we visited the factory of Sunaga, where we met a young man, Yasuhiro Sashimura. We again bade our farewells, not knowing that we would meet again. We have had several happy meetings with Dr. Shimizu at the various worldwide conferences and, to our surprise, we were subsequently asked if Yasu could come to work in our factory. At an earlier time we had been asked to accept another young Japanese man who spoke no English at all, but we were fearful of the responsibility of not being able to communicate in our country area, and I was still attempting to teach at school and run a business. On the other hand, Yasu's command of English is good, and we were sad to see him return to Japan after two and a half years, but a firm friendship has been forged between our families. Those who have some understanding of Japanese know that it is not easy for them to distinguish between "L" and "R," so when we are speaking about pattern shells, we have one type that we still call "Clab" to remember Yasu.

The huge displays continued in 1984 for the opening of the Thames Barrier by the Queen. This vital device prevents an irregular kind of flooding caused by winds and spring tides. It was such an occasion that caused the flooding of the Wells factory on the Dartford marshes, which is mentioned elsewhere.

In 1985 the Bicentenary of the *Times* newspaper was the occasion for a large display at Hampton Court, a great setting for a display with that lovely house as a backdrop. This time Mr. Rupert Murdoch and the Prince of Wales were in attendance.

When Mrs. Thatcher removed the Greater London Council in 1986, we were called up to provide a large barge display on the Thames at the side of Westminster Bridge with the old GLC building in the background. It was also the year of a special 350 celebration at St. John's College, Oxford, where they have beautiful gardens, immaculate lawns, and slightly anxious gardeners. We have done another display there since that time and happily we have overcome the problems. In fact, we have done displays for special occasions at most of the Oxbridge colleges.

Fog and rain are the curse of firework shows, but the rain certainly spoiled a centenary display for an independent school in Cheltenham, Dean Close. The school had produced a wonderful outdoor show with special lighting, but by the evening it was clear that we were in trouble. In the pouring rain, high-powered lights were exploding. However, the fireworks were set up and covered in polythene and we decided to go ahead while everyone watched from indoors. At the beginning my colleague rushed out with the coat of arms and ignited it before the lances got wet; it had been destined for the end. Then at the end he did the same with the "Good Night," but unfortunately in the rush he placed it upside down.

It is worth mentioning one or two memorable smaller displays toward the end of the 1980s where things did not quite go as planned. Greenwich is south of the Thames, but the favorite site for the display is on the opposite north bank. We had to get police protection as a group of thugs had called to threaten us if we went ahead, on the grounds that the local old people objected to the noise. However, noise in those days was not the problem it has become 10 years later.

A National Trust house with a very special herd of cattle asked us if we could put on a quiet display to stop the herd stampeding. We agreed to do this, but were horrified to see the display advertised as a "Silent" display. In the event it was not easy as we found that the cattle started running as soon as they heard the piped match cracking.

At yet another National Trust property near Heathrow Airport, concern had been expressed not only about the height of the shells and incoming aircraft, but also about the fact that the display by the lake would upset the ducks. As we waited for the start of the display, a Royal Society for the Prevention of Cruelty to Animals van arrived. Expecting trouble, we watched a man take cages out of the van and place more ducks on the water, so we asked him if he was worried about the fireworks. He replied not at all, adding that they fly off for a while, get used to it and then return. Maybe not all, but most.

At two other National Trust properties the display takes place on an island, and all operators have their own stories about tackle falling into the water, particularly after the show in the dark when you are tired, and being confronted by those who complain about water pollution, which is more or less nonexistent.

These were the days before the modern obsession with risk assessments. Life has become so crazy in the West that it is surprising that anyone is willing to do anything. The cost of a display is increased by such assessments. A display in the City of London absorbed much of the budget with as many as 10 visits with the various authorities who had an interest in the event.

The actual venues and people vary enormously, but I recall one U.S. Air Force July 4th display where we were placed well away from any building on a field at the end of a runway. The officer concerned assured me that they had five fire tenders ready. With a straight face, I asked him if he could manage another one? However, it is getting bad in the United Kingdom, where the new brand of safety officer can make life very difficult. Frequently they know little about fireworks, and I wonder if they are picked because many of them are so humorless.

The 800th anniversary of the City of London took place in 1989 and was the occasion for an enormous display near Tower Bridge and the Tower of London. It was again a tremendous success apart from a complaint from a lady who had used her penthouse flat to host a party for all her friends to watch the show. It did, however, cost us a lot of money to remove, at her insistence, a small blob of dross from a very high roof light.

The large display for the 50th anniversary of the Battle of Britain in the presence of Prince Charles also took place on the Thames in 1990, followed by the 550th anniversary of Eton College in 1991, which was attended by H.M. the Queen Mother. Over recent years we have staged a number of displays for Eton, latterly in connection with military pageants and the millennium celebrations.

Corby used to be a steel town and is not very far away from Kimbolton and a regular venue for displays in November and the summer. One summer display was memorable. The preparation work was complete and we were sitting in the van eating our supper. There was much alcohol about the place and suddenly two youths, a plastic pint of beer in hand, started to have a drunken argument. One got hold of the back of the other's trousers at the top and pulled hard, whereupon the trousers split completely at both sides and the youth was left standing there without any pants. After all the excitement, we found that one or two smaller youths had stolen a few rockets while we had been distracted.

Over the years I have been involved in the various competitions that take place in Europe, and it can be very difficult to venture an opinion about them in quite a short space. Inevitably, those who win are delighted and those who lose tend to whine. In the early days, while doing consultant work, I had been involved in Cannes, San Sebastian, and Monaco, but with retirement from teaching in 1988 and the building of the new factory in 1990, we have been more involved in Knokke, Hannover, Barcelona, Montreal, and Cannes.

In 1972 I was asked to act as technical organizer of a BBC firework competition for TV. The first problem was how to make sure that the TV was never blank for the six-minute show. One Spanish manufacturer complained that his wonderful flight rockets appeared as a blank screen, and only then did he realize that in those days a TV camera could not take low- and high-level light emission at the same time. Later we arranged TV displays on this basis. But the biggest problem in competitions has been value for money. It will never be solved. Is it breaking evens or selling price? A manufacturer can always do better than someone who buys. Some competitors go over the top and put the whole value of the show in the finale. This means that the only answer is to have a maximum weight of material allowed in the show, but a kilo of Chinese fireworks may be better value than a kilo of European material!

Although degree standards seem to have gone down in recent times, it was generally agreed that a first-class degree required an overall performance rather than just facts. This is often the case with a firework competition, but awarding places can also be difficult. There is the additional problem that shows can take place one after the other and sometimes on different days; there can be a psychological advantage to the last performer.

Over the years I have come to believe that the best fireworks should be seen to win, but this can be a disadvantage where there are professional jurors and amateur ones where the latter are likely to respond to gimmicks. At Barcelona we considered it a real honor to receive both the professional and the amateur jury firsts; it does not always happen.

In both respects it is important that professional judges be of high caliber and that some also be available to guide the amateur jurors. There are competitions in which jurors and organizers are also allowed to take part in the competition; this is disgraceful and will always lead to accusations of a "fix."

People do like to win prizes, but there is a lot to be said for simply having a festival without any winners or losers.

After 25 years as a school chaplain and chemistry teacher, I decided to retire in 1988 at the age of at 57. Mark Lancaster also was leaving school to read civil engineering but decided to enter the army on a short service commission where he ended up in Hong Kong's New Territories with the Royal Engineers. One day a call from Hong Kong broke the news that he also wanted to carry on the firework business. The existing laboratories and magazines were quite inadequate, and I made it clear that building a new plant would require a half million Sterling just for a start. This was not such a problem as it seemed since we had been highly profitable from the importation business and with a small labor force; there was no need to borrow so much.

In 2005, it is exactly the opposite: We have a larger labor force, much higher overheads, expensive bureaucracy, and small profitability because of too many importers.

For most people it is very hard to get a site and even more difficult to get planning permission, but we were lucky, and people on the whole were glad to see us moving to the periphery of the community.

At this time we were joined by Tom Smith, whom I had met while he was doing a chemistry D. Phil at Oxford. Tom worked with us for a number of years in order to gain experience of the explosives industry until he left to set up his own consultant explosives business, still in Kimbolton.

There can be no doubt that the industry has changed dramatically in these past 14 years, and it is with great regret that I have to say that there is no serious future for the firework industry as we have known it. Manufacture may last another generation and individuals may well carry on, but we shall never again see the huge quantities produced in the past by Brocks, Standard, and the old producers. It is easy to blame the economics of the Pacific Rim, but European bureaucracy is not far behind.

The Health and Safety Acts in the United Kingdom were laudable in their lofty aims, but there can be no doubt that the open-ended nature of this type of legislation also has been very destructive. Individuals seem to have polarized so far as their attitude to legislation is concerned. Needless to say, those who have suffered injury through the foolishness or the selfishness of others may well have cause to be grateful. Others are more concerned about the enormous and expensive bureaucracy that has slowly been built up. In the early days some bureaucrats were arrogant and even boasted about the Draconian powers that they possessed, but time has indicated that the system has proved to be inefficient, lacking in business acumen, and consequently destructive of it. Kimbolton cannot afford fine buildings, canteens, and gymnasiums.

Since the mid-1980s a whole series of pieces of legislation have led to the view that there are too many problems for the explosive industry; indeed, there is not very much of it left now. We all joke that health and safety is the only growth industry we have. If it were now to slow down and take stock of what it has gained, all might be well, but, alas, those in its employment cannot remain static, with the result that they are kept well occupied by constantly moving goalposts or thinking of new projects. They are becoming a big source of government revenue.

A look at the chapter on legislation in this book well illustrates these issues, but discussions in the EEC about differences in attitudes to fireworks indicate how difficult it can be to find unity. In Sweden fireworks may well be used at –30°C but in Greece at 50°C. Some countries want the maximum net explosive content of a rocket to be 20 grams, while others need very much more. Aerial shells are forbidden in some places but not in others. It can be very depressing, and it was not an easy decision to build a new factory.

The inspector we consulted about the new factory was one of the "old hands" and could not have been more helpful. Today it is different, where the attitude is much more like "tell us what you propose and we will tell you whether we agree with it." The "old hands" would relish the prospect of planning such a thing; these days officials do not want the responsibility, and their eyes glaze over when you ask them for a decision.

We decided to build the factory on the German cell block system pioneered between the wars and have been happy with it. When it was complete, we were told that it was "overkill" for fireworks and Mark Lancaster's army chart indicated that about 13 kilos of Semtex would be required to crack the steel reinforced walls. They would not regard it as "overkill" today.

Since moving the factory and with increasing age, it has been quite pleasant to concentrate on research and development and leave some aspects of the business to the next generation. It also is clear that fireworks displays have been taking on the extra dimensions of entertainment, which the public expects with the addition of music, lasers, and complex firing systems. I do not even miss getting home in the early hours of the morning cold and damp after spending the whole of a November day in a field. I never expected that I would feel like that—*sic transit gloria mundi*. Nevertheless, the company has been involved in some very fine displays in the years since 1990.

Two of the largest undertakings at the end of the century were in 1995 to celebrate 50 years after the war in the Far East and in 1997 for the Hong Kong handover.

As always, there are organizational problems for huge national events in the United Kingdom. There is the issue of who might have the responsibility for sponsoring it, but, unlike Paris or Sydney, we do not have any grand sites where huge numbers of people can gather. While the Thames might be the best place where barges can be used, it does meander somewhat and the buildings along each side hinder the viewing.

The Ministry of Defense decided that the 1995 display would consist of three barges on the Thames and that the displays should be accompanied by music of the war period. This was not all that easy. Each arm of the service had its own evocative melodies such as *Hearts of Oak, The Dambusters March,* and *Pomp and Circumstance*, but we found ourselves wondering what kind of fireworks to use with "The White Cliffs of Dover" or "Coming in on a Wing and a Prayer."

As we discussed the display content, allowing for three barges, it moved to four barges and ultimately five. These barges would be spaced between Tower Bridge and Westminster Bridge, a distance of about 3 kilometers. One barge was quite near the rebuilt Globe Theatre with its thatched roof, but of course, being new, it had a sprinkler system. In all there were ten barges as five were needed for the huge sound system. However, we had to accept that the average viewer would see and hear only two barges at the most and, needless to say, sound and light do not travel at the same speed. There is now a great dearth of large barge platforms on the Thames and, as it happens, the tenth had to be raised from the bottom and dried out.

The tide was not very helpful at the time of the display and, with a lack of suitable moorings, the barges had to be held by tugboats. When the time for firing arrived, Mark Lancaster's barge was still being turned 90 degrees in the river, but this was somewhat minor as we had never tried to do something like this before—over 3 kilometers. Dame Vera Lynn, who is generally associated with the wartime songs, was asked to press the button to start the show, and I had the privilege of guiding the proceedings on the terrace of the Tower of London with the Prime Minister (John Major) and some of the cabinet. As it happens, John Major was our constituency Member of Parliament, and our respective children had all been taught together in Kimbolton School.

Just before the display Lord Cranborne asked me if I was nervous. I replied that I was very nervous. It was not the fireworks that caused the anxiety but rather that we had never before done such an extended display to music. However, we were using DAT tapes and manual firing systems, and I was not to know how bad I was to feel in 2004 in Dublin, where an even more complex display was to depend entirely on electronics.

Over the years, the senior boys in the school had worked with us on shows and later came for university vacation jobs. On the three days before the 1995 show, about 30 of them came back to work with us—a pedriatic surgeon, medical students, teachers, architects, and so on. They all did a marvelous job under the direction of the display manager at that time, Tom Smith.

As you can never predict the weather in the United Kingdom, we were blessed with wonderful weather and a highly successful display. As the barges were turned and towed back down the river, the one that had been under water happened to be the farthest away at Westminster. Some of its rotten boards had caught sparks and now the gentle breeze brought them into flame. Attempts to put out the flames were not successful and there was a fear that the odd unexploded shell might take off, and what if it was under one of the bridges? At last the huge fire barge appeared and not only put out the flames but also was so powerful that it washed a goodly amount of tackle into the river.

There were ideas that the millennium display on the Thames would emulate this concentrated pyromusical with 18 tons powder weight on five barges. Politics deemed that Syd Howard from Australia was to be responsible for this, but a lesser powder weight was spread over 16 barges and over a greater distance.

The Hong Kong handover was equally complex in terms of politics and available money. Many events were planned for the actual handover, and while the foreign office was keen to have fireworks, in the end the display was the main British piece of entertainment and we had to find our own sponsorship. Mark Lancaster had served in the army in Hong Kong, and we were fortunate to gain sponsorship from Ing Barings Bank.

Bearing in mind the old expression "carrying coals to Newcastle," we decided to make as much of the display as we could with our own material. The whole display was shipped out in polythene tubing, but much of the finale was obtained locally, and inevitably this was not so carefully protected. Ing Barings asked for their name to be included, and because of the distances it was to be burned out with railway fusees. As it was June 30, 1997, and the busiest time at home, I stayed at home and the boys went to Hong Kong assuming that they would work three days and rest for two. Alas, it also was monsoon time and they had to work five days, one minute "apple bobbing" on the deck and frying eggs the next. By the time of the display it was raining "stair rods" as we say in the United Kingdom. Leading up to the display, watching TV was a most depressing experience, but feelings of empathy were not very helpful.

A small percentage of the finale was lost, as was the Ing Barings name, but the rest of the show was magnificent, and anyone watching would not have been aware of any of the enormous problems. What could Mark Lancaster have said just before the show as a member of the foreign office said, "Don't let us down, the whole world is watching you."

It was also during these years that we made a brave attempt to participate in the competitions held at La Ronde in Montreal. As with all such competitions abroad, the challenge is great and the preparations take up an enormous amount of time in the factory. In 1993 we were happy to receive the Special Silver Jupiter, which was awarded for "Pyrotechnic Excellence, Dynamism and Professionalism." Attempts to improve on this more recently were thwarted by the most appalling wind and rain.

It is inevitable that technology has to move along and the presentation of fireworks has to go with it. It is interesting to look back and see how the shows have gradually been able to produce spectacles that would have been impossible in the past. The first time I went to Cannes to attempt a coordinated display in the sea off La Crossette, our primitive firing system was a disaster and we were beaten by Dynamit Nobel using suitable tapes. With a modern firing system, three barges, and musical accompaniment, we produced magnificent results and the winning prize, the Vestale D'Argent. We did not realize that we would have to return in 2003 with the winners of the last years, but we were blessed to win again to obtain the Vestale D'Or. This was very much to the credit of our boys, to Darryl Fleming and Michael Lakin.

But at the time of writing, the St. Patrick's Day display in Dublin presented the greatest challenge yet. The harbor site is flat and a number of buildings restrict the view, but the display has grown in size and popularity over the past decade. It was decided to make 30 floating pontoons, to be widely spread and leading up to the three main display points. To an old man unskilled in computers, it was a very frightening experience wondering what might happen if the system failed. Actually, the sequencing over the pontoons worked like a dream. The 2005 display celebrating Nelson's bicenenary of the Battle of Trafalgar in Portsmouth was a special triumph seen on television all over the world. The new firing systems also produced an award-winning show on the harbor wall at Funchel, Madeira.

Looking back to when I arrived in Kimbolton in 1963, I thought that I might stay for three to five years and never dreamed of the existence of Kimbolton Fireworks. It was never planned either; it has all just happened as one door after another has opened.

Although Kimbolton has only just over 1,000 electors and would be classified as a village, it is an ancient township with a castle and a thirteenth century church. It also is famous for the fact that Queen Katherine of Aragon, the first wife of Henry VIII, lived in the castle and died there. It also has schools with in excess of 800 pupils of all ages. However, more than this, in World War II it was surrounded by airfields, with the nearest one being manned by the U.S. Air Force.

A schoolmaster/poet colleague, John Greening, recognizing all these facts and knowing that Mark Lancaster is also a Royal Engineers Bomb Disposal Major, composed the following poem:

Fireworks this village is famous for. They come for miles to witness
the Reverend's all-star variety performance, where the judge
lived who once condemned Guy Fawkes to death.

Catholicism broods here, like the one that wouldn't go off, lurking
in its priest-hole, its side-alley churches. Even our spire flares
a Roman Candle to guide us, as it guided the B-17s

back from their attacks on V2 launch sites, or bombing
a ball-bearing factory. Remember the war. The blackout.
The bagatelle-board rumble from Warren Hill. And the formations...

I met the heir to the fireworks empire at the post-office counter.
He's in the Territorials. Expecting to be called up
to Kosovo quite soon. Bomb disposal. Once November

and the Millennium are safely over, he'll be out there, clearing
mines and reminding us that war can never be cleared, however
we celebrate our VEs and our VJs with elaborate triumphant displays.

keeping his eyes on the ground, on the patterns the soil makes.

John Greening
September 1999

It can truthfully be said that it had never occurred to me that it was possible to gain an income from fireworks until there was a need for capital expenditure in the mid-1960s. Once we had expanded, it gave way to the internal joke that I knew nothing about business but a little bit about how to make money. It was inevitable that when Mark went to the business school, I asked him to come up with a mission statement, which turned out to be "Unrivaled in Quality and Service." However, I had studied classical languages in my youth, so with the help of a colleague I came up with "Ex Luce Lucellum," which we translate as "Out of light, a little profit." On the other hand, I am confident that those who know me know that my love of fireworks is far greater than material gain.

1

THE HISTORY OF FIREWORKS

R. E. A. Butler

Firework manufacture has a long history, but the development of the pyrotechnic art has been remarkably slow. The Chinese may have made fireworks of sorts more than 1,000 years ago; displays have been fired at public and private celebrations for 500 years, and their popularity, now worldwide, seems undiminished. Nevertheless, firework displays basically have changed little over the centuries, with rockets, shells, and Roman candles in various forms remaining the main display components. Certainly, the quality and range of colors have been improved, shells are more spectacular, rockets are propelled higher, the use of new materials has brought some original effects, and fashions in set pieces and in the style of displays have changed. Although modern technology, in the form of sophisticated electronic firing boards and musical accompaniments, now enhances the spectacle, the essential ingredients of the firework exhibition do not alter. The fireworker still strives to excite and delight with a combination of color and noise. He creates patterns of beauty and brilliance using natural materials and employing a knowledge of chemical reaction, together with the benefits of experience, and often much patience, dedication, and intuition. The invention of gunpowder heralded the beginning of the pyrotechnic art, and this dark mixture is still the firework maker's principal material. Thus, in this capacity as a bringer of pleasure and beauty, gunpowder makes some amends for its evil reputation as a source of death and destruction.

It is probable that the first gunpowder was formed when, quite by chance, charcoal, saltpeter, and sulfur were brought together. The result of this accident must have been obvious if the mixture was exposed to some means of ignition, and the potential use of this new explosive material must soon have become apparent. Traditionally, the Chinese are credited with the discovery at a time well before historical records.

Certainly, the evidence suggests that gunpowder originated in the East, with China or India being the likely source, although the Arabs and Greeks have certain claims. Tradition also credits Roger Bacon, an English friar of the thirteenth century, with the invention of a gunpowder mixture. Michael Swisher, however, argues conclusively in *Pyrotechnica* (November 1997) that:

> Roger Bacon did not invent gunpowder. He knew of it, and described it in several of his works with varying degrees of detail. The only application he describes is in a firecracker, a child's toy made in diverse parts of the world.

Swisher also explodes the popular myth that Marco Polo brought knowledge of fireworks back from China in 1295. Polo makes no mention of fireworks in his account of his exploits, and the writings of Bacon and others show that gunpowder and fireworks were known in Europe well before that time.

As for the application of gunpowder, the invention of the gun usually is attributed to a Franciscan monk named Berthold Schwarz, and the town of Freiburg has erected a statue in the town square in his honor. Attractive as the image of the "Powder Monk" might be, Professor J. R. Partington, in his meticulously researched *History of Greek Fire and Gunpowder* (1960), finds no evidence to suggest that Schwarz even existed. It is thought that the invention could well have had Asian origins, although it was more than two centuries later that the first artillery was reported in China, and that was on Portuguese and Dutch ships.

The Chinese, however, had employed pyrotechnic mixtures long before this date. Ancient manuscripts describe explosive bombs, which were fired from giant catapults and burst on landing or in the air. Similar missiles were merely dropped on the enemy from fortress walls. Firecrackers were used in early times, just as they are now, to scare away evil spirits from wedding and birth celebrations and from funerals, and they also were much in evidence at various religious festivals. These crackers often were made by packing gunpowder into bamboo cases or rolled paper tubes, thus laying the foundations of modern firework manufacture. They exploded when thrown onto the fire; hence the origin of the name *firecracker*.

An encyclopedia by Fang I Chih, dated around 1630, refers to "fire trees and silver trees" used in the Tang dynasty (seventh to tenth centuries) in which gunpowder was thought by the author to have been used. These fireworks may have been the forerunners of those used in big displays that frequently were put on in China in the seventeenth and eighteenth centuries, and which were described in various writings by travelers returning to Europe. Apparently the development of Chinese fireworks proceeded very slowly, and in 1821 Claude-Fortune Ruggieri, the French pyrotechnist, remarked that the "Chinese fireworks . . . were no different from what the Chinese have been making for three or four centuries; this convinced me that we in Europe are far superior to the Chinese." But, of course, this could have been wishful thinking.

In India, too, progress appears to have been slow, for war rockets were in use at a very early time. Here, as in China, fireworks of sorts frequently were seen at celebrations and public festivals, and fifteenth and sixteenth century writings, such as the Marathi poem of Saint Ekanatha, describe displays and mention rockets and fireworks producing garlands of flowers, a moonlight effect, and hissing noises. By the eighteenth century, displays were organized on a lavish scale. The first English display in India was in 1790 near Lucknow and was said to have taken six months to prepare.

In Europe, pyrotechnics for military purposes saw an early peak of achievement in the form of Greek fire. Highly combustible material, including sulfur, resin, camphor, and pitch, was blown by a bellows device out of copper or iron tubes, or even hand pumps, and was almost inextinguishable. Old manuscripts suggest several ways of attempting to combat the fire, especially the application of wine, vinegar, sand, and even urine. For 400 years, the Greeks guarded the secret of their devastating weapon and used it with spectacular effect on land and sea; but by the tenth century, the Saracens had learned the formula and used it against the Crusaders. By the fourteenth century, gunpowder appeared in European warfare and made the short-ranged Greek fire powerless against far-flung missiles.

In the wake of gunpowder came the arrival of firework mixtures, both of them appearing in Europe, probably as a result of information on their manufacture being brought from the East. Italy seems to have been the first area in Europe to make fireworks, as opposed to military pyrotechnics, and put on displays. It is clear that before 1500 fireworks were employed extensively at religious festivals and public events, as frequent displays were becoming popular entertainments. Florence probably was the center of an expanding manufacturing industry, as demand for the new spectacle increased. Before this period, fireworks had been used as scenic effects at theatrical productions. In fact, fiery torches and the like had been added embellishments in the amphitheaters of classical Roman times. Now the fireworks became the main concern, although elaborate scenic sets and buildings were to form backgrounds to displays for many years to come.

Firework displays were seldom seen in England before the end of the sixteenth century. Shakespeare refers to "fireworks" on several occasions in his plays, suggesting that the term was in general usage in England in Tudor times. Other literature of the period often mentions the "Green Man," whose function was to walk at the head of processions carrying "fire clubs" and scattering "fireworks" (in this case, probably meaning sparks) to clear the way. The origin of the Green Man is a mystery, but the character probably evolved from the wild men, satyrs, monsters, and so on. A sixteenth century Danish print might provide a clue, since it shows a floating firework device, figures holding rockets, and two figures wearing leaves and carrying fire clubs, resembling the Green Man of a later date.

The earliest record of a firework display in England was in 1572, when a large show was put on at Warwick Castle to mark the visit of Queen Elizabeth I. The Queen is said to have enjoyed the spectacle immensely, and this approval served to encourage the organization of many more displays, including two shows fired at Kenilworth Castle, Warwickshire, to entertain Her Majesty during a visit there in 1575. The first of the displays on the River Thames was in 1613 to celebrate the marriage of King James's daughter Elizabeth. The site has been used with great regularity ever since.

The early displays in England were mainly the work of firework makers from France and Italy, especially the latter, who seem to have been supreme in Europe until the end of the seventeenth century. It was not until considerably later that English pyrotechnists began to challenge the continental lead. Responsibility for the provision of fireworks and the organization of displays was put in the hands of the military, and Ordnance officers, ranked Firemasters, were appointed to take charge.

While the English lagged behind, two distinct schools of firework making appeared in Europe. In the northern area, such states as Poland, Sweden, Denmark, and the German states were developing new methods of firework presentation, which differed markedly from the traditional style of the Mediterranean countries. Brock considers that the split was closely related to religious matters, and the intense feelings that the Reformation aroused found outlet in more sectarian spheres, including pyrotechnics. In fact, the fireworks made in the north and the south remained very similar in effects; the divergence occurs more often in the presentation of displays.

The Italian style, illustrated especially by the Ruggieri brothers of Bologna, and followed by the manufacturers of France (who were joined by the Ruggieri family at a later date), had grown from the early ceremonial displays in Florence at the Feast of Saint Peter and Saint Paul. Invariably, collections of small fireworks were arranged on, and in front of, huge, elaborate structures, built in the form of castles, temples, and classical edifices, and known as "machines" or "temples." The imposing frontages were lavishly adorned with rich decorations, and the whole was illuminated from within and without. The audience thus was entertained before the actual display began, and when the fireworks were lit, they tended to heighten the general spectacle of the "machine" rather than provide purely pyrotechnical amusement.

The breakaway northern school took their lead from Nürnberg, where experts such as Hoch, Muller, Clarmer, and Miller challenged the masters of Florence and Bologna. The displays in the north gave the fireworks the prime importance and diminished the role of the "machine." The fireworks were set out in neat rows on the ground for all to inspect before the display was fired. If the "machine" was used, it was of a much less elaborate construction than in the south, although sometimes real buildings or landscapes were used to add atmosphere to the shows. The effect achieved by firing displays behind a foreground of water was realized in this period, and engravings of displays at Stockholm, Paris, Versailles, and on the Thames illustrate the early beginnings of this still popular practice.

Spectators accustomed to either northern or southern types of displays were scornful of the attempts of the rival school, as can be clearly discerned from contemporary publications. The most authoritative was *The Great Art of Artillery*, (46) penned by Casimir Simienowicz, the Lieutenant General of the Ordnance to the King of Poland, in 1650. His displays, although following the techniques of the north, included some features of the south, thus giving his shows decorative effect before firing time, yet concentrating on pure firework amusement during the performance. This kind of compromise display often included figures and architectural structures, smaller and less intricate than the "machine" and made of a wooden frame, over which was papier-mâché, which concealed fireworks. At a certain point in the show, sparks and stars would be seen to issue from the model with spectacular effect. Various figures made their appearance in the different shows, although the Cupid was perhaps the most popular, and the tall obelisk was a regular feature at displays for many decades.

In their various styles, displays increased enormously all over Europe. As far back as 1532, Charles V, the Holy Roman Emperor, celebrated his military victories with displays. In 1690, Peter the Great of Russia put on a five-hour display to celebrate the birth of his son, Alexis. Louis XIV and Louis XV enjoyed numerous shows in Paris and at Versailles in celebration of royal births and weddings, state occasions, and victory or peace festivals. In almost every European country, visiting royalty were treated to huge displays. Peace treaties, such as that signed at Aix-la-Chapelle in 1742, were excuses for expensive performances in many European capitals. Unfortunately, according to reports at the time, the show staged in Paris resulted in "forty killed and nearly three hundred wounded by a dispute between the French and the Italians, who, quarreling for precedence in lighting the fires, both lighted at once and blew up the whole." Numerous prints and engravings of the time undoubtedly flatter many of these shows by always depicting them in full, extravagant splendor.

In fact, not all displays were the spectacular successes their advance publicity proclaimed. The pyrotechnic celebration planned to take place in London's Green Park in 1742 was to have been the greatest display of all time. An official estimate of the cost was more than £14,500, and Ruggieri and other notable Italian manufacturers were brought over especially for the occasion. Nearly six months were spent in erecting huge temples and various ornate machines of elaborate design. One machine, designed by Cavalieri Servadoni, was 410 feet long and 114 feet high. Eleven thousand fireworks were prepared to accompany the "Music for the Royal Fireworks," which Handel had composed especially for the occasion.

At the appointed hour, King George II, accompanied by an impressive array of aristocracy, paraded to his seat past the huge, excited crowd. However, all was not well behind the scenes, for violent arguments had arisen between the English and Italian fireworkers. These disagreements were brought to a dramatic end as an explosion rent the North Pavilion, which burst into flames. The fire caused widespread confusion and alarm, but it eventually was brought under control so that the planned fireworks could begin. Judging by eyewitness reports, however, the display was anything but the memorable spectacle that had been promised. Such descriptions as "pitiful and ill-conducted," "the Grand Whim for posterity to laugh at," and "the machine was very beautiful and was all that was worth seeing" were just some of the less abusive comments. Certainly, it was the last big display London was to see for many years.

Private firework companies had long been operating on the continent of Europe, but in England artillery officers were still in charge of displays, although the actual arrangements probably were under the control of civilians. Small English companies undoubtedly made fireworks for the shows, and large quantities were regularly imported from France and Italy. It is recorded that a Swede, Martin Beckman, made fireworks for the celebrations that marked the coronations of Charles II (1660) and James II (1685), both on the Thames, as well as that of William of Orange (1689). The eighteenth century, the "Age of Elegance," gave the English manufacturers the opportunity to show their skills and to increase their sales and production.

It was during this period that the "Pleasure Garden" became, for the respectable townsmen and their ladies, the fashionable place at which to be seen. Taking a lead from London, most towns of note established these exclusive resorts, with their concerts and tea parties, opportunities to exchange gossip and to be sociable, not to mention the availability of medicinal waters in such towns as Bath, Harrogate, and Leamington Spa. Soon other entertainments were added to amuse and excite the clientele. Male and female bare-knuckle fights, dog- and cockfights, and bear- and bullbaiting were all popular attractions, and firework displays eventually became regular items on the programs. Many small manufacturers found this new and expanding market just the incentive they needed to develop their businesses and make their reputations. The Brock family business, in particular, made great strides during this period, with impressive displays at the famous Marylebone Gardens, and later at Ranelagh, Vauxhall, and the Spa Gardens, Bermondsey. London boasted scores of resorts, ranging from the most fashionable and exclusive to others that were rather less salubrious and often short-lived.

Outside London, the most outstanding resort was the Belle Vue Gardens in Manchester. It started as an extension of a public house that exhibited a few animals for the amusement of customers and developed into a sporting and amusement center, a zoological garden, and, after 1852, a site for regular firework displays. Spectacular recreations of famous battles using pyrotechnic effects were a specialty that continued until 1939. The fireworks were provided by the Belle Vue Fireworks Company until 1926, when Brock took over. Immediately after World War II, shows were fired every night throughout the autumn, with a huge display on November 5th as a finale. Rising costs later necessitated the reduction of the number of shows and the budgets until the Gardens closed in the 1970s.

The other English pleasure gardens had long since disappeared, victims of the vast and rapid expansion of London and other cities during the nineteenth century, although some of them were converted to public houses and still serve, if for a wider clientele. The Tivoli Gardens in Copenhagen remain, complete with regular firework displays. The numerous "theme parks," epitomized by the Disney enterprises, are modern equivalents of "pleasure gardens" of the past.

Events of national importance continued to provide income for the companies through the nineteenth century. Notable among these were the Peace Treaty of 1814, the Jubilee of George III, and the coronations of George IV in 1821 and Victoria in 1838. The expensive Hyde Park displays that marked these occasions rivaled in effect the many held in France at that time. (Napoleon, in particular, was a great firework enthusiast and had much to celebrate before his ultimate defeat.)

As the nineteenth century wore on, the techniques of firework-making, a process that mirrored the vast development that was proceeding in all fields of scientific research and means of communication in Europe, took a dramatic leap forward. In time, the better understanding of chemical reactions produced new pyrotechnic effects, especially a wider range and greater intensity of colors. A better use of propellants and more efficient methods of firing led to ever-improving pieces and consequently to better displays. Newly discovered knowledge was circulated by the numerous scientific books that were being published, encouraging a large and new generation of amateur pyrotechnists. The expanding sales of newspapers and journals such as *The Illustrated London Times* and *The Illustrated News of the World* publicized and popularized displays with frequent and large illustrations. Better means of transport enabled manufacturers to put on displays in areas previously regarded as inaccessible and allowed people from increasingly wide circles to travel to watch the exhibitions. Moreover, all over Europe, royalty and other dignitaries were able to move around with much greater facility and often were treated to a spectacular firework display in each city they visited. In this way, Victoria and Albert were entertained during their tour of 1845; and 60,000 people assembled in the Crystal Palace grounds to enjoy the display given in honor of the visit of the Grand Duke Vladimir of Russia to London in 1871.

There was, of course, an ever-increasing number of small and public displays, but the contracts for the huge and expensive shows to celebrate national events were heaven-sent opportunities to the manufacturers, not only to increase their profits considerably, but also to publicize their wares, as indeed such events are today. Hence they greeted the victories during the Crimean War with much enthusiasm, and the Peace Treaty of 1856 was the signal for a glut of displays in every sizable town in Britain, including no fewer than four in London, as well as in most French cities. For English manufacturers, Queen Victoria proved to be an even greater asset, since the celebrations that accompanied Her Majesty's Silver and Golden Jubilees required numerous and lavish displays, not only in Great Britain, but also throughout the Empire.

Apart from the occasional spectacles marking national events, the most important displays in Britain certainly were those put on at the Crystal Palace by the Brock company. Charles T. Brock inaugurated the series at the new and popular resort at Sydenham, London, in 1865. The idea was such a success that the company gave regular displays every year until 1910, and then again after World War I from 1920 to 1936, when fire destroyed the building. Thus a series of nearly 2,000 displays was brought to an end. They had come to be known as "Brock's Benefits," after an explosion nearly ruined Brock's business in the 1820s and, as we are told on a contemporary poster, a sympathizer gave Brock "the gratuitous use of his commodious Ground (in the City Road) to display an exhibition of fireworks for his benefit." In an effort to popularize the newly opened Alexandra Palace, a firework competition was held at the resort in 1876. Among the prize-winners were Pain's, Wells, and Wilder, as well as Brock's—company names still familiar until a few years ago.

The displays of the period, as typified by the early "Benefits" at the Crystal Palace, bore a significant resemblance to the modern displays in respect to type of firework and general display programs and underlined the changes and improvements made in the previous years. Gone were the elaborate scenery and decorated buildings, and gone too was the obelisk or similar central model. Instead, the spectacle was provided by a greater effectiveness from the fireworks displayed and an altogether wider range of items. The discovery of potassium chlorate by Berthollet at the end of the eighteenth century allowed the firework industry to produce colors that had never been seen before. The addition of metals such as aluminum and magnesium, the latter in around 1865 and the former in about 1894, gave fireworks an increased brilliance. Rockets of 1/2 pound and 1 pound caliber soared higher, and shells increased in diameter to 200 mm, 300 mm, and 400 mm. (It was Ruggieri who first wrote of a shell with lifting charge and projectile contained in one unit in 1812.) The potassium picrate "whistlers" made their first appearance at the Crystal Palace.

Yet, it was the set pieces that were the main attractions of the time. The Crystal Palace with its shrubberies and fountains presented a splendid backdrop for a firework display, but the sheer size and magnificence of the setting demanded a performance on the same grand scale; hence Brock's introduction of the huge pictorial set piece. Using thousands of colored lances on frames and more than seven miles of quickmatch, "fire pictures" up to 800 feet long and 90 feet high were produced, depicting such epic events as the Battle of Trafalgar, the Siege of Gibraltar, and the Eruption of Vesuvius. After 1897, portraits in fire of royal personages, visiting dignitaries, politicians, and, later even film personalities became popular. A spectacular variation was the transformation set piece in which a design made up of quick-burning colored lances gave way to a picture of slow-burning white lances, the white having been obscured by the brighter colors until that point. By very careful choice of compositions and using lances of different lengths, a transformation piece could be presented in which a rural scene changed from spring to summer, then to autumn, and finally to winter. The largest set piece ever constructed was Brock's spectacular depiction in 1898 of the destruction of the Spanish fleet in Manila Bay in the same year. The structure covered 60,000 square feet.

Various other set pieces, such as firewheels up to 100 feet in diameter, lattice poles, flights of 5,000 or more rockets, bouquets, and the famous Niagara of fire, a magnificent cascade that covered an area of 25,000 square feet and burned a ton of iron filings, became regular items. So too did humorous lancework pieces, such as the seal that seemed to balance a ball on its nose, donkeys that threw off their riders, elephants that blew showers of sparks from their trunks, and the British lion that lashed its tail and winked at the delighted crowd. Innovations at the "Benefits," which have disappeared with the very large set pieces, were the "living fireworks," so popular at the time. In these, live actors, dressed in asbestos and outlined in lancework, wrestled, boxed, enacted scenes, and even walked the tightrope. On one famous occasion, an actor named Bill Gregory, who was listed in the program as Signor Gregorini, was to slide down a wire from the tower to the terrace. At his first attempt, he became stuck halfway down and had to stay there suspended for the remainder of the display. As Brock notes in his book, ". . . his remarks left no doubt as to the country of his origin." It is interesting to find that a member of the Gregory family was still working for the Brock Company in the 1960s (Figs. 1.1 and 1.2).

While the Brock Company gained remarkable publicity and deserved acclaim for their big displays at Crystal Palace and elsewhere,

Figure 1.1 Brock's Firework displays at the Crystal Palace, London. Early twentieth century (Brock's Fireworks)

other companies in England also were extending their business and influence. Wells (established 1837), Pain's (1860s), and Wilder's (1876) were among the firms sharing an expanding market in this country and throughout the Empire. On the continent of Europe, many of the famous companies were well established by the turn of the century and were rapidly extending their arts, techniques, and businesses. For example, the Lacroix company was established in 1848 to rival Ruggieri in France, and Hansson of Gothenberg dates back to 1888.

The number of displays did not decrease, and the big national events, such as the Peace Celebrations of 1946, the coronation of Queen Elizabeth in 1953, various independence celebrations around the Commonwealth, the Silver Jubilee of H.M. Queen Elizabeth in 1977, the marriage of the Prince of Wales and Lady Diana Spencer, and the fiftieth anniversary of VE Day, still demand a pyrotechnic show. The millennium celebrations in 2000 provided the companies with plentiful opportunities. Moreover, displays are becoming increasingly popular at carnivals, regattas, and fund-raising events, and as accompaniments to

Figure 1.2 Brock's Fireworks at the Crystal Palace, London. Early twentieth century (Brock's Fireworks)

outdoor concerts. Traditional English occasions, such as the Henley and Cowes Regattas or the Shrewsbury Flower Show, would be incomplete without their displays, and fireworks are regular items on the programs of county shows and town carnivals throughout the summer. In the United States, state fairs usually boast expensive shows; the spectacular "Setting the Rhine on Fire" show is a highlight of the German fireworks calendar, although the Moog company, which for many years provided the show, no longer exists; the French resorts regularly entertain their visitors with displays; and in Italy and Spain numerous religious festivals have been celebrated with fireworks for centuries.

The English Bonfire Societies, especially at Lewes and Bridgwater, are interesting traditions. Antipapal feelings were aroused in Lewes in 1554 when, during Queen Mary's persecutions, 17 Protestants were burned at the stake in the town. Then, after the failure of the Catholic Gunpowder Plot in 1605, King James ordered celebrations throughout the country. In the 1660s diarists Pepys and Evelyn described the "great bonfires and fireworks" on November 5th. Unfortunately, the celebrations got out of control in the eighteenth century, and drunken riots caused much fire damage. To control the situation in 1853, the Lewes Bonfire Council coordinated the activities in the town, and eventually six bonfire societies were formed in different parts of the town. Stunning costumed processions featured traditional effigies of Guy Fawkes and the Pope, sometimes joined by other perceived "enemies" of Britain and contemporary celebrities. Firework tableaux depicting local and political events have become increasingly spectacular, and recently large aerial fireworks have enhanced the effect of the displays, which now take place in fields outside the town rather than in the street. The Bridgwater Carnival dates back to the Gunpowder Plot and features a unique squibbing display. More than a hundred 38-mm-diameter squibs are let off the length of the High Street.

Twenty years ago it could have been said that, as exciting and spectacular as the big displays undoubtedly were for the spectators, to the manufacturers they often were more useful as a means of publicity than as great profit-making items. At that time, most firms depended on sales of small fireworks through retail shops to accrue their main profits, and their factories were engaged mainly in making fountains, small rockets and candles, "volcanoes," and "bangers" for the general public rather than large shells and set pieces for exhibition purposes. In Britain, more than 90 percent of all fireworks were sold within the six weeks preceding Guy Fawkes Day on November 5th. Similarly, the peak sales period in the United States. was just before the July 4th Independence Day festivities. In France, the main demand comes before July 14th, Bastille Day, and in most European countries, firework parties on New Year's Eve are an old-established tradition. A combination of ever-stricter regulations governing the sale of many types of shop goods, higher production costs, and the increasing importation of inexpensive Chinese fireworks has led to a ravaging of firework manufacture in many countries. Factories have been forced to close, turn to military pyrotechnics for survival, or rely on importation to furnish their display needs. If family firework parties are less popular than in the past, the number of organized displays fired by enthusiastic amateurs has proliferated. There is now a large and growing market for "do-it-yourself" display packs. A further result of the changes is that the old established companies are having to compete with a rash of small businesses, which can buy goods from home or foreign producers, for displays or the "packs" trade, without needing to have the expertise or the plant and machinery for manufacturing. This is taking place throughout the European Economic Community (EEC), and attempts are being made to produce a unified standard. It is not as easy as it seems: Stability tests might require –30°C in Sweden or +50°C in Greece.

These trends are unlikely to change in the near future. Fireworks are potentially dangerous objects if they are handled negligently. When the instructions are not followed, accidents inevitably happen and lead to much negative publicity. Considerable opposition has been directed against the sale of fireworks, or even their manufacture, in almost every country where they are made. Certainly in the United States and the European Union, governments often have reacted vigorously, and sometimes with almost hysterical haste, in response to campaigns for greater control. In the United States, different legislation applies in the separate states. Some administrations ban the sale of all types of firework, while others allow only "safe and sane" types, which include flares, candles, fountains, wheels, sparklers, and so forth, but not those with explosive elements.

In Britain, ever-stricter laws have been imposed in recent years. The sale of shells to the public was banned in 1996, and "bangers" were banned in 1997, when it also became unlawful for those under the age of 18 to purchase any sort of firework. There are rigid laws governing the transportation of explosive materials and regulations concerning the amount of fireworks that retailers may store, the storage conditions, and the manner in which pieces may be displayed in shops. The controls within which firework factories must operate have been very severe, but those covering importation have become very relaxed since the opening of the borders in the EEC. The current regulations in Britain have come about after many centuries of change, but once again changes are taking place in 2004. At last, the 1875 Act and various other pieces of legislation have been gathered together in the Manufacture and Storage of Explosives Regulations (MSER) of 2005. As has already been seen, there were scores of small factories established by the eighteenth century, ready to supply the needs of the Pleasure Gardens as well as a growing clientele elsewhere. Long before that time squibs and crackers had been made in considerable quantities for the local general public to celebrate the traditional festivals, such as Guy Fawkes Day, Queen Elizabeth Day, or Saint John's Eve. The diarist Samuel Pepys mentioned a firework party he attended with his family after the victory over the Dutch in 1666. Many of the manufacturers came from the Continent as religious refugees, although the making of fireworks usually was a spare time occupation, carried out in their homes after a day's work in the silk or woolen industries. It was in these conditions of manufacture that the main dangers of explosions lay.

The situation was made worse after 1695, when the making of fireworks was completely banned following the antigovernment riots that flared up during the November 5th celebrations of that year. In fact, the Act made clear that "if any Person shall make or cause to be made, or sell, give, or offer, or expose to sale any Squibs, Rockets, Serpents, or other Fireworks, he shall forfeit Five Pound." Unfortunately, the legislation was not, nor could be, adequately enforced and only served to drive manufacturers to work in increasingly dangerous conditions in back kitchens, with no opportunity to extend and improve their technically illegal businesses. There was no diminution in the supply of fireworks for Guy Fawkes Night or Pleasure Gardens requirements. Gunpowder, compositions, and stars were stored around kitchens near open fires. There was no official control or inspection, since theoretically no fireworks were being made. The ludicrousness of the situation was highlighted when the authorities engaged people to break the law if fireworks were needed for public displays. The addition of potassium chlorate to the list of ingredients in firework mixtures added greatly to the potential risks involved.

Inevitably, many accidents occurred. In 1839, three people were lucky to escape death when a spark from an open fire fell on a pile of gunpowder lying on the table, and this in turn set off a barrel of powder nearby. It was recorded that an explosion shook the premises of Mr. Mortram, when some rocket stars, which were drying in front of an open fire, burst into flame and exploded. The *Weekly Messenger* of September 4, 1825, vividly describes the "dreadful explosion" in Brock's own factory in Whitechapel, London, when ". . . ten houses were seriously damaged, and over sixty had their windows broken from top to bottom," and "one poor woman . . . was so dreadfully injured by the broken glass that she lies in London Hospital without hopes of recovery." Two boys apparently had been ramming rockets when the ramrod struck against the funnel. The friction caused a spark, which fired some nearby gunpowder. Other fireworks were set off, and eventually the fire reached the gunpowder magazines. This caused the real disaster and for a considerable distance was heard "a sort of rumbling noise as if of distant thunder, and the next moment a tremendous and deafening explosion followed, and the air was illuminated with lights of various descriptions, and accompanied by continued reports."

Firework accidents occurred with frightening regularity, and the damage to property was great. Moreover, the lists of persons badly injured and killed increased continually, and these included "third parties" who were unconnected with firework manufacture. The authorities finally decided to act. To try to implement the old legislation would have been very difficult and undesirable because many people were employed in the industry and fireworks were in great demand. So the "Gunpowder Act of 1860" was enforced after 165 years of illegal manufacture. The new law sensibly laid down regulations concerning the making and storing of fireworks and the preparation of compositions. Justices of the Peace were empowered to grant licenses to those who wished to make or sell fireworks, and inspectors were appointed to make regular visits to ensure that the new laws were enforced. The Act was not perfect, but it did encourage many manufacturers to move their businesses out of the back streets and to set up bigger and safer factories.

C. T. Brock, who started his Crystal Palace displays in 1866, found that the large quantities of fireworks that he now required necessitated the building of a large new factory at Nunhead. This was built with special regard to safety, and over the next few years Brock carried out a series of experiments concerning this subject. Many basic conclusions were arrived at, including findings on such important issues as "the liability of fireworks to ignite by concussion or friction," "whether twenty yards is ample distance to prevent an explosion in one shed communicating to other sheds situated at the statuary distance," or "the liability of fireworks to explode en masse if from any cause they should be accidentally ignited." The experiments were witnessed with great interest by a Royal Commission, and the results formed the basis for the Explosives Act of 1875 insofar as it relates to fireworks.

The results of the Act were immediately seen, as the number of firework accidents decreased dramatically. Yet, one important clause had been omitted from the regulations, and this concerned potassium chlorate and sulfur mixtures. Sulfur by itself is not very reactive, and even in a finely divided state it is difficult to ignite. Similarly, potassium chlorate is a relatively stable compound at ordinary temperature and decomposes only when heated almost to its melting point of 368°C. Yet, when mixed together, sulfur-potassium chlorate is unstable and is liable to ignite spontaneously or to detonate at the slightest frictional provocation. Accidents from these causes had been prevalent before the 1875 Act and continued to spoil the new safety records thereafter. No fewer than 28 accidents and 11 deaths related to these causes occurred in the 28 years after 1875. Ironically, the last of these accidents was recorded in Brock's factory, when a man emptying crimson stars from a canvas tray into an earthenware jar suddenly found that the slight friction involved had been enough to cause ignition. Soon the whole building was alight, and the man suffered severe burning, from which he subsequently died. The authorities at last saw the necessity for a complete ban on all sulfur-chlorate mixtures. Consequently, legislation to this effect was introduced by an Order of Council in April 1894.

The Acts of 1860, 1875, and 1894 were so successful in reducing accidents that they have been the basis of the laws relating to the manufacture, storage, and sale of fireworks in Britain. Other countries have, of course, made their own laws and, in general, most have regulations similar to those in Britain as far as the more vital issues are concerned. Despite the obvious potential danger in working with explosives, the firework industry is now a remarkably safe one in most countries. Accidents do inevitably occur from time to time, and when they do, they tend to be rather spectacular ones, especially when in the form of explosions, sometimes with loss of life. In the East, where regulations sometimes are lax, until recent times explosions were a frequent occurrence, often with tragic consequences. The Madras accident of 1936 accounted for the deaths of 39 people, and the explosions in Macao in the same year claimed 23 more lives. Italy, Spain, and Portugal are now tightening their precautionary measures after a bad history of accidents. In northern Europe, the record has been good, with little loss of life. The United States, despite all its precautions where the public is concerned, until recently had a factory explosion rate that was much higher than average, although certainly not as high as that of Latin America, where regulations still seem to be liberal or are flouted by manufacturers. A number of recent accidents in China have caused quite a serious disruption to trade. One county in the south stopped all manufacture for many months and forced the closure of many factories.

Firework manufacture is an art, but for the manufacturing companies it also is a business. As such, it faces the multitude of problems that confront all other businesses in the modern, competitive commercial world. Above all, the work must show a profit, and with rising costs of materials and transport, increasing wage bills, higher insurance premiums, escalating costs, expenses of various overheads, expensive safety regulations imposed by the authorities, and the invasion of inexpensive competition from the Far East, the fireworks business in many countries is facing difficult times. In recent years in many countries, small and large firms have been finding it increasingly difficult to fight against mounting economic pressures and cut-throat competition. Takeovers and mergers often have been answers to problems in an age when the big combine can find the capital necessary for mechanization and organization in order to hold prices down to competitive levels. Sadly, many of the great pioneering companies no longer manufacture; even Brock's has gone (Fig. 1.3) and Standard has transferred production to China.

Brock's Fireworks was founded around 1700 by John Brock in the Islington Road section of London. Ironically, but appropriately, he was buried on November 5, 1720, but not before he had established the reputation of his company with major displays at the fashionable Pleasure Gardens at Vauxhall and Bermondsey. Later, regular shows at the Ranelagh Gardens further enhanced the firm's stature, and the aforementioned "Brock's Benefits" and Crystal Palace displays after 1865 continued its rise to prominence. "Crystal Palace," as they became known, moved to new premises in South Norwood in 1877, and such was the need for bigger and better works that they again moved in 1901 to Sutton, Surrey. Brock's fireworks were constantly in great demand for public and private shows. Their archive records, dating back to the Aix-la-Chapelle Peace display at Kenwood in 1749, list an illumination of the Taj Mahal during the visit of the Prince of Wales in 1875; gold and silver medals won at firework festivals in Cape Town, Brussels, and Barcelona in the 1880s; and a display from the Brooklyn Bridge in New York in 1892 to celebrate the 400th anniversary of the arrival of Columbus in the Americas. The catalogue is long and varied and includes the Dusseldorf Exhibition display on the Rhine in 1906, the Stockholm Olympics in 1912, and the peace celebrations in 1919 in all parts of the Empire. The success again had resulted in a need to expand, and a new factory was built at Hemel Hempstead in Hertfordshire in 1933. This was said to be the largest fireworks factory in the world, covering 207 acres, with 200 buildings, 60 storage magazines, housing, a social club, and playing fields. The Jubilee of King George V was celebrated with huge displays in all the colonies, and this highly profitable event was quickly followed in 1937 by the need for scores of shows for

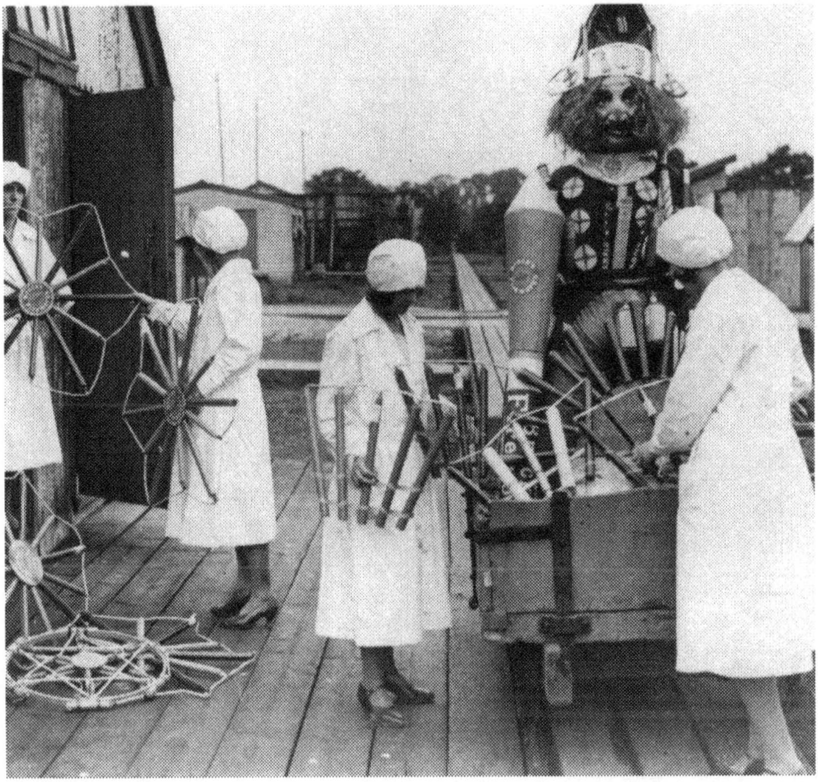

Figure 1.3 Display goods and Guy Fawkes effigy. Early twentieth century (Brock's Fireworks)

the coronation of King George VI. After World War II ended, the list continued with peace celebrations (1946), the Festival of Britain (1951), the Empire Games at Cardiff (1951), the Prince of Wales's 21st birthday (1969), and numerous independence celebrations around the Commonwealth. Meanwhile, Brock's continued to benefit from a substantial share of Britain's "shop goods" market (Fig. 1.4 and Fig. 1.5).

In 1956, a subsidiary factory was built at Swaffham in Norfolk, another was later developed at Skelmersdale, near Liverpool, and Wilder's Fireworks were taken over in 1971. But the great company was feeling the effects of economic pressures and foreign competition, as well as, it must be said, a decline in quality. Brock's was obliged to import fireworks from China and Malaysia; their display business declined. The Hemel Hempstead factory was closed in 1973, and a new one was built at Sanquhar in Scotland. By 1987, both Swaffham and Skelmersdale had gone, and Brock's was depending on the manufac-

Figure 1.4 Large rockets as sold before 1939 (Brock's Fireworks)

ture of commercial explosives, especially signals for the army, in order to remain solvent. In 1987, Brock's merged with Standard Fireworks to end an era in fireworks history and share a fate suffered by other celebrated British firms such as Pain, Wells, and Wilder.

James Pain founded his company in the 1860s, assisted by the advice and experience of his uncle, a member of a well-known family of pyrotechnists called Mortram. After manufacturing on a limited scale (and suffering several accidents) at Walworth in London, near the Surrey Gardens, where regular shows boosted their reputation, they soon moved to Brixton and in 1876 were winners of a prestigious prize for manufacturers at the Alexandra Palace. The requirements of the 1875 Explosives Act caused a move to large and modern premises at Mitcham in 1877. At about this time, Pains also built a factory at Parkville, Long Island, New York, and competed successfully in the American market, producing much-appreciated displays at such venues as Manhattan Beach and the Chicago World's Fair. Like Brock's, Pains traveled the world with their typically British, part-scenic, part-fire-

Figure 1.5 Large shell. Early twentieth century (Brock's Fireworks)

works displays, especially for royal visits to various countries of the Empire and eventually for independence celebrations. In 1898, Pains had fired a display in Lisbon to commemorate the 400th anniversary of Vasco da Gama's voyage to India. Long-standing shows in England included those for the Cowes Regatta and the Alexandra Palace, while smaller displays at various seaside resorts were regular dates for decades.

The progress of the company, however, was not without financial and organizational problems. During World War I, the manufacture of military and naval pyrotechnics and Very lights acted as a stimulus to growth, but the 1920s and 1930s were lean years, with only the coronation of King George VI to provide substantial earnings. The last of the Pain family died in 1926, and his nephews, Philip and Arthur Milholland, took over to steer the business through those difficult times. World War II again saw Pains helping the war effort with the production of Very cartridges and wing-tip flares. Peacetime brought a return to normal firework production and increased prosperity for the firm. But this was not to last, and in 1963 Pains were amalgamated with the Wessex Aircraft Engineering Company Limited (WAECO) under the direction of the Bryant and May's British Match Company. The takeover meant the closure of the Mitcham plant in 1965 and the transference of manufacture to a new, extensive, and modern factory at High Post near Salisbury.

WAECO was founded in 1933 to make wind direction smoke generators for use in civil light aircraft. As a result of the trade depression, the firm took up experimental pyrotechnic work with the encouragement of the nearby Porton Governmental Experimental Establishment. The company prospered, especially under the direction of E. H. Wheelwright, after 1937. During the war, smokes and other items were made for the government, but later pesticidal smokes (known as Fumite) and fuses for stage and film effects were added to the list of products. In 1947, Wheelwright launched into fireworks under the Wessex brand name and, with new buildings and modern equipment, made large strides. The range of products was completed in 1952 when the Wessex marine signal made its appearance.

Unfortunately, the British Match Company was more interested in military pyrotechnics than the seasonal fireworks trade, and Pains-Wessex ceased to retail fireworks in 1979. Under new owners called Chemring, and incorporating the firm of Schermuly, which already had taken over Wells's Fireworks, the business continues to manufacture signals. In 1980, John Deeker, an ex-director of Pains-Wessex, reestablished the old name of "Pains Fireworks" at a site in Whiteparish, near Salisbury, and produces displays using material purchased from other sources in England and abroad.

A third long-established English firework firm was Joseph Wells and Sons Limited. The original company was established in 1837 at Earlsfield and Camberwell in the City of London. Before this, Joseph Wells carried on the business of Public Decorators and Limelight Contractors. This consisted of decorating the streets, houses, and gardens with flags, festoons of bunting, ornamental shields and designs, little candle bucket lamps, and Japanese lanterns. Such service was in great demand for the celebration of weddings and births, at religious festivals, or for the passage of royalty or other dignitaries through towns. Before the introduction of electric lamps, gas limelights were in constant demand in theaters and concert halls, as well as for outside floodlighting and illuminations. Early in the nineteenth century, fireworks became allied to the decoration business, as they were needed to give realistic representation in mystery plays and were employed in pageants and processions held to celebrate great occasions. The invention of colored fireworks widened the possibilities of the pyrotechnic art and thus increased its popularity.

The company's expertise in firework manufacture increased rapidly, so that by 1873 Wells won the gold medal for "the superiority of signal rockets" at the Hyde Park International Expedition. In 1875, they were awarded first place for Roman candles at the international competition at Alexandra Palace. At this time, they were supplying regular displays at both the Cremorne and North Woolwich Gardens. To meet the increasing demand and to comply with the requirements of the 1875 Explosives Act, a new factory was opened at Honor Oak Park in 1878. By 1883, these premises rivaled those of Brock and Pain in their modernity. Such was the reputation of the Wells Company under the direction of Joseph Richard in the years leading up to World War I that they were known in the trade as the "Rolls Royce" of fireworks. A huge overseas business had been built up in Australia, South Africa, New Zealand, and other countries of the Empire. Wells fired many shows in Europe, and only Brock and Pain exceeded their sales in Britain.

Delight after a series of fine displays for the Earls Court Exposition in 1913 turned to dismay the following year when war broke out and all firework manufacturers were ordered to cease sales. Fortunately, Wells was able to remain profitable through government contracts for signal cartridges, parachute rockets, and smoke flares for the navy. The Peace Day celebrations in Hyde Park featured a huge display to which all the British companies contributed. After a few good years, during which the new factory at Colchester (opened in 1915) supplemented the Honor Oak production, the business suffered during the years of the Depression, along with the whole of British industry. The Jubilee of King George V in 1935, and the accession of Edward VIII, to be quickly followed by the coronation of George VI in 1937, were a "godsend" to the firework companies. In 1938, a new factory was built at Dartford, Kent, and the Colchester works was closed. Incidentally, at this time, Wells signed a deal with the Disney organization to produce a special box of fireworks with Disney characters depicted on the labels. These Mickey Mouse fireworks proved popular, but they—and all fireworks—disappeared for several years when World War II broke out. The firm again turned to the manufacture of military signals. (The Honor Oak works closed in 1947.)

The postwar years were boom times for the firework trade, and 1953 was all set to be the best of all, with the prospect of a host of lucrative public and private shows during Queen Elizabeth II's coronation year.

But, for Wells, tragedy struck in the form of coastal floods, which swept through the Dartford factory. The firework stocks (which were not insured), materials, chemicals, and paper were all destroyed. The other firework companies assisted as best they could with stock, materials, and machinery, and Wells was back in business before the end of the year. But the company never really recovered. In 1968, Joseph Wells and Sons was bought out by Schermuly of Dorking; the manufacture of fireworks ceased in 1971. In 1973 Schermuly was taken over by British Match, which already owned Pains-Wessex.

The Dartford factory was bought by John Deeker, who, in association with Unwin Pyrotechnics, manufactured display fireworks under the old Pains name. When he and Wilf Wells, the last Wells director, moved to Hampshire, the old buildings were used by Unwin's and, finally, Astra Fireworks. Thus, sadly, ended another era in the fireworks story.

In addition to Brock's, Pains-Wessex, and Wells, several other companies were established rather later in Britain, the best-known being Standard Fireworks and Lion Fireworks, both based in Huddersfield, Yorkshire. It is said that the Huddersfield trade came about through the use of blackpowder by coal miners in the previous century. Certainly, where the Yorkshire fireworks industry was concerned, the town of Lepton played a vital part. It was there that Allen Jessop, as early as 1847, enjoyed a hobby of making squibs in his cellar to sell locally for the November 5th celebrations. After passage of the 1875 Explosives Act, records refer to the factory as No. 55. After the death of Allen Jessop in 1880, it appears that the family business was continued under his son Elliott until it was taken over by the Parrett family, who in turn sold out to Standard Fireworks in the 1930s. Two other sons, Humphrey and Eli, started a new business close by, and yet another son, Ben, went into partnership with Harry Kilner, a member of another local firework family, whose mother was, in fact, the daughter of Allen Jessop. The records show that they were taken over by Standard Fireworks in 1917. The Kilners probably made the greatest contribution to the development of the industry in Lepton. Originally called the Yorkshire Fireworks Company, they became Lion Fireworks in the 1930s, folding in about 1970. The Shaws were another family manufacturing fireworks in 1865 in the area. They were listed as factory No. 32 in 1876 and traded as Globe Fireworks until the 1940s. It is unfortunate that so many of the records of these companies have been lost. The story evolves like a jigsaw puzzle with many pieces missing. Yet the scraps of information culled from existing documents and the memories of the decreasing number of people who were directly involved provide an insight into the beginnings of the firework industry in Britain. Several areas probably spawned similar groups of families involved in firework making, but those of London and Yorkshire seem to have enjoyed greater long-term success.

Yet to be discussed is the one Yorkshire company that earned the greatest acclaim—Standard Fireworks. In 1891; James Greenhalgh started selling fireworks in his drapery business in Huddersfield, obtaining the goods from the Shaw, Jessop, and Kilner factories, and having some of these made up with his own labels. He also imported fireworks from China, transporting them to Huddersfield on barges. Almost inevitably, manufacturing began around 1910, when a factory was built in an old stone quarry at Crosland Hill, on the outskirts of the town. Edward and Richard Greenhalgh, together with their sister Kate, were the driving force of the business, which became the largest manufacturer in Europe, employing more than 600 people. The Greenhalgh brothers were dedicated pyrotechnists who knew what they wanted and, like all pioneers, were not afraid to get their hands dirty. Standard Fireworks competed well with the other main companies for big and small display business, but they probably were more renowned for their "shop goods." Until the 1980s, Standard dominated the sales of small fireworks, especially for the November 5th celebrations. Ironically, Chinese fireworks, which had helped the business to become established in the beginning, were a major factor in its decline. Inexpensive imports swamped the market and sales slumped. Standard became a wholly owned subsidiary of the Yorkshire-based industrial and property group Scottish Heritable Trust in 1986. A year later, Standard merged with Brocks. Mel Barker bought the management in 1992, but manufacturing ceased in 1998. The company is now owned by Black Cat of Hong Kong.

Alongside the "main players," many small companies appear in British pyrotechnic history. We know that Frederick Copley, proprietor of a hairdressing business, was manufacturing in Sheffield in 1841. In 1881, a separate factory was built, and his advertisements included the boast that he was "pyrotechnist to their Royal Highnesses the Prince and Princess of Wales."

In the 1790s, Michael and Sarah Hengler were making fireworks to give more dramatic effect to circus acts. Soon they were supplying shows for several London Pleasure Gardens, including Ranelagh and Vauxhall. After Michael's death, "Signora" Hengler continued to manufacture, as well as perform on the tightrope "amidst a shower of fire." She died dramatically in a fire at her factory in 1845.

Hammond's Fireworks, founded by Thomas Hammond, is known to have existed between 1873 and 1974. In 1890, the company prided itself on being "the cheapest house for good fireworks" and was advertising packs at various prices from 10s 6d to £20, which were "ideal for garden parties and children's fetes."

Michael Riley established Riley's Fireworks in 1844 at Ossett, Yorkshire. The business flourished, and the name became well known in the north of England, to both retail buyers, to whom they offered "between 250 and 300 different kinds of fireworks," and the display business. Many northern towns were thrilled by regular displays at galas and seaside resorts, like Bridlington, and for special events such as the 1918 peace celebrations and the coronation of George V. It seems that the company did not survive the economic depressions of the inter-war years.

Wilder's Fireworks was a well-recognized label before being bought out by Brock's in 1971, after more than 100 years of production in Birmingham. Excelsior Fireworks was founded by Oswald Bradley near Ripon in Yorkshire at the turn of the century but moved to Southport in Lancashire in 1911. The company ceased trading in 1971. Guy's Fireworks began life as Comet Fireworks at Calverly near Leeds in 1946, the brainchild of a chemistry teacher, Hugh Allen. It changed its name in 1948 and was dissolved in 1957. Wizard was founded by Philip Rose, previously of Phoenix Fireworks, on the disused aerodrome at Chedburgh, near Bury St. Edmunds in Suffolk. It prospered with the sale of a broad range of popular retail fireworks, but it also branched out into agricultural devices and distress flares. Wizard special effects were used in major films including *Cleopatra*, *Bridge on the River Kwai*, *Battle of the River Plate*, and *The Yangtse Incident*. Financial difficulties, particularly affecting their backers, Coopal of Belgium, caused the last Wizard fireworks to be sold in 1962, the factory being taken over by Haley and Weller for the manufacture of Benwell fireworks.

The Benwell label made its appearance in 1949, when Ben Weller of the Haley and Weller toy-making company set up a firework factory at Wilne, near Draycott, in Derbyshire. In 1952, the manufacture of signals, flares, and smokes was added to the fireworks production. The transfer of the fireworks division from Draycott to Chedburgh in 1965 was followed in 1970 by Haley and Weller, becoming part of MY Dart Plc, which in 1987 became MY Holdings. A decision was made to cease fireworks production, but signals and other military pyrotechnics are still being made as part of the Chemring Group.

Astra Fireworks was another company that was born and thrived in the boom period for fireworks immediately after World War II. Established in 1947 at Bromley in Kent to make sparklers, it quickly gained a reputation and supplied fireworks to both Guy's and Benwell's. By the 1970s, Astra had taken over Rainbow Fireworks, which had been established in Berkshire in 1949, and acquired a bigger factory at Sandwich. The company was purchased from its founders in 1980, and experienced personnel from Brock's were brought in. Yet, despite improved production methods, bright new label designs, and attractive packaging, firework sales dropped and Astra moved into military pyrotechnics. All production ceased in 1992. Astra continued as a trading name, selling imported material until 1996.

These brief histories have many common threads. The old pioneering companies, established by families of enthusiasts, have all suffered and died in the modern era of high overhead costs, "big business," and inexpensive foreign competition. Moreover, the industry now lacks the "characters," such as Willy Stott of Standard Fireworks and Jack Lineham of Pains, who learned their trade through years of experience at the mixing bench, in the filling sheds, and on the display sites. An example of those who have spent a lifetime in the business is Malcolm Armstrong. He learned his trade under Edward Greenhalgh at Standard. Starting in 1946 by doing jobs around the factory on Saturday mornings, he was put on the payroll in 1950. He became factory manager at Guy's Fireworks in 1956 at the age of 21, and when that company failed, he moved to Haley and Weller's Benwell factory, eventually becoming works manager. By the time Malcolm was 26, he had moved to Astra, where, as general manager, he modernized the company and made it competitive. In the 1970s, he became interested in theatrical pyrotechnics and realized that there was a growing market in this field. With Reg Harden, an ex-Brock employee, he set up Theatrical Pyrotechnics Limited (TPL), at first in Astra sheds and later in a new factory at Manston Airport in 1984. The company is now internationally renowned and has provided material for countless stage productions and a formidable list of epic films, including several James Bond spectaculars, the *Star Wars* trilogy, *Alien*, and *Judge Dredd*. (TPL's main rival in manufacturing effects for the entertainment industry is LeMaitre Fireworks, established in 1978.)

During a period when so many British companies have been forced into mergers, takeovers, or extinction, one shining success story is that of Kimbolton Fireworks. The Reverend Ron Lancaster developed his interest in fireworks while a schoolboy in Huddersfield in the 1940s and gained early experience at the local factories. After originally intending to study medicine, he was ordained into the Anglican Church, and, after six years serving as a parish priest, he spent the rest of his career teaching divinity and chemistry as a chaplain at Kimbolton School in Huntingdonshire (now Cambridgeshire). In 1964, he set up a couple of laboratory workshops in a nearby garden in which to pursue consultancy research for Pains, and to finance his fascination for fireworks, he started to put on displays at local fetes and carnivals. The business grew quickly, and soon scores of displays were being fired all over the country, including prestige shows for the Henley Regatta, Shrewsbury Flower Show, National Trust and English Heritage Events, and royal functions. Ron Lancaster was the firework advisor to the London Celebrations Committee, which organized the Silver Jubilee celebrations in 1977. In 1978, Kimbolton Fireworks put on the display outside Buckingham Palace to mark the twenty-fifth anniversary of the Coronation of Queen Elizabeth II.

In 1988, Ron Lancaster retired from teaching but did not cease his involvement in the fireworks business. A new factory with modern technology was built, and the workforce was increased. All types of fireworks are manufactured (except rockets), and candles and shells are, as always, Kimbolton's specialties. Taking pride of place among the many awards in the trophy cabinet is the Silver Jupiter won at the Montreal International Competition in 1993. In 1995, the company staged the massive VJ 50th Anniversary display on the River Thames, when 18 tons of explosives were fired from five barges, synchronized to the same music over a distance of two miles. To add spectacle to the Handover Ceremony in Hong Kong in 1997, Kimbolton Fireworks fired a huge display over Victoria Harbour. Ron Lancaster was made a Member of The Most Excellent Order of the British Empire in 1992 for his services to the fireworks industry. He still keeps a vigilant eye over the progress of the company, but much of the day-to-day running of the business is in the hands of his son Mark, who particularly works in the retail sector; son-in-law, Stuart, the commercial director; and display manager, Darryl Fleming, while father attends to production and development (Fig. 1.6).

The rise and fall of Wallop Fireworks is an interesting story. Set up by businessman Robert Kime Shaw Porter at Middle Wallop, Hampshire, in 1946, the company owed its initial success to managing director Jack Browning. After early problems with the building program, obtaining materials in postwar England, a lack of experienced pyrotechnists, and competition from established companies, the busi-

Figure 1.6 A barge being towed up the River Thames. 50th anniversary of the end of the war in the Far East (Courtesy Leo Kennedy)

ness prospered, especially aided by a sales contract with WHSmith. Bangers were a specialty, selling under provocative labels such as "Wallop Whambang" and three-star "Biff Bang Wallop." Other intriguing titles included "Moonrakers Curse," "Titania's Treasure," and "Golden Gremlin." In association with Schermuly, the manufacturer of distress flares and smoke generators proved very profitable. However, cash flow problems eventually forced the company to abandon firework manufacture in 1954 and move exclusively into military pyrotechnics.

Before we leave this survey of British pyrotechnic companies, mention must be made of Octavius Hunt. Hunt lived in a cottage at the end of Dove Lane at Redfield, Bristol. His original occupation is not known, but as early as 1851 he had experimented with the manufacture of matches to the extent that he was able to tour the local area selling them from a wheelbarrow. The business thrived, and Hunt gradually bought up the other properties in Dove Lane so that a factory could be built. The business became a subsidiary of the Bryant and May Match Company in 1912 and turned to producing special, handmade matches, including windproof "survival" matches for the forces, fuses, and Bengal matches. The latter, often sold under the brand name of Bronco, were supplied in quantity to the firework companies, since they remained alight in windy or damp weather. In 1936, Octavius Hunt Ltd.

started making sparklers, which were sold under the Stella brand name. In 1991, the company was taken over by the Chemring Group. Besides special matches, it is the only match factory remaining in England, and sparklers, military smokes, and pesticidal smokes are the main business. In 1998, a management purchase was made by its managing director, Jan Reynolds.

The Italians rightly take a great deal of the credit for pioneering the pyrotechnic art in Europe. One of the first books on fireworks was *De Pirotecnia*, which was written by Vannuccio Biringuccio in 1540. He described the huge Renaissance feasts at the courts of popes and princes in the rich cities of Rome, Florence, Naples, and Venice and the enormous "machines," ornamented with a variety of fireworks, which have already been discussed. During the seventeenth century, famous architects such as Bernini and Buontalenti designed "machines" representing palaces, gardens, and mythical scenes as the bases for pyrotechnic displays. Few names of the fireworkers have been handed down, but in the early 1700s, the five Ruggieri brothers became famous for their shows in many European cities and began a tradition that continues in France today. By the end of the eighteenth century, fireworks were being made by hundreds of small groups of workers in villages and towns all over Italy in preparation for a few displays on saints' days and other religious feasts throughout the year. Aerial shells were used in the 1861 celebrations to mark the end of wars that preceded the unification of Italy. These became the specialty of Italian manufacturers. Some of the families of artisans developed into bigger commercial concerns and remain to this day. Viola of Sicily, Pagano of Naples, Parente and Martarello of Rovigo, Panzera of Turin, and Soldi of Florence all derived from a family tradition and represent scores of others. In the south, in Sicily or Naples, for example, there may be as many as 200 tiny enterprises. The larger, better organized, capital-intensive factories are now in the north. Most converted temporarily to military pyrotechnics during World War II, and some, like Coccia of Rome, continue in that field. Because of the strong local traditions, the displays differ widely in their content and style in different parts of the country. In the north, the wide-fronted, electrically fired shows are popular, while in the south, leisurely, but rhythmical and very noisy, daylight shows are traditional. Many small businesses are under increasing pressure. They lack the finances and technical ability to comply with the stricter laws being introduced, especially by the European Union, to reduce the frequent accidents, both in manufacturing and at public displays. Moreover, it is an increasing trend for young people to seek better paid, cleaner, and safer employment.

Figure 1.7 A plastic dove fixed to a line rocket (Courtesy Pirotecnia Soldi)

A pyrotechnic tradition in Italy is the Florence Dove. On the morning of Easter Day, the "Dove" (a line rocket) is positioned inside the Duomo in Florence. A wire, 4 m above the ground, runs to the Chariot of Fire filled with firecrackers. As the Gloria part of the Mass begins, the priest ignites the blackmatch and the Dove travels along the wire to light the chariot. The Soldi factory produces 40 to 50 such Doves each year to be used in numerous churches at Easter in the Florence area.

A large number of well-established firework companies operate in Germany, many of them making military pyrotechnics as well as display and retail fireworks. As already discussed, the German states were among the leaders of the northern tradition as early as the sixteenth century. Scores of small concerns undoubtedly thrived in the nineteenth century, some of them surviving to become the large commercial companies of today. Wolfgang Buchwald, in his superb publication Feuerzauber (1996) (41) comments that there were 54 companies in existence in Germany in 1934. Inevitably, World War II and the subsequent partition of the country resulted in the demise of several of these, together with the movement of key technicians from one firm to another. Certain areas of the country stand out as important, and it is interesting to note how the availability of raw materials, such as charcoal and potassium nitrate, had a profound effect on the original location of many companies.

The Hamburg area is a prolific pyrotechnics quarter. The firm of J. G. W. Berckholtz was established there in 1838 and was fortunate to survive wartime bombing. After the war, Wilhelm Berkholtz still owned the company, but it was leased to a Czech technician, Dr. Kustynowitsch, who was able to secure business by putting on displays for the British army. Gunpowder was even brought from England for the purpose. The doctor was ambitious and rented factory space with the Burmester company in Trittau and Moog at Wuppertal. Financial troubles led to an amalgamation with Moog-Nicholaus in 1949, and by 1964 a new plant had been built at Quickborn. Unfortunately, Wilhelm and his son, Horst, were to become bankrupt in 1975, thus ending a long company history.

The Carl Flemming firm was founded in 1910 as a fireworks company. It turned to signals manufacture during wartime and has continued in that line since, now at a factory at Hornbek. Nico Pyrotechnik was founded by Hanns-Jurgen Diederichs, formerly a colleague of Moog and Nicholaus, in 1949, on the site of the Burmester factory at Trittau. The company has flourished, with the purchase of the old Lünig firm and the securing of a major shareholding in Pyrotechnik Silberhütte. Comet Fireworks of Bremerhaven began in 1953 and now specializes in signals as well as fireworks. It absorbed Rudolf Brandt's Gebrüder Bock company at Trappenkamp.

Further south, in the Rhine Valley, perhaps the best-known German company, Moog, had been in business at Wuppertal since 1916. The present plant was established in 1924 by Wilhelm Moog, in collaboration with his stepson, Dr. Fritz Feistel, who later set up laboratories in Berlin, which grew to become the Deutsche Leucht und Signalmittelwerk Berlin. After the war, business gradually recovered, and, after the brief liaison with Berckholtz, Moog displays soon were being acclaimed all over the world. Moog was joined by Heinrich Nicholaus and Dr Feistel in 1949. The latter also helped the Meissner company to found Standard Pyrotechnik Mannheim, which later moved into a new factory at Schwegenheim, with Wilhelm Schübel as manager. The factory closed in 1982. (Dr. Feistel's son, Fritz, who also was a chemist, set up the Feistel factory at Göllheim in 1960, producing all types of fireworks and signals.)

Heinrich Nicholaus was the son of Adolf Nicholaus, whose firm, founded in 1921, had made toy fireworks and sparklers at Meiningen. They manufactured signals during the war and, with the departure of the Americans from Türingen in 1945, the family had moved to a refugee camp at Aalen. When Adolf died in 1949, Heinrich transferred to Moog, but the Nicholaus logo, a red cockerel, Nico-Hahn, became a part of the new firm Pyrotechnische Fabrik Hans Moog-H. Nicolaus.

The great annual pyrotechnical event in Germany is known as the "Rhine in Flames." Since 1955 a series of displays and other events have taken place along the river near Koblenz in August, in front of huge crowds. As the country's leading fireworks manufacturer, Moog invariably fired these shows. Klaus Moog took over the company in 1971, but financial worries forced him to sell out to the Piepenbrock Group of Osnabruck in 1981. The latter also absorbed the Feistel concern at the same time. The Feistel and Moog factories are now closed.

The Weco company, in Eitorf, prospered under the ambitious policies of their leaders, and Hermann Weber, originally from Nicolaus of Meiningen, was honored by the state for his contribution to postwar industry. He died in 1993, but the firm is buoyant with Frank Weber-Pickard at the helm. Weco was an early importer of Chinese fireworks, has interests in the United Kingdom, and in 1989 bought the firm of Wicke of Hattingen, which had manufactured since 1866. In 1990, with the cooperation of Zink Fireworks, Weco secured a major investment in Sachsenfeuerwerk, located near Freiberg in Saxony.

The Harz area has long been associated with blackpowder and fireworks. Gunpowder was made in the area in the eighteenth century, and the firm of Eisfeld was established near Harzgerode by 1855. Fireworks of all sorts were made over the years, and great efforts were made between World War I and World War II to improve the technical ability of the manufacturers. The company became famous for its fine displays at Nürnberg. As many as 700 workers were said to have been employed in military pyrotechnic manufacture during World War II. By 1973, fireworks were again in production under the name VEB Pyrotechnische Fabrik Silberhütte.

Louis Kleinknecht, who kept a grocery store, opened a firework trade business at Brackenheim in 1852. In 1884, Wilhelm Fischer, who had worked in Kleinknecht's store as a young man, took over the firework business and started a very simple production in Cleebronn. The business grew very strong in the 1920s. Shares were sold to finance this growth, but Fischer finally lost the majority of the shares and was pushed out. Fischer continued in Brackenheim but set up his own factory on Lake Constance. Certainly, at one stage early in the century, money was injected by the Eisfeld company, and in 1929 it became part of a group that changed its name to Depyfag-Deutschen Pyrotechnische Fabriken. The factory at Cleebronn must have been one of the most picturesque anywhere. Situated on a hillside among vineyards, it was greatly extended during World War II, and more than 600 workers were employed. Special lifts and conveyors had to be installed because of the steepness of the slope. It was purchased after the war, first by Dynamit

Nobel AG and later by Buck Werke in 1992; the manufacturing was transferred to Uftrungen.

This was not the end of the fireworks story in Cleebronn, for a new company had been created in 1949 by Paul Zink, who had been a foreman with the Fischer concern at Lake Constance. Walter Zink, the son of Paul, now runs the company and is a very practical man, with a high reputation for making high-quality fireworks, especially rockets.

The Lünig family was making fireworks at Stuttgart at the beginning of the twentieth century. Artur Lünig founded the firm, which has built an international reputation for big, spectacular displays, the most recent of which was designed by Matthias, a member of the fourth generation of the family. The firm went into partnership with Nico Pyrotechnik in 1992 and was closed at the end of 2003.

Clearly, as in all countries, the old family firms in Germany often have been forced by financial circumstances to close or to surrender their independence to survive. It is therefore refreshing to record the firm of Fritz Sauer of Gersthofen, north of Augsburg. Founded in 1854, the firm has seen good and bad times, but it is still under the direction of the original family.

The history of fireworks in Spain and Portugal is still largely unexplored. It appears that, as in so many other aspects of technology and philosophy, Iberia benefited from both the Christian and the Muslim traditions during the Middle Ages and the Renaissance. Gunpowder and its uses were introduced by the Arabs at an early date, but fireworks as we know them probably arrived from Italy in the sixteenth century and were used extensively for civic and religious ceremonies. At that time, guilds of *coheteros* (rocket makers) are often referred to and there are records of examinations to become a master of the trade.

Three firework traditions can be recognized today. In the Galicia Province of northwestern Spain and northern Portugal, pyrotechnists have developed rocket-making to perfection. The large examples (*cohetones* in Spanish, *foguetöes* in Portuguese) are propelled by two motors tied to a 4-meter long bamboo cane and can lift a kilogram of stars to an enormous height. They are fired singly with flights of several dozen smaller rockets interspersed. The almost silent launching of the rockets and the release of shooting stars covering the sky give a unique effect. Unfortunately, the rising cost of skilled labor, more stringent safety regulations, and the availability of shells from other parts of Spain and abroad have caused a dilution of the foguetaö tradition, and many factories in the area are now manufacturing a range of fireworks. Gabriel Rocha, the leading Galician pyrotechnist, still sells large quantities of his famous big rockets, including some with parachute effects, to other Spanish companies. In Portugal, the Silvas at Viana do Castelo and the Fernandes at Lanhelas, Minho, were the leading manufacturers. The former company is no longer in business, but the Fernandes are thriving and are famous for their intricate set pieces.

On the Mediterranean coast, in Valencia and Catalonia, a large number of companies produce very spectacular shows using the whole range of firework types and effects. The displays are called *castells*, or castles, perhaps as a relic of the "machines" of the past. Igual is the foremost Catalonian firm and won the contract to fire the display that opened the 1992 Olympic Games in Barcelona. In Valencia, the Caballer family, whose members run three competing firms (Pirotechnia Caballer, Ricardo Caballer, and Miguel Zamorano Caballer), and the Brunchu concern, founded in 1809, are leading companies. "Pirofantasia," recently formed by Bernado Sanchiz and Miguel Gonzalez, presents displays in a modern, innovative style. Also in Valencia, the Jurado family companies (Garuda and Pirotechnia Arnal) and Pirotechnia Turis, under the direction of Gaspar Lopez and Vincente Domingo, are thriving concerns.

In central Spain, displays show elements derived from both the Western and Eastern traditions. Pirotechnia Zaragozana from Saragossa, run by the Perez Sanz family, had early beginnings and is a major exporter.

The regular international competitions are important to the progress of Iberian companies because they introduce new ideas to the industry. The San Sebastian contest, started in 1964, is the most prestigious, but others are staged at Alicante, Blanes, and Bilbao, while the more local competition at the Valencia Fair in July has been an attraction for more than a century.

Fireworks manufacture in Portugal has followed a familiar trend. Family concerns supplying fireworks for local festivals gave way to large companies staging national displays, exporting abroad, and adopting modern methods. The Expo '98 show in Lisbon and the huge millennium display in Funchal, Maderia, were typical digitally controlled performances. Portugal has become renowned for the sheer number, variety, and complexity of rockets in its displays. Some have sticks more than 5 meters long. Manufacturers include the Luso Pirotecnia Group, which includes GJR, Pirotecnia Oleirense, and PB Pirotecnia Debarbeita; Pirotecnia Minhota; Viana and Filhos Brothers; Pirotecnia das Beiras; and Pirotecnia Bombarraalance.

France has a very long fireworks tradition. The Ruggieri family emigrated from Italy in the seventeenth century, and the company known as Ruggieri Artificier, manufactured in Monteux. Lacroix of St. Etienne, established in 1848, absorbed Ruggieri in 1997 and dominates the industry in France. This is especially so since old-established firms such as Berastegui and Rollet of Angouleme, Legoux of Tourville-en-Auge, Maurel of Pertuis, and several other small concerns have ceased to exist. Inexpensive foreign imports were the main cause. Seventy percent of sales are made for Bastille Day on July 14, the biggest displays being at Paris and Carcassonne. Fifteen percent to 20 percent are fired at the tourist resorts in the summer and about 10 percent are made for New Year's Eve celebrations. As in most advanced countries, shows are increasingly accompanied by music, lasers, and water fountains, and electronic firing devices aid synchronization.

In 1895, legislation was passed in Sweden to make the manufacture and displaying of fireworks safer. This effectively eliminated the long-standing and widespread practice of casual firework-making in kitchens and sheds around the country. Thus, at the end of the nineteenth century, there were six firework factories in Sweden, all of them small family concerns. C. R. Hansson had been making fireworks for a couple of years when, after three months experience at the Pains factory at London, he started his own factory at Göthenburg in 1888. Hansson's displays were famous around Scandinavia until they were forced to cease fireworks production in 1983. The company still makes distress flares under the name of Nammo LIAB. The old trading name Hansson Pyro Tech still exists for imported Chinese material.

The Hammargrens, starting as tobacconists in Göthenburg in 1879, began to import German and Italian fireworks at the beginning of the twentieth century. The business expanded, and they started their own fireworks factory in the 1920s. Financial problems caused it to close in 1992, but the company survives as an importer and wholesale dealer of "shop" goods.

Sadly, the third major Swedish company, Torner's, which was started in Stockholm in 1880 by A. R. Torner, an officer and teacher at the artillery fireworks corps, was forced to close in 1972. Thus, the recurring theme seen throughout Europe is suffered in Sweden. After the lucrative, palmy days of the 1950s and 1960s, craftmanship and skill have given way to mass, inexpensive imports and commercialism.

Gunpowder came to Denmark in the middle of the fourteenth century. The first recorded display took place in 1559 for the coronation of King Frederick II. He was so much a fireworks enthusiast that he had a book on the subject printed especially and exclusively for his son, who became Christian IV at the age of 11, when his father died in 1588. The book remains at the Royal Museum in Copenhagen, complete with notes in the margin written by Christian himself. Under the building that now houses the library is a building called the "Svovlhuset" (Sulphur House), where Christian often made fireworks himself, a hobby that was referred to in his funeral address. For his coronation, Christian had a massive display at Castle Kronborg, Elsinore, consisting of 80,000 pieces, and another huge show was arranged for the wedding of his son in 1634.

On August 15, 1843, a fireworks display helped to open the famous Tivoli Gardens in Copenhagen, and regular shows have been fired ever since. At present there are two each week, with a special celebration on the opening night anniversary. The first Tivoli pyrotechnist was C. Hoegh-Guldberg, an amateur craftsman who later, in 1865 and 1867, recorded his skills and experience in a two-volume publication. The Hoffman-Barfod family had learned their fireworks skills in Austria around 1800, before moving to Denmark. They started producing the Tivoli displays in 1938 and still make all the fireworks at their factory near Roskilde. Happily, because of the regular business, Lars Hoffman-Barfod, has been able to withstand the competition from the East, which has ruined other Danish firms. Verner Poulsen, a former employee of the Hoffmann-Barfod company, has founded a successful factory near Ringsted to manufacture indoor fireworks.

In the eighteenth and early nineteenth centuries, chemists and pharmacists in the Netherlands earned extra income by using their knowledge to make and display fireworks despite frequent accidents. Among them were the founders of companies still operating today: Schuurmans of Leeuwarden, founded in 1821, "Kat" of Leiden (1823), and Ruysch of Utrecht.

Early in the nineteenth century, Mr. Schuurmans kept a small general store and made squibs, fountains, and firecrackers for local sale. His son set up a small fireworks factory in 1821 and supplied the neighborhood with material for celebrations on New Year's Eve and for private weddings and parties. The reputation of the business grew so that in 1896 Schuurmans obtained the right to supply fireworks to Queen Mother Emma and later Queen Wilhelmina. Despite several factory explosions and fatalities and the problems posed by World War I and World War II, the company survived until the boom years after 1946, although the Schuurmans family had given up the management of the firm. (Mr. P. A. de Bruin took control between 1937 and 1998.) Following a familiar pattern, signal cartridges for the Ministry of Defence became important in the 1940s, and in the 1960s large-scale importation from China became more profitable than manufacturing in the Netherlands. The business is now a public liability company, a section of which, called JNS Pyrotechniek and directed by Hein Hofmeester, recently has set off many spectacular displays, with musical accompaniment, including the show for the fifth anniversary of the European Technology Programme in Rotterdam in 1991 and award-winning spectacles in Canada in 1991 and 1995. The accident at Enschede in 2000 has caused many problems for Dutch manufacturing. New, more restrictive, rural sites are being discussed with the industry.

The Hendrickx brothers have long dominated the fireworks trade in Belgium, although the spread of Antwerp is threatening their factory site. Van Cleemput of Brussels has ceased to manufacture. In Switzerland, Hamberger of Oberried, Bugano of Neudorf, and Muller of Kreuzlingen are well-respected names, as are Leibenwein and Pokorny in Austria. Jakob Hatteland operates in Norway.

Details on fireworks history in Eastern Europe are hard to come by. Certainly, blackpowder was known in Russia in the fifteenth century, and in 1674 rockets and firecrackers were used in a display at Ustyug. Peter the Great was an enthusiast, and displays were made obligatory at every state occasion and festival during his reign. His fascination led him to set up his own workshop to make pinwheels, rockets, and lancework. In 1693, introduced by a 56-cannon salute, he is said to have put on a huge display that included a portrait of Hercules prizing open the jaws of a lion, all in lancework. Much is written about the advances in firework technology brought about by Professor Lomonosov in the 1750s, including the means to make green stars. This period saw great advances in Russian pyrotechnics. Set pieces on the vintage Brock scale, measuring 300 feet long by 150 feet high, are described. In 1855, the Mikhaylovskaya Artillery Academy was founded as a scientific center for research into military technology, and pyrotechnic research also was performed there. Private companies and displays were not allowed in Estonia and Latvia while they remained Soviet Republics. Twice a year on New Year's Eve and the Soviet Union Army Day (November 7), large displays were organized by the Soviet army. After independence, Mihkel Paarde, a former military officer, and Aarne Uustal, who had contacts in Finland, introduced techniques from outside. The first company to manufacture was RUF Estonia Ltd., created by Ivo Melder and Andrus Kaen in 1992. There now are 10 factories in Latvia and Estonia producing fireworks for displays and selling to the public.

In Russia after the Revolution of 1917, fireworks were produced in handicraft workshops at the parks of culture with the use of old recipes.

During World War II, these workshops were closed, and beginning on August 5, 1943, the great victories were celebrated with the 107-mm "Salut-signal" round shells containing 14 large red and green stars produced in munition factories. After the war, the workshops resumed their activity in Moscow, Leningrad, Kiev, Odessa, and other cities. However, since 1957 development and manufacture of modern fireworks were concentrated on defense industry enterprises. In 1968, the manufacture of fireworks in park workshops was forbidden, and they were all closed except for the Kharkov workshop, which exists to the present day. From then until 1991, all firework displays in the former Soviet Union were conducted by the Soviet Army or by the Kharkov pyrotechnists. As a result of the conversion of the defense industry and the liberalization of the economy, private display companies became possible. At present, fireworks in Russia are manufactured by a few state enterprises. The main developer and producer of Russian pyrotechnics is the Scientific Research Institute of Applied Chemistry established in 1945 at Sergiev Posad (Moscow region). Also, there are munition factories "Signal" in Chelyabinsk, "Chapaev Production Association" in Cheboksary, Krasnozavodsk chemical factory, and Leningrad factory "Sokol" in Saint Petersburg. They produce aerial round shells of 60 mm, 100 mm, 105 mm, 195 mm, and 310 mm; star mines of 45 mm, 100 mm, 125 mm, and 195 mm; Roman candles of 33 mm and 45 mm; along with consumer fireworks. Round shells and cakes are imported from China. The formation of the Russian Association of the Manufacturers and Displayers of Fireworks, "RAPID-FIREWORKS," in 1997 has developed a code of best practice for the Russian pyrotechnists.

Fireworks are known to be freely available again in the shops of the former Czech Republic, especially for "Silvestr" New Year's Eve celebrations. This tradition was suppressed during the communist regime, when shows were allowed to commemorate only socialist anniversaries. The state-operated Zeveta Bojkovice concern had the monopoly of sales. This business now trades as an independent company known as Pyros Zeveta Bojkovice, alongside others such as Konstructa Trencin and ZVS Dubnica. The companies have invested in publicity campaigns inviting people to buy fireworks throughout the year to celebrate festivals and personal anniversaries in an attempt to boost sales and to meet the competition from the familiar Chinese imports.

Pyros Praha was, until 1975, part of Konstrukta Trencin and specialized in research and production for the army. It was then created into a separate state-run enterprise and branched out into manufacturing special explosive effects for films and television. With the removal of communist control, Pyros Praha, like other former army products manufacturers, including Explosia Semtin, Policske Strojirny Policka, and Zbrojovka Jablunka, had to solve the problems involved in the transition from military products to civil products, and to secure markets in a competitive commercial world. Pyros Praha started to manufacture fireworks. Sadly, they were forced into liquidation. The biggest firework company in Slovakia is Privatex-Pyro Ltd. It was founded in 1991 in Dubnica nad Vahom, a town with a long-standing pyrotechnical tradition, by Lubomir Masar, whose skill and enthusiasm ensured success. It has achieved a sound reputation for both shop and display goods, imports from China, and exports to EU countries.

Fireworks were introduced into the United States by immigrants. As early as 1608, Captain John Smith remarked in his history of Virginia that rockets had been let off, terrifying the Indians. In 1731, regulations to restrict their use in Rhode Island were imposed after one Benjamin Thompson was injured making fireworks to mark the repeal of the Stamp Act. In 1783, a large display in New York, viewed by George Washington, marked the celebration of independence from Britain. The Fourth of July became, of course, the country's chief firework day. The first American book on fireworks, *A System of Pyrotechny*, by Professor James Cutbush of the U.S. Military Academy at West Point, was much influenced by the British army officer Captain Robert Jones, who masterminded many early displays, and by the French experts such as Ruggieri. The earliest known manufacturer of fireworks for pleasure in the United States was the du Pont family near Wilmington, Delaware, in 1809. The business thrived until the 1940s and continued to supply powder for the industry until the 1970s.

In the nineteenth century, many American manufacturers were branches of European companies. Masten and Wells of Boston and Manhattan Beach Fireworks were established by the English firms Wells and Pain, respectively. Later in the century, H. P. Diehl and A. L. Due thrived close to the King Powder mills in Cincinnati. At this time, Italian immigrants brought new technology, especially for the making of multibreak shells. The Gruccis from Bari, based in New York, and the Zambellis (now in New Castle, Pennsylvania) and Rozzis (now in Loveland, Ohio) both from Caserta, are still major companies. Chinese immigrants organized imports from China, especially firecrackers, at an early date. The Communist authorities broke off trade in 1948, but fireworks continued to move through Hong Kong and Macau, and from firms in Taiwan. In 1972, new trade links were established between the United States and the People's Republic, of China and vast quantities of fireworks of low price, but often dubious quality, began to flood the market. Zambelli Internationale claims to fire nearly 1,500 shows around the country on each July 4th. Almost every major city puts on a spectacular show on that day, the most notable being around the Washington Memorial in Washington, D.C., off the Esplanade at Boston, on the lakefront at Milwaukee, and at the Macy's show in New York City. New Year's Eve is the second most popular time for displays, and then there are innumerable shows for state fairs, feast days, and regular or one-off local celebrations. The Venetian Night at Chicago, the Cincinnati "Riverfest," and the "Harborfest" at Newport News, Virginia, are notable displays.

Regulations have inhibited the U.S. fireworks industry since the beginning of the twentieth century when, after a series of accidents, some of them fatal and some concerning children, vigorous anti-fireworks campaigns ensued. A combination of restrictive legislation and economic depression in the 1930s caused many firework companies to close. Some companies turned to making military pyrotechnics during World War II, but afterward even more restrictive laws destroyed much of the market. Moreover, the regulations differ from state to state. Added to this burden is the increasing amount of litigation when accidents occur. Ridiculously high monetary awards are made to successful claimants, no award of costs is made against unsuccessful plaintiffs, attorneys' fees are huge, and liability insurance premiums have escalated and sometimes are unobtainable. The manufacturers' plight is deepened by the increasing invasion of both "shop" goods and display fireworks from China, Taiwan, and Japan. Probably as much as 85 percent of fireworks fired in the United States come from Asia.

As one would expect, displays have benefited from innovative technology. Music has become a familiar part of shows, with companies such as Austin Fireworks becoming famous for its "Concert in the Sky" performances. Computerized multiswitch firing boards are commonplace. The $2 million 1986 Statue of Liberty display in New York used 40,000 shells fired electronically from barges. The Disney theme parks in Los Angeles, California, and Orlando, Florida, probably are the largest users of fireworks in the United States and have invested in equipment that propels shells by compressed air rather than gunpowder.

There is increasing demand for fireworks in the United States, but high production costs, together with the problems outlined previously, have caused many firms to go bankrupt and others to cease manufacturing in order to organize shows using imported materials. Some of the best known companies, apart from those already mentioned, are

Precocious Pyrotechnics of Belgrade, Minnesota, directed by Garry Hanson, and the Rozzi family business in Loveland, Ohio.

Hands Fireworks is the leading manufacturers and display company in Canada. In 1873, William Hand emigrated from England, with pyrotechnic knowledge acquired in employment at Woolwich Arsenal, and set up a firework factory in Thorold, Ontario. His first public show was for the "Whit Monday" public holiday at St. Catherines on May 25, 1874. This was the beginning of a catalogue of "Pyro-Spectaculars" that were produced almost every year for half a century. The shows were massive, involving not only fireworks on a grand scale, but also large casts of actors, acrobats, jugglers, and marching bands, performing in front of colossal sets more than 500 feet long and 50 feet high. The epic themes included "The Siege of Sebastopol," "The Last Days of Pompeii," "The Taking of the Bastille," "Cleopatra," and many commemorating events in Canadian and world history. The final items included a barrage of high-level fireworks and huge lancework set pieces depicting "The Mapleleaf" and "God Save The King." The displays took place regularly at Toronto, where more than 120,000 people would watch the 10 or 12 sold-out performances in any one year, but also at large festivals in the United States and other major Canadian cities, including Montreal, Quebec, Hamilton, Winnipeg, Ottawa, and Halifax.

In 1875, Hand moved from Thorold to Hamilton to satisfy the need for larger premises. A partner, Walter Teale, died in a fatal explosion in 1900, and a year later William Hand failed to recover from burns in another factory accident. His son, who was known as TW, took over the running of the business. In 1919, the company bought up a small concern called Great West Fireworks in Brandon. By 1930, the Hamilton urban area had grown so much that the fireworks factory had to move. By purchasing the small Dominion company in Cooksville, Hands acquired a large rural site for a new factory and the considerable technical experience of Dominion's former owner, Dominic Carson. TW died in 1931 and his sons, Hugh and Bill, guided the firm through the war years, when military pyrotechnics were manufactured, and into the boom time afterward. Urban sprawl caused yet another move in 1963, this time to Papineauville. Although the Hand family sold the business in 1977, the company is still called Hands Fireworks Inc. Under the leadership of Richard Brown, the firm makes military pyrotechnics, smoke signals and rescue flares, theatre maroons, and unique "polar bear scaring devices," as well as a wide range of fireworks. Government legislation, which has virtually prohibited the importation of Chinese material, has helped Hands and other Canadian companies to maintain sales and quality.

Other Canadian manufacturers include Ampleman Pyrotechnics of Montreal; Blue Smoke Fireworks of Calgary; GSH Pyrotechnics; and Specialised Pyrotechnics; which acquired the factory at Kenilworth when Canadian Fireworks expired in 1997. The Butchart Gardens in Victoria are now famous for their use of fireworks, and the prestigious international competition is organized by La Ronde in Montreal each year.

There must be hundreds of fireworks factories in Latin America. Mexico certainly has a strong fireworks tradition. It is estimated that more than 3,000 festivals each year use fireworks there, many of them featuring the *castells* of eastern Spain. Besides the innumerable local fiestas, the two national festivals—Independence Day in September and the day of the Virgin of Guadeloupe in December—are celebrated across Mexico. Articulos Pirotécnicos Mexicanos of Mexico City, Fuegos Artificiales La Union/Azteca Fireworks of Guadalajara, and Lux Pirotecnia are well-known company names. In Brazil, Caramuru of Sao Paulo are the only manufacturers left after the effects of the Chinese "invasion," both in Brazil and in terms of reduced exports to Canada and other South American countries.

The best-known name in fireworks in Australia is Howard's. Sydney Howard experimented with compositions in his spare time and, after firing small shows for the local community, set up a factory at Eastwood, New South Wales, in 1922. Five years later they moved to bigger premises at Waterloo Road, Epping, New South Wales. Son Harry joined the business, and soon the company earned a high reputation all around Australia, displaying at agricultural shows and in the big cities for major celebrations. Following the usual practice, they made military pyrotechnics during World War II, after which they returned to the manufacture of commercial fireworks, with Harry's sons, Leslie and Sydney, joining the firm. In 1965, a modern factory was built at Box Hill, New South Wales, and this site doubled in size by 1980. A successful excursion by Leslie and his wife into the importation and sale of Chinese "shop" goods was thwarted by government regulations banning the sale of "shop" goods in Victoria in 1982 and in New South Wales five years later. This legislation also put an end to that area of Howard's business.

In 1985, Sydney left the company to set up his own firework business called Syd Howard Fireworks International, in competition with the established company, which then became known as The Original Howard and Sons Pyrotechnics. The latter has enhanced its reputation with displays for independence celebrations of Nauru, Fiji, Soloman Islands, Vanuatu, and Papua New Guinea. Scores of displays marked the bicentennial of European settlement in Australia. Prestigious awards were won at the international competitions in Toronto and Vancouver in 1991 and 1992. Howard's fireworks are sold widely in the United States and are used extensively at the Disney theme parks, and the firm is rightly proud of its traditions. Also in New South Wales, Foti's International Fireworks of Leppington are successful manufacturers and displayers.

The first firework manufacturing in Japan probably started about 1620, although displays were performed there before that date. It is said that the word *Hanabi*, the Japanese word for firework, was used in 1585 for the first time. "Hana" means "flowers" and "Bi" is a softened sound for "Hi," which means "fire," and "Flower of Fire" is the Japanese name that corresponds to the word "fireworks." It is recorded that in 1613 a messenger from James I of England put on a fireworks display for Ieyasu Tokugawa, the founder of the Tokugawa government, although it is uncertain whether the fireworks were introduced into Japan by the Dutch or the Portuguese, together with firearms, at about this time.

In 1659, Kagiya, the famous fireworker, began his work in Edo (now Tokyo). In 1733, the renowned displays on the River Ryogoku in Edo were originated in commemoration of a Buddhist service for the many victims of a cholera epidemic in 1732. The fireworks were fired from ships, and the displays took place more or less annually until 1963.

In 1810, another famous fireworker, Tamaya, left the Kagiya concern and set up his own factory. However, in 1843, the Tokugawa government ordered him to remove his factory from Edo after the outbreak of fire for which he was responsible. Firework factories developed in many other parts of Japan under the patronage of feudal lords.

For a long time Japanese fireworks were made only from blackpowder, but in about 1880 the introduction of potassium chlorate from Europe allowed colored flame compositions to be used. Another milestone in Japanese firework history came when Gisaku Aoki introduced the double-petaled Chrysanthemum in 1926, and two years later, in the fireworks display to celebrate the enthronement of the Emperor, Aoki displayed a triple-petaled Chrysanthemum, with the center of the flower in red, the inner petals in blue, and amber around the edge. After this, multipetaled Chrysanthemums were widely developed and became the country's most representative firework. Japanese manufacturers also are noted for the very large shells they produce. A 36-inch diameter shell takes many weeks to make and may contain several layers of stars and smaller shells.

A large percentage of the fireworks produced in Japan used to be exported, but this is no longer the case; the bulk of them are used internally. This is because they require so much expensive hand labor that the finished products are very costly. There are many small manufacturers of shells that sell their products to the larger companies or to exporters. Marutamaya Ogatsu possibly is the largest company with an international reputation for exporting and displays. The original factory at Fuchu near Tokyo may date from the seventeenth century, but the present site is in the Yamanashi prefecture.

Hosoya used to be a name respected for shell production, but the company now concentrates on signals and commercial items such as golf balls that issue a stream of smoke when struck. The Koa company is another well-known signals manufacturer, having dropped its fireworks division. As with Europe and America, Japan imports considerably from China and Taiwan, but, unlike Europe, there are still more than 100 large and small companies, such as Koa, Sunaga, Mantsuna, and Aoki (Fig. 1.8).

China usually is considered the birthplace of blackpowder and firework making. Hundreds of factories supplied material for displays celebrating religious festivals and public events, and firecrackers, fountains, and candles marked family births, marriages, and funerals. Ironically, all large displays were banned during the years of Communist austerity, although exports were encouraged, especially through Hong Kong. Recent political changes in China have meant that many manufacturers are now allowed to invest private capital and trade unhindered across the world. Certainly, China leads the world in the volume of production and export. A large number of small companies, especially in the Guangdong, Guanxi, Hunan, Jiangxi, and Hebei Provinces, feed the large exporters and display companies, which operate under such group titles as Temple of Heaven of Beijing or Red Lantern, and brand names such as Sunny, Vulcan, and An Ping. As has been discussed, the invasion of cheap Chinese fireworks into almost every firework manufacturing country has had profound effects on domestic production, often to the extent that little remains of the list of original companies.

Taiwanese shells are now respected and are familiar additions to many displays. San Tai Fireworks, Mantsuna, and Wan Dar supply many of them. Displays take place during the Moon Festival in August, during the Chinese New Year in February, and for the birthday of a Buddhist deity called Ma Cha in May.

Figure 1.8 Ron Lancaster and Takeo Shimizu at the Hosoya factory in Tokyo

Early records in India refer to a display in 1443 in the Mahanavami Festival. Today fireworks feature in the numerous Hindu ceremonials, especially The Festival of Lights in October, and at Hindu weddings. There are scores of small producers across the country and a few large factories, such as Standard Fireworks, Sri Kaliswari Fireworks, Ravindra Fireworks, and The Arasan Fireworks Factory, which was established in 1937 and boasts that it is "one of the leading manufacturers and exporters of KUIL Brand Fireworks Varieties to South Africa, Finland, Italy, etc."

There are a multitude of small differences between the products of manufacturers in different countries, and indeed within the same country, which probably few but the experts would notice. The vast majority of spectators at displays will be thrilled by certain effects and take away a general impression of the presentation as a whole. The expert will see much more. He will know through experience what the manufacturer was attempting to achieve with a particular firework, what some of the problems would be, and to what extent success has been achieved. The experienced worker would be able to compare objectively the quality of the colors of stars, the efficiency of the rockets, or the spread of a shell burst. He would notice whether rockets and shells had burst before the zenith had been reached or after, and whether candle stars had been blown high enough or landed burning on the floor. Pyrotechny obviously is a potentially dangerous pursuit in which, unlike most other hobbies, few can "dabble." Hence the number of enthusiasts and experts is small. Better communications have led to an easier exchange of ideas and information between them, and international festivals and competitions, such as those at Montreal, Bilbao, Cannes, San Sebastian, and Hanover, encourage this interchange. Where insularity may have perpetuated localized characteristic effects and styles of presentation in the past, intercommunication tends to diminish this tendency.

Sadly, the number of true experts and enthusiasts has been eroded dramatically over the past few decades. This chapter has traced the demise of the bulk of the old-established companies in almost every country. The effects of mounting production costs, government safety regulations, and especially the relentless "invasion" of inexpensive goods from China and Taiwan have caused business after business to turn to military pyrotechnics for economic salvation or to become importers and distributors of material from the Far East. Moreover, scores of concerns, often run by individuals who have no expertise or background in firework manufacture, have sprung up to share in the increasing market for "do-it-yourself" packs and organized displays. Indeed, the escalating cost of fireworks and deeper concern over safety have brought greater demand for large shows.

In fireworking, progress used to be thought of in terms of creating new effects. By experimentation, a new chemical composition might produce an original star color or noise effect, or a new shell burst or set piece pattern might be achieved. Today progress is measured mainly in terms of the spectacle that can be accomplished with the use of vast quantities of fireworks, sophisticated electronic firing devices, and the enhancement that musical accompaniment and lasers can give to a show. With so few manufacturers left in many countries, fewer young people are learning the skills of a successful pyrotechnist. There are, therefore, many question marks over the future of the industry, but, without doubt, while fireworks remain popular as the perfect spectacle for celebration all over the world, displays will continue to be fired.

REFERENCES

1, 2, 3, 24, 29, 37, 39, 41, 46, 51 and grateful thanks for special details from Sergio Soldi, Dr. Franco Baudacco, Walter Zink, Dr. Uwe Krone, Dr. Felix Gõni, Björn Söderberg, Lars Barfod, Hein Hofmeester, Michael Swisher, Robert Cardwell, Ettore Constabile, Campbell Innes, Yasuhiro Sashimura, Dr. Roland Wharton, Dr. Ivan Oleynik, Ivo Melder, Antonio Rodrigues, Laurie Jackson, and L'ubomir Masár.

2

FIREWORKS DISPLAYS EARLY EVOLUTION TO THE MODERN APPROACH: A PERSONAL VIEW

Mark Lancaster

Much of this book is dedicated to the principle and practice of the manufacture of fireworks, which generally remains much as it was 50 years ago. Perhaps the only exceptions to this would be the introduction of limited automation, together with the use of some of the more modern chemicals, such as the chlorine donors helping to improve color. Despite this, fireworks manufacture can be considered a mature process that has experienced few fundamental changes in recent years. The same cannot be said for the manner in which displays are fired, the methods of which have advanced significantly over the same period with the introduction of electronic firing systems and new lighter weight firing equipment.

In this chapter, initially using the United Kingdom as the example, I examine how the method of firing displays has changed in recent years and how to some extent, especially in larger displays, traditional methods have been abandoned to be replaced by a more generic global approach. Finally, almost as exceptions to prove the rule, four countries, which to some extent have resisted this pressure of change and stuck to their traditional way, are studied. These countries provide a refreshing change to much of what can be seen today.

EARLY DEVELOPMENT

In the modern age, entertainment is piped to us in our own homes and the available options available make leisure time relatively easy to fill.

This was not the case 100 years ago, and certainly in the United Kingdom, much time was spent at the "pleasure gardens," the earliest of which was founded as long ago as 1750. There were several in London, principally at Vauxhall and Ranelagh, then later and more famously at Crystal Palace and Belle Vue in Manchester. The pleasure gardens are long gone, the only traditional one remaining being the Tivoli Gardens in Copenhagen. It was, however, these gardens that provided the first permanent sites for frequent firework displays that became fashionable in London around 1870, when the Brock family first seized the opportunity presented by the increased patronage at the Crystal Palace site.

The advantage of the permanence of these sites did much to establish the early character of British firework displays. By modern standards they were quite incredible in their extravagance. At the core of each display were elaborate set pieces, the likes of which are never seen today. These could be 28 meters high and run to an average length of 60 meters, although this could be extended to more than 180 meters. Considering the unpredictability of British weather, I marvel at the thought of how exciting it must have been to see on these great frames the reenactment of the Battle of Trafalgar, the Eruption of Vesuvius, or indeed the great transformation pieces. The work involved to make these pieces must have been immense, the manufacture of thousands of color-changing lances, together with half-length lances, and the enormous time it must have taken to assembly such pieces. I envy my father, who perhaps saw the very last display of this type when he was a boy, performed by the Brock company at the Crystal Palace shortly after World War II (Fig. 2.1). Perhaps the only modern equivalent, and much smaller in scale, would be some of the transformation pieces used as part of the Mexican *castillo*.

The displays held at Belle Vue in Manchester immediately after the war captured much of the style of the time. On the site itself was laid a set of steel rails, similar to a train track, which led at either end to massive wooden sheds. Here the various lancework pieces, such as the typical Wild West bar scene, or replica locomotive, could be prepared and stored, protected from the elements, ready to be rolled out immediately before their use. Another characteristic of these shows was the "Weird White Waterfall." The Belle Vue site had a wire cable from one end to the other, along the length of which could be strung a waterfall. Each stick had its own igniter, and at either end were two mighty maroons and two enormous benzoate whistles of approximately 30 mm bore. On ignition, everyone jumped with the bang and then squirmed from the inter-

Figure 2.1 Erecting lancework at the Crystal Palace, London (Brock's Fireworks)

action of the two powerful whistles. The whole display had the degree of care and preparation that we sometimes lack today. It was not uncommon for even small displays to take up to three days to prepare on-site, with a special marquee being erected for this purpose. Much of the work that we now do before we leave the factory took place on-site, with many of the lancework frames being prepared from scratch.

Rockets were among the more frequently used traditional British fireworks that accompanied the magnificent lancework. Indeed, if truth be known, Brock's, for one, were not very skilled at shell making. Rockets seem to have been frequent troublemakers over the history of displays. Perhaps this is one reason why, apart from Spain and the magnificent examples found in Portugal, they are infrequently used in modern displays. (As I write, following this year's ban on the sale of star shells to the general public in the United Kingdom, their popularity has staged something of a comeback.) In earlier times, rockets were a very important feature of any display in much the same way that they are used in Spain. English rockets were big, usually with a bore of either 18 mm or 24 mm. These rockets were fired in groups but, unfortunately, as we know, out of all fireworks that are projected into the sky, the rocket provides the greatest of dead weights to fall back to land. When fired in such volume, this inevitably presented a problem. This partly explains the demise of rockets in large public displays; another reason is perhaps the relative inaccuracy of timing when firing a large number of rockets at once. A mass of flight rockets is a unique effect and, where we can, we do still use them, although, alas, the opportunities are becoming ever scarce. Roman candles also were a feature of many displays; they provided careful tapestries and mosaics of color, which could be done by the use of long-tailed comets and the careful placing of the candles along the full width of the site.

Another traditional British firework that has lost favor in recent years are set pieces. Normally stuck on top of an 18-foot wooden pole, they comprise just about every static ground piece imaginable, and their use overlaps with the magnificent lancework devices mentioned earlier. Wheels, saxons, and gerbs—all forming differing mosaic patterns—have always been a firm favorite with children and offer an interesting and different alternative to much of the aerial effects of displays. They are, however, both susceptible to the weather and difficult to waterproof, as well as being laborious to construct. They also suffer from being difficult to see by all but those in the first few rows of the audience. They are another traditional firework of which we see less and less in all but the more traditional country fair–type show. The main exception to this rule, and one of the reasons for its inclusion in this chapter, is their use in Malta, where the variety and quality of the examples produced must be among the best in the world.

THE MODERN DISPLAY

Having identified the traditional elements of a British firework display—intricate lancework devices, set pieces, rockets, Roman candles, and star shells—it is interesting that perhaps only the last two are still extensively used and can be seen at all current displays. The other types are still used, depending on the site. The modern display often must perform to audiences of much greater size than those of the past, which also can restrict choices in the use of some types of firework.

Perhaps the biggest factor in shaping modern displays has been time. Despite an ever-greater increase in facilities designed to make better use of time and cut the length of time needed to achieve various tasks, it seems to become an ever more precious commodity. The majority of our displays are now rigged and fired on the same day. The majority of the fusing and preparation takes place in the factory, long before the display site is ever reached, leaving only the final assembly and placing of Roman candle barrages and racks of mortars into their positions on the display site itself. Gone are the days when marquees would be erected on-site and great teams of men would spend several days in preparation. Even a relatively large show now can be prepared by just two or three men in a day. Certainly, the electrical firing of displays helps, but even now it may come as a surprise to some that we still fire smaller shows by hand. (As it happens, electrical ignition is not new; there are records of its being used to initiate the "Royal Portraits" that were fired from the royal box at the Crystal Palace as early as 1879.) The great long trenches that once were dug to bury the mortar tubes are no longer required, and we recently replaced all of our firing stock with new, lighter, easy-to-handle, fiberglass tubes. The new working practice is now so accepted that groans are heard from display operators if they are asked to erect simple lancework saying "GOOD NIGHT."

The basic formula for our modern displays relies on Roman candles and star shells. In its simplest terms, the Roman candles are set out along the width of the display site and are fired continuously to form the basis of the display. Star shells are then fired over the top of the Roman candles to add power and variety.

Roman candles invariably burn for approximately 30 to 45 seconds. (Naturally, if necessary, they can be made to burn for a shorter period by varying the number of shots. It is, however, more difficult to have them burn longer because of the limitations of the tube.) The length of the duration of the candles is used to stipulate the length of each of the different "sequences" of the show. Imagining the area of the sky where the display will be fired as a picture frame, the Roman candles will tend to occupy the full width of the frame and the bottom third of its height. The nature of Roman candles allows the creation of an enormous range of patterns in this area by carefully positioning frames and using long burning stars that leave trails. The mosaics or patterns are perhaps the greatest characteristic of modern British style shows. Probably the biggest mistake is the overuse of candles where the sky is too full, and the result is a mass of stars forming no pattern at all. Comet candles probably are used more because of the precise effect they produce rather than the slightly more random results of a bombette candle.

Except in the largest shows, star shells are not fired continuously, as seems to be the pattern for typical modern displays. Clearly, it is a matter of taste, but using star shells only seems restrictive, as one is forced to rely purely on the variety between the effect of the shells, and lose the added dimension that the use of Roman candles can give. I accept, of course, that in some situations the site and audience dictate that it is almost impossible to use anything other than star shells (e.g., Hong Kong Harbour). Star shells normally are used in the last 20 seconds of a 45-second sequence, color-coordinated with the candles. We start with smaller caliber shells, say, 3 inches in diameter, in large numbers, slowly building up to the larger caliber shells, although we rarely fire anything larger than a 10-inch shell. Often the last shell of a sequence, timed to fire as the Roman candles finish, is a long-burning shell, acting as a bridge to start off the next sequence. (It is interesting that many Japanese believe that the optimum size for a star shell is 8 inches.)

Back to our imaginary picture frame, it is important to try to vary the show as much as possible. By using the Roman candles, the base third is always full, but with differing patterns. The top two thirds are full sporadically, the central third being filled first by smaller caliber shells, and then slowly the upper third is filled as large shells follow. It is always possible to elevate people's feelings to great heights in a show, but it is very difficult to keep them there, which is why this variety in effect is so important.

From a color perspective, the show is split into the same 30 to 45 second sequences, each one of which is color-coded, using some of the traditional combinations—violet and green, red and silver, blue and yellow. More recently, however, we have begun trying to use some of the newer in-between colors such as turquoise, lilac, and citrons. Sequences may be overlapped with a transition along the spectrum (i.e., from citron to green). The rule we do try to enforce is that we will succumb to the dreaded multicolor sequence only at the finale. In our experience, it is all too easy to end up with a messy "salad" of colors that allows you to show off nothing to its best effect. Another mortal sin is to have a wonderful golden willow shell completely obscured by brighter low-level fireworks. By the end of the show, we aim to have used each of our effects once, but only once.

Perhaps the greatest trend of recent years has been the introduction of the pyromusical. I would estimate that 65 percent of our total shows and nearly all of our summer shows are now accompanied by music, which offers many advantages and adds an extra dimension to the show. When attempting to play with the audiences' emotions, much can be gained with the addition of music. There is now a growing culture, in Britain at least, of audiences eating a picnic on the grounds outside a stately home or castle in England, listening to a classical orchestral concert live. Then at dusk, the last piece, normally something "original" such as the *1812 Overture*, is accompanied by fireworks. The potential of this style of show is enormous; quieter moments accompany Roman candles, and as the music grows, so too do the star shells, ending in a tremendous cacophony of sound, light, and color. Performed well, it can be a wonderful sight.

Anyone will tell you that the synchronization of a firework display to a live orchestra is not easy. A conductor never plays the piece at exactly the same rate (often I am convinced this is out of spite, as they normally resent the presence of fireworks distracting the audience from their beautiful music). It is not surprising, then, that most of us prefer to fire to a DAT recording, which at least can be relied on to be consistent in its timing. By some, I am sure that we would be considered antiquated, but for the average show we still rely on a manually operating electrical firing system. Notwithstanding this, we spend considerable time working out our firing cues, for which we do use a computer, but the use of the millisecond timing computers are reserved for the large elaborate displays. Damp November nights do strange things to computers and pyrotechnical delays alike, and, apart from enjoying the sanctity of ultimate control, sound and light do not travel at the same speed. We still rely on the audience to do 50 percent of the coordination for us. This is, of course, helped by the initial careful selection of music, which ideally is popular, patriotic, or emotive.

THE MODERN GLOBAL DISPLAY

If people read the preceding and wonder why I believe it uniquely describes a typically British display, when in its broad form, it could also describe a typically French, German, Italian, Dutch, Canadian, or American display, this is precisely my point. There is little doubt that nearly all of the large important international displays held around the world are very similar in style, whoever performs them. This was a point drummed home to me in 1997, when I saw three of the largest displays I have ever seen all performed in Hong Kong over a period of three months—one by Kimbolton Fireworks, one by an Australian company, and one by an American company. While different in detail, they were the same in style. The site had something to do with this, but not everything. It clearly exemplified the ever-increasing generic style of firing firework displays, particularly with the heavy, some might argue, overreliance on star shells.

Firework makers have always been guarded over what they consider to be their own formulas and production method. Some even delude themselves that they are unique in what they do. The reality, of course, is that individual development has been replicated simultaneously in different factories all other the world. Many of the skills and production methods individuals think are unique to them are in fact duplicated by others. It is true that formulas and working practice are not identical everywhere, but based on the guiding principles that we are all forced to follow in fireworks, strong similarities are inevitable. That said, there is little doubt that if you travel the world, you will see an enormous variety of quality in the fireworks produced. It is, after all, much easier to produce a bad blue star than a good one. Indeed, not every factory has the advantage of a hydraulic press and a carefully gauged increment board enabling it to produce consistently performing fountains, but clearly some practices are better followed than others.

The exception always proves the rule, and perhaps one area where development has genuinely continued independently until recent years with interesting results is China. Here, after so many years of isolation, some unique and different effects have been developed, and it is only in recent years that we in the West have been exposed to them. The intricate toy fireworks are good examples, as I doubt if there is any factory in the Western world that would have the time or the economic inclination to make such pieces.

The modern world no longer allows people to work in isolation. People are exposed more and more to other people's products, and the exchange of knowledge probably is greater now than it ever has been. In the past 20 years, we have begun to attend conferences, such as those organized by CANMET in Canada, where knowledge has been shared by those around the world. From a safety point of view, this exchange of information must be a good thing, as unsafe working practices, such as the chlorate-sulfur admixture problem, are slowly being superseded by safer practices. It would be wrong of me to suggest that the first edition of this book helped precipitate this process, but I know that it was believed by the author at the time that the greater the common knowledge, the potentially safer our industry would be. He always added with a wink, though, that he had not put all his best formulas in print.

In addition to the production process, the same evolution has occurred in public display of fireworks to a greater or lesser degree. With some noticeable exceptions, which we will look at shortly, we are slowly moving away from national characteristics in the performance of firework displays to a more generic approach. The exchange of production methods has been mirrored by an exchange in display methods. The growth of international firework competitions, such as those held at Montreal, Toronto, Stockholm, Cannes, and San Sebastian, have given a much greater opportunity to display operators around the world to examine the various techniques available to them. In the early days, this must have been a good thing, and the contrasting styles were often the source of the genuine interest in the competitions. I think to some extent that this is no longer the case, and we are in danger of adopting a more prosaic generic approach. This is partly an understandable consequence of competition organizers insisting on a basic format for displays (in their duration and the site available for the display to be fired), but factors channel styles into an acceptable generic format. I doubt, for instance, if the site or the audience at La Ronde in Montreal would really allow the intricacies of a Mexican *castillo* to shine.

A second factor that has led to this convergence of styles is the products themselves. The economic situation in the world has slowly forced more and more firework factories around the world to close and has resulted in an overreliance on Chinese material. (Certainly, in the United Kingdom, firework factories after the war could be counted in the teens; now Kimbolton Fireworks is the only one still manufacturing display fireworks.) With the production of fireworks so heavily dependent on manual labor, it is an understandable consequence, but it also is a mistake. One day the economic advantage enjoyed by China will be eroded and the existing market segment for distinctive national fireworks will grow. The concern, of course, is common to all craft industries. Once a factory closes and the knowledge is lost, it can never be regained without an enormous degree of effort. This overreliance on Chinese material naturally has resulted in shows looking the same. Unfortunately, a Chinese 3-inch peony shell is going to look the same wherever it is fired in the world. It is no coincidence that, in the examples at the end of this chapter, those countries that have resisted this globalization of style have all retained a healthy national firework manufacturing base.

Traditionally, firework displays were performed by the manufacturers of fireworks themselves. Clearly, it is not the case today, as by far and away the majority of "firework companies" to be found in the telephone book have never even added a lifting charge to a shell, let alone manufactured one in the strictest sense of the word. In their defense, with the dominance of Chinese fireworks, this is inevitable. Since the majority of advancement in the firework world has been in the display side of the business, it can be argued that the displaying of fireworks is now an entirely separate skill. It is, however, a contributing factor to the slow erosion of individual national style. I believe that it is still an advantage to understand and participate in the manufacturing process when it comes to designing displays, for only with the thorough understanding of a firework that comes from its manufacture can it be put to its best use in a display. There is little use in designing a show that relies on split-second electrical ignition if you fail to appreciate that a cold damp November 5th night is likely to add several hundredths, if not tenths, of a second to the delay fuse of a large star shell. On the basis that they are now separate skills, I hope people will forgive me if I now draw a slender line between a true pyrotechnist and a display operator.

The majority of firework displays around the world were traditionally linked to religious festivals. Indeed, in all of the examples cited later where there is still a strong pyrotechnical style, this is still the case. In recent times, of course, certainly within Western Europe and the United States, this is no longer the case. Countries still have their national days, but it is corporate clients who now dominate and with them a certain style of show. While doubtless November 5th is still our busiest single day for firework displays in the United Kingdom, the volume of business is dwarfed by the demand over the summer for corporate product launches, balls, pyromusicals, and other such events. Any corporately driven event must conform to various requirements, normally for it to be short and sweet, resulting in an all-aerial display of relatively short duration where it can be difficult to impress the audience with anything less than a massive array of color and noise. This requirement contributes to further eroding that distinctive national style of event.

To a degree, it is at the smaller displays where national style can still be easily displayed, and I would not wish to be accused of painting with too broad a brush. I cite here four examples of countries that still retain a strong independent identity in the techniques they use when displaying their fireworks. These are by no means the only examples; Portugal, with its enormous rockets, could just as well be used as an example, although at this time I have chosen just these four. However, it is interesting to note that, where distinct character has remained, it often is due to the close ties these events still have to religious festivals. It seems that the modern corporate dollar has to take much of the blame for the erosion of tradition.

SPAIN

The typical Spanish firework is perhaps the easiest of all European fireworks to identify. What it used to lack in finesse, it made up in power. Star shells are a good example. Unlike the Japanese, who rely on layer upon layer of paper on the outside of the shell to provide the resistance before breaking, the Spanish rely on brute force, literally exploding the case and forcing the stars out at high speed. The result is characterized by a slightly ragged burst, but one of intense power and tremendous noise. Spanish shells have always appealed to me, not least because they invariably are of a single color and provide an interesting, if more expensive, alternative to Chinese products. We have had a long association with Spain, and I remember well the passion that seems to go into every Spanish display I have seen. The brutal power of a display is matched by the fast and furious pace with which it is fired. If the French are the figure skaters of the firework world, then the Spanish are the ice hockey players, as their displays rely on tremendous volume and power and are designed to literally stun the audience. Displays are primarily aerial-based and, with the exception of the mascleta, less use is made of ground effects, although Roman candles and mines are used to supplement the large spherical shells. The Spanish also are masters at making interesting cylindrical effects shells, many of which seem to culminate in some form of noise!

The Spanish also make considerable use of small flight rockets. Normally with a diameter of about 14 mm, these are fired en masse, from either large metal cones or bins. The effect is stunning, but the flight of the rocket tends to be erratic, with an odd one shooting off on the horizontal toward the audience. It perhaps is not surprising then that the Spanish audiences seem to treat the potential risk of injury of going to a firework display as one of the hazards that you accept by your attendance. One only has to witness the melee in any Valencian square at the mascletas to realize that the principal reason for going is a love of fireworks, rather than to make a claim against the display companies' insurance.

The majority of Spanish displays at home revolve around religious festivals. Perhaps the best known of all the Spanish festivals is the Fallas de Valencia, held in the city over six days, normally during the Feast of Saint Joseph on March 19th. In an age when our bureaucrats seem determined to legislate the British firework industry out of existence, this festival makes our Guy Fawkes Day appear like a public safety demonstration. My first experience of the festival was in the early 1990s, when I went with my father. My first impression was one of astonishment, as the Valencians simply seemed to be prepared to set fire to their own city. From the first night, while fire crews stood by, flames sprouted upward and the ancient streets disappeared into clouds of sparks and smoke, signifying that the start of perhaps Europe's greatest single firework festival had flared into action.

The origins of the festival are blurred. The word itself appears to stem back to the period when Valencia was occupied by the Moors, having its origin in the word *facula*, meaning firestick or torch. Records dating back to King James I mention "fallas" or bonfires to mark civic occasions, such as a royal birth or a visit by a neighboring monarch. Dating from the sixteenth century, the region surrounding Valencia also is famed for its celebration in honor of Saint John and Saint Anthony, by the building and consequent burning of high pyres or "fogueres i fogates" in January and June of each year. It is known that these "fallas" were simply barrels filled with firewood and tar, placed on top of a pole and painted, at first with grotesque designs, and later with a hunch or "buito," the epitome of the modern papier-mâché figurines that can be seen in the modern festival (Figs. 2.2 and 2.3),

The modern festival within Valencia itself seems to owe its origin to Valencia's Guild of Carpenters. At the end of winter, cleaning up their workshops of shavings, useless pieces of wood and other leftovers, they choose to celebrate on the eve of the day of their patron saint, Saint Joseph. Over the winter months, the carpenters worked using their

Figure 2.2 March fallas, Valencia, Spain

Figure 2.3 March fallas, Valencia, Spain

lanterns, or "cresols," suspended from wooden candelabras to light their work during long winter nights. This item was then burned on the bonfire on Saint Joseph's Day, as it was no longer required. The festival grew and the community added all of their own unwanted items to the carpenters' bonfires to celebrate the "Falla de Saint Joseph." In time, the satyrical spirit of the Valencians resulted in the dressing up of the torches with a jacket or a pair of trousers. Slowly the grotesque effigies began to form.

Each district of the town now produces its own "fallas." Initially each was made spontaneously by the community, but as competition between districts to produce the best effigy has increased, a more professional approach was adopted. As early as 1903, districts had advanced to using wax to sculpt the hands and face rather than simply using gloves and a mask. Now artists especially employed by a district will first make a sketch, followed by detailed models of the proposed work. Once the idea is approved by the district committee, proper work begins with the making of plaster of Paris molds for the various figurines. Still very much a community job, work on each of the fallas can take up to a year, with up to 200 artisans involved, not to mention the countless painters and casual laborers. The results can be enormous. The statue in Figure 2.4 is over 80 feet high. Traditionally, all the fallas were burned at the end of the festival, but since 1935, by popular demand, the fallas voted the best each year are saved from the fires and exhibited year-round in the Monteolivete Museum.

Figure 2.4 Paper sculptures ready to be burned, Valencia fallas, Spain

Considered a working class festival, and despite being originally shunned by the middle classes, it was almost inevitable that the effigies would soon be used as a medium to poke fun at those in authority. The winning effigy in 1996 came from the Velluters quarter, the old silk district. Their model, at 105 feet, apparently was one of the largest ever seen. Entitled "Outlaws," it ridiculed Spanish political life, corruption, and injustice. Desperate Dan, cowboys, a pig, and a vulture all played their part in conveying the message. It was said to have cost the district more than £70,000 to complete.

No one sleeps during the week of the fallas. In the city, streets are packed with hundreds of small girls and women processing toward the Plaza de la Virgen and a 30-foot Virgin Mary, to whom they deliver flowers. Leading each district is their "fallera," the queen of the district, who will compete for the title of fallera mayor, the queen of the fallas. The processions are impressive, with everyone dressed in traditional

Figure 2.5 Paper sculptures ready to be burned, Valencia fallas, Spain

Valencian costumes passed down through generations. But the real stars must be the effigies themselves—remarkable towering papier-mâché political statements, often with a heavy dose of satire and rich in sexual innuendo (Figs. 2.4, 2.5, and 2.6).

As the sun sets, the marching bands heat up and drums become even louder, echoing around the city. The climax of each evening is the firework display fired promptly at midnight on the banks of the river in the center of the city. Traditionally, a different company fires each display over four evenings. While not technically a competition, there is little doubt that rivalry is fierce and little heed is paid to budgets. Historically, each of the displays was fired by a Spanish company, although on occasion an overseas company has been invited to fire one of the displays. It is a relief that, despite economic pressure, the majority of the fireworks used are still Spanish in origin. Heavy emphasis is placed on single color sequences and novel effect shells, invariably

Figure 2.6 Paper sculptures ready to be burned, Valencia fallas, Spain

ending with gratuitous noise. They are capable of making excellent bright color stars, but both their Roman candles and star shells are characterized by the golden trail left behind in both the break and the black-powder delays of Roman candles. The Valencians are a knowledgeable audience, and you are left in little doubt as to their views on any particular show. It is not that they cannot appreciate some of the subtleties in color and effect that fireworks can offer, but one feels their appreciation of a display is directly related to the noise it produces.

MASCLETA

As if to confirm a love for noise, every morning of the festival one's slumbers are interrupted not by a chorus of chirping birds, but by the "despierta" (a chorus of firecrackers proclaiming that the day's festivities are about to begin). As if unable to wait until midnight for their next dose of noise, crowds flock to the town squares for the daily "mascleta" at 2 o'clock in the afternoon. Probably the most famous mascleta occurs in the Plaza del Ayuntamiento. Mascletas are simply strings of firecrackers of all shapes and sizes, woven on wires stretched across the square. The power of this event is indescribable. Pregnant women are advised not to attend. This event undoubtedly is the scariest sight I have ever experienced. More than 200,000 people crammed into that square to watch thousands of these mascletas fired around them. The crowds love it, and burns seem to be taken for granted. Once started, you feel for 10 minutes as if the world is about to come to an end. It took four days for my hearing to recover.

The origin of mascleta is thought to have been in the use of gun salutes honoring the start of civic or religious celebrations. The "masclet" itself was a small tube approximately 4 inches long with a diameter of 1 inch. The tube was baseless, being fired vertically on its bottom using the ground to provide resistance. The loud noise was produced by the heavy confinement of the blackpowder using plaster plugs. These devices were relatively easy to make, and their sequential firing gave birth to the modern mascleta.

Further development occurred with the introduction of potassium chlorate flash mixtures, allowing for large bangs to be produced without the confinement, resulting in modern mascleta being prepared in flimsy cardboard tubes. Old gunpowder trails have been replaced by modern blackmatch connecting each banger (Fig. 2.7).

The basic principle of performing a mascleta display has remained unchanged through the years. It is born through a combination of sounds through time, attempting to keep a distinctive rhythm. The show itself can be split into three sections: the beginning, the main body, and the end.

Beginning

Traditionally, the beginning consisted of at least 10 aerial maroons to mark the start of the display. In modern times, smoke, whistles and hummers have also been added.

Main Body

The main body consists principally of the ground-based elements of the show, the long strings of mascleta held on wires at head height, allow-

Figure 2.7 Mascleta—March fallas, Valencia, Spain

ing their sound to reverberate off the floor. The fast tempo gradually increases to the "traca final" or "earthquake."

End

Coinciding with the last effects of the main body of the show, the display heads to the air again, with a series of explosions that are of greater frequency than those on the ground. The frequency gradually slows, but as it does, the power of the explosion increases until the deafening finale.

Of all the European countries, it is perhaps Spain that has maintained much of its unique character in its use of fireworks.

MEXICO

Mexico is obsessed with fireworks. It has a strong tradition, with estimates of over 3,000 festivals using fireworks every year. Probably the largest of these fall on September 15th, or Independence Day, and December 12th, which is the day of the Virgin of Guadeloupe. Not surprisingly, some of the largest displays are held in Mexico City, where thousands of spectators gather in the central square for the annual display. Away from the city itself, one of the towns best known for its pyrotechnical attraction probably is Zumpango, about 70 kilometers north of Mexico City, where every year on December 8th, the feast of the Immaculate Conception is held. The event takes the form of a fiesta in which several different local companies take part. While not a competition in the strict sense of the word, it nonetheless is highly competitive, with the honor of producing the best display highly sought after. The Mexicans share a love of noise with their Spanish cousins. Their displays, however, take a considerably different form. Mexico is renowned for its ability to produce some of the most delicate and intricate fireworks to be found anywhere in the world. At a time when labor costs are soaring, it is to their credit that they have continued to dedicate the many hours necessary to produce the delicate and highly engineered pieces that adorn their "castillos."

The most important and central part of any Mexican display undoubtedly is the "castillo," the Spanish word for castle. The exact date this unique pyrotechnical tradition was started is not known. It is believed that the Spanish brought it with them around the middle of the sixteenth century. With this in mind, it is not surprising that the castillo holds much in common with the Spanish "fallas"; both share a similar heritage and role in acting as the focal point for the festivities and subsequent display (Fig. 2.8).

The castillo is a structure between 15 and 20 meters in height, and it resembles an obelisk. Other types based on poles and triangular structures can be found around Mexico, but the obelisk is the most common. Traditionally, it is built of wood, but in recent years aluminum has begun to be used to aid the construction process. The castillo is laborious to construct. Up to 10 people performing different tasks are needed in its construction, with the average assembly time being three to four hours for factory prefabricated examples. The construction sequence begins with the transportation of the pieces to the place of the show. Larger communities have their own dedicated firework teams called "coheteros," and local semipermanent castillos. This feature alleviates the need for transportation and allows the building of more complex pieces, some of which may take several days to construct. The various elements of the castillo are assembled, and anchors are placed at the four corners to stabilize it. A complex series of ropes is used to place the various wheels on the frame, and each segment is slowly added from below, forcing the structure ever higher. Each segment is traditionally secured together by "lazos," a special tarred string. At the top of the frame, and perhaps the most important element of the castillo, is the crown, which acts as the finale to the display. Once the main body is in place, the various lateral or side wheels are put into position.

Figure 2.8 Mexican castillo (Courtesy David Hall)

Finally, the most delicate part, the massive wheels, which can be up to 12 meters in diameter, are placed in front of the whole structure. The wheels are incredibly elaborate and can burn for several minutes. They are masterpieces of pyrotechnical art that rely heavily on pyrotechnical delays to transport their fire from one effect to the next, and the centers bear some resemblance to the old English transformation pieces.

The modern aluminum castillos have enabled the Mexicans to manufacture bigger wheels, still the core of many of their displays. Another advancement in recent years has been the use of electrical ignition. Negating the reliance on pyrotechnical delays, as has been the practice of the past, electrical ignition has helped synchronization, but it also has eroded some of the skills necessary in controlling the display through the ingenious use of these natural timing devices. The "castillo" is formed by an assembly of wheels of different arrangements, which are then set off symmetrically, together with arrangements of several types of gerbs, fountains, pinwheels, cascades, mines, Roman candles, and girandoles. Perhaps some of the cleverest and most unique fireworks are the "tenedores." Best described as a collapsible string ladder, they consist of lances secured to each end of a small tube. Each tube forms the step of the ladder, and the strands are then gathered together in a bundle and secured to the castillos. Once lit, the lances ignite, burning their tethering and allowing the ladder to drop down to form a row of lights. These often change color, and by careful placing and separation of these lights, different patterns can be formed on the "castillo."

Tradition dictates the order of the display, which normally starts at the base of the tower. Sometimes more than one tower is used, with two smaller structures on either side of the taller central structure. Initially, the lower lateral wheels on either side ignite and slowly the fire spreads upward to other elements. Next, the larger central castillo wheels ignite. These often have color-changing lances to add to the effect. The cores of the wheels contain differing images of, say a lancework Virgin, which will then be replaced by a second image as the time fuse passes to another set of lances. The climax to each castillo is the lighting of the corona on top, which takes off into the air. After a short pause, the finale moves away from the castillo with the firing of star shells of differing calibers to bring the display to a dramatic and noisy conclusion.

A traditional effect at many Mexican displays is the "lluvia," or rain. Similar to, but not quite the same as, a long burning glitter, these stars are transported skyward either by the large "voladora" wheels or rockets. At their apex, the lluvia is ignited and proceeds to float earthward with its distinctive trail. It is common for this rain to be accompanied by a mass of maroons, giving a heart-stopping experience common to many a Mexican display.

The castillo is the core of all Mexican displays and normally is accompanied by "toritos," or bulls. Toritos are wooden constructions embedded with a variety of small wheels and crackers often carried by children. They are highly dangerous by our standards. The children proceed to roar about the square dancing and weaving through the spectators leaving a trail of chaos behind them. More affluent villages also may have a massive waterfall strung along structures in the town square. Most festivals invariably are concluded with an aerial barrage of large-caliber star shells (Fig. 2.9).

While all Mexicans seem to share a love of fireworks, no two Mexican festivals are identical; many are based on castillos, but there also is considerable regional variation. Far from Zumpango, there is a festival at San Miguel in the center of the country, where up to 10 castillos are built and are purported to be the largest seen in the country. In the south, there is a unique festival at the village of Chiapa de Corzo, where mock Spanish galleons carry out sea battles with a blaze of fireworks.

Figure 2.9 Mexican fireworks (Courtesy David Hall)

JAPAN

Japanese star shells generally are considered to be among the best in the world. The perfectly spherical bursts, bright colors, and accuracy of the color change stars are the hallmarks of a high-quality Japanese shell. The manufacture of these shells is discussed in detail in Chapter 23.

Japanese displays are unusually long by Western standards, with shows taking between one and two hours, but apart from the shells themselves, probably their most unique feature is the manner in which they are fired.

In principle, there are two different firing methods for star shells: ordinary shooting and quick shooting. We shall look at each in turn.

Ordinary Shooting

This basic shooting method relies on a mortar tube very similar to the one used in the West, but differs in the system of ignition. In the West, we are accustomed to a bag of gunpowder lifting charge to be secured to the base of the shell, from which a leader of piped match fuse extends over the brow of the mortar. To initiate the shell, this fuse is lit by a portfire, which in turn transfers the fire to the lifting charge. Alternatively, a squib or electric ignitor may be placed directly into the bag of lifting charge.

This is not the system used for the ordinary firing technique in Japan. Here a measured quantity of gunpowder is dropped directly into the mortar. The shell is placed on top of this. Once the shell has been carefully lowered in the tube, a further small quantity of powder called "powder spray" is dropped on top. The powder spray helps to transfer the fire down the side of the shell. A special slow-burning star, called "sindoro" (a rough translation of its name would be "cinder"), approximately 3 mm × 3 mm × 30 mm in size is pressed. This is then lit and dropped down the tube, its fire being transferred to the lifting charge by the powder spray. After firing, the mortar is cleaned of any remaining debris and the process is repeated.

Quick Shooting

Quick shooting is a different format. A bag of gunpowder lifting charge is attached to the base of the shell, as in the Western method, but no leader is added. The top of the shell is connected to a short wooden handle, so the result looks very similar to the World War II German stick grenade (Fig. 2.10).

To fire the shell, first small coils of iron are heated in a charcoal fire until red-hot. These coils are then placed at the base of the mortar. The shells are then dropped into the mortar by use of the handle, obviously

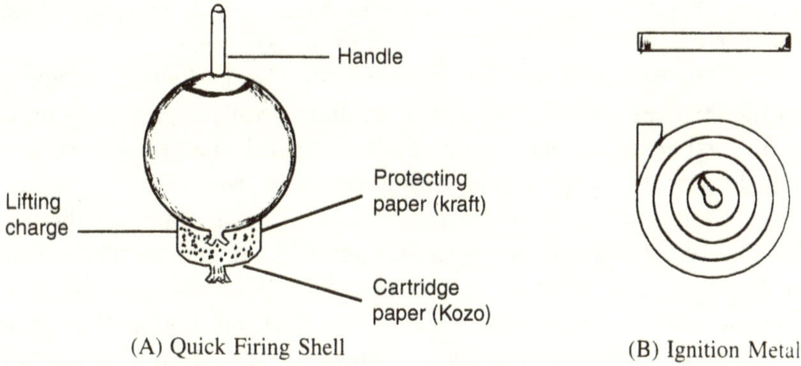

Figure 2.10 Quick firing shell and ignition metal

on contact with the hot coils, the lifting charge is ignited, and the shell rises into the air. It is possible to light and fire about 10 shells in quick succession using this method (Fig. 2.11).

The mortar used for this method of firing is rather larger than that used for other methods in an effort to ensure that the shell does not stick to the sides when lowered. This is especially important because cinder is liable to collect on the walls. Obviously, this system of firing is more dangerous than the ordinary system, but it does have the advantage that it enables shells to be fired relatively quickly from a small number of tubes. In addition, it prevents the premature ignition that can be caused by long lengths of piped match connecting series of shells.

MALTA

The association between Malta and fireworks is well documented. The majority of displays almost exclusively revolve around festivals held by individual villages in celebration of their own patron saint. For example, the village of Luqa celebrates the festival of Saint Andrew, and Kirkop celebrates that of Saint Leonard. In a staunchly Catholic country, the church is still the cornerstone of the community, and hence these festivals are viewed with great importance. Matters can be complicated when individual villages have two patron saints, invariably resulting in two sets of followers and their associated musical bands. Many of the firework factories (of which there are approximately 35 licensed in Malta) are based on these band clubs, and enormous rivalry is all but guaranteed. The factories traditionally manufacture solely for the festival, and all funding comes from donations collected at the previous year's festival. Such is the importance of the festivals to the community that they are run as nonprofit-making enterprises and the workforce is made up of volunteers from the community.

We have been to Malta around August to see the fantastic displays for the festivals, and I have little doubt that, as well as providing some of my first experiences of fireworks outside the United Kingdom, they inevitably will be some of my lasting memories. My family's long and close relationship with the firework makers of Malta started in the late 1950s. My father's early visits to St. Michael's firework factory in Lija evoked suspicion because he was perceived as a young cleric from England who was obsessed by fireworks. A strong relationship grew throughout the years. Kimbolton Fireworks is one of the few companies outside Malta that has imported the wonderful fireworks made on this

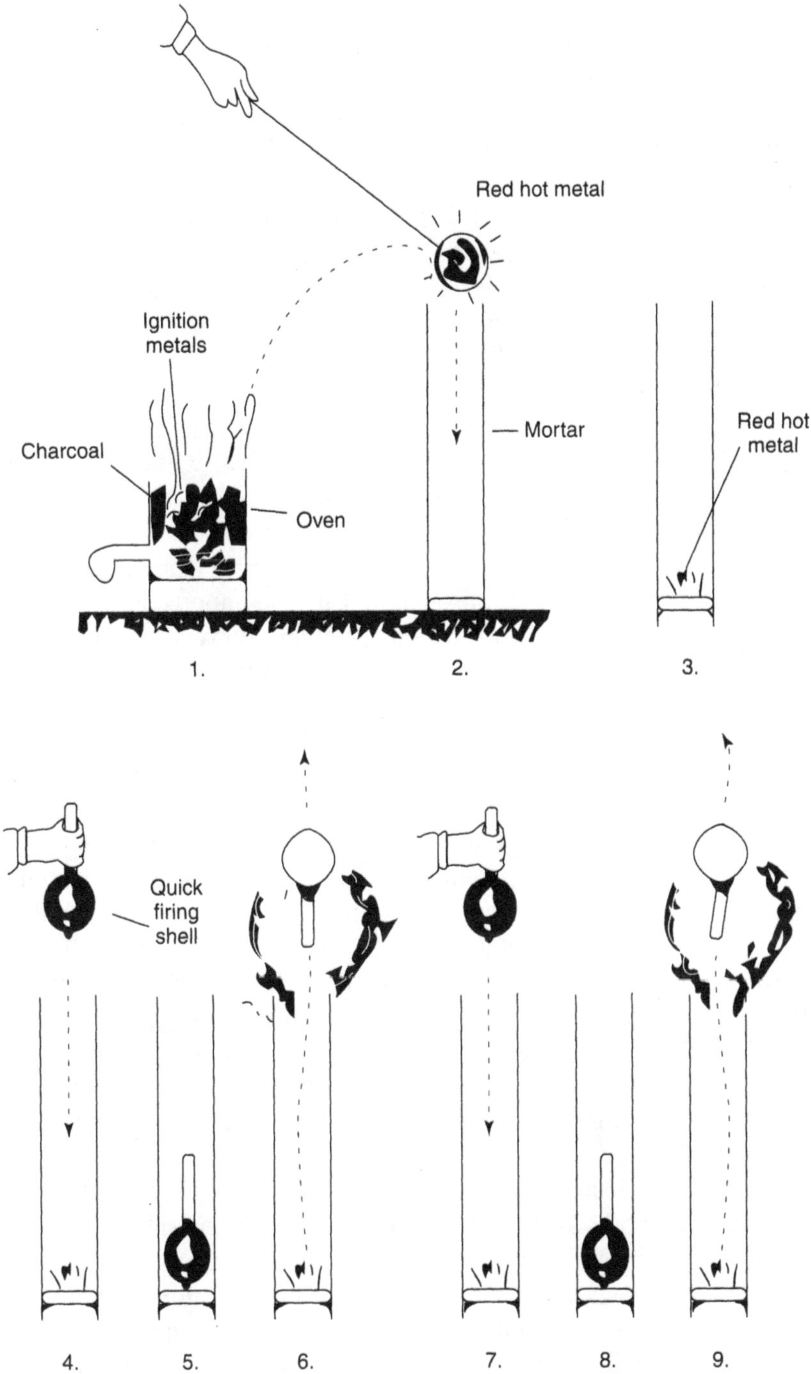

Figure 2.11 Quick shooting operations (Courtesy Dr. T. Shimizu)

small island. (Odd as it may seem, for many years no money ever exchanged hands as Kimbolton Fireworks supplied some of the chemicals used in manufacture.)

I have many happy memories of being packed off on my own to Malta for my summer holiday to take part in the festivities. (As fireworks seemed to revolve solely around alcohol, this was something of an adventure in itself for a 14-year-old.) In later years I have almost begun to feel embarrassed that I use fireworks as a source of income, as at St. Michael, like many other factories in Malta, the majority of time dedicated to manufacture is voluntary and unpaid, stemming from a passion and selfless dedication to the community festival perhaps unequaled in any similar event anywhere else in the world. There is little doubt that this dedication is fueled partly by the strong sense of competition—mostly friendly—between the rival villages and, in extreme cases, rival groups within villages.

Maltese fireworks rely heavily on this volunteer labor. By modern standards, at a time when traditional manufacture in many countries is failing under pressure from inexpensive Chinese imports, Maltese fireworks are uneconomical to produce. They are, however, some of the most intricate and wonderful fireworks to be found anywhere in the world. Owing much to a strong Italian connection, the majority of fireworks produced are complex multibreak cylinder shells. The majority of the production seems to be 6 inches in diameter, although both smaller and larger calibers are made. Each break of the shell often is only a single color, although wonderful married stars and splitting stars also are made. A Maltese shell is characterized by a relatively tight burst when compared to a Japanese shell, but it has very intense color. The Italian influence continues as the last break of any shell is a maroon. In common with many others, my favorite Maltese shells are the repeating maroons in which breaks of smaller multiple maroons, called crackers, are repeated with a final enormous single maroon to finish. These cracker maroon shells, anywhere from two breaks upward, never fail to stun audiences in both Malta and the United Kingdom alike.

For people accustomed to the more traditional American or European fireworks display of relatively intense but short duration, Malta offers something refreshingly different. Festivals traditionally last up to three days and can be split into different sections. While fireworks accompany the buildup to the festival, the first major event is the removal on the first day of the statue of the patron saint from the church, accompanied by fireworks. Unlike in the United Kingdom, daylight shows are common and play an import part of the daytime activities, the show is noise-based with ample use made of the cracker maroon shells mentioned earlier, and smoke shells are used to add color to the proceedings. What makes Malta different is the manner in which the displays are fired. Apart from the finale to the display, fireworks are fired singularly, with a considerable gap of several seconds between firings. To the eye accustomed to the more manic overlapping way that shows are fired elsewhere, this appears almost amateurish. This is far from the case, but it certainly allows each shell to be appreciated without its beauty being marred by an infringement of color by another firework. Almost as beautiful as the fireworks themselves are the amazing snaking lines of mortars, laid one after another in a never-ending line slowly winding its way up the hillside forming a continuous stream of fire, interrupted only by a darting figure, cigarette in hand, entering or leaving one of the several bunkers acting as a sanctuary for the firers. (I always recall Joe Portelli telling me how difficult it was to get the firers to use these sanctuaries, as each was determined to see the product of his year's work. This I sense, however, is a problem of firework men the world over) (Fig. 2.12).

In firework terms, a world away from the raw power of multibreak cylinder shells are the rows of intricate lancework devices that line the streets of Lija. Their closest firework cousins would perhaps be the Mexican castillos, but unlike the Mexican castillos, whose partial charm lies in their totality, each of the lancework devices in Malta is a masterpiece in its own right. Traditionally associated with the second day of the festival, each is an example of firework engineering at its best, as even in daylight when the various pulleys, wheels, and lattices that are obscured at night are seen, it is almost impossible to decipher how each works. My personal favorite is perhaps the first I ever saw. Simple in effect, it consisted of three different colored circles that shrunk in turn and then grew in diameter, each inexplicably forming from the inside of a larger circle only to take its place in size. Difficult to describe in words, it is best imagined as the firework version of one of Escher's drawings (Fig. 2.13).

The display itself seems never-ending. The crowd stands in awe as each is fired, only to rush down the street past the freshly extinguished piece to be in the best position to see the next. As a child, I was secretly pleased at the gradual demise of lancework devices. Those were always a feature of English displays, forced out of the display by the ever frenzied requirements of busy corporate launches where shows lasted only three minutes in a blaze of corporate color. But having

Figure 2.12 Finale shell battery, Malta

experienced the set pieces in Lija, I am determined and convinced that they do have a place in many of the modern shows at more traditional venues, demonstrating how simple engineering can be just as impressive and dramatic as that hidden away inside black firing boxes.

The highlight of the festival is the third day, normally during the late afternoon, when the statue is paraded through the streets for several hours accompanied by enormous amounts of noise. Finally, it is returned to the church, where it stays for another year accompanied by an overwhelming finale of fireworks.

There is little doubt that Malta is a unique place, where fireworks in their purist form are enjoyed for their own sake, by some of the friendliest and skilled pyrotechnists I have ever had the pleasure to meet. Certainly, one of my dying memories will be parading behind the Statue of Our Lady, around the streets of Lija, the sounds of the village band deafening in my ear, with a bottle of beer in one hand and a bundle of donations for next year's display in the other. At the time, as a 14-year-old, it was my idea of heaven!

Figure 2.13 Set pieces in the street, Malta

Figure 2.13 *(continued)*

3

GUNPOWDER

It would be fair to say that the use of blackpowder in the firework industry is a splendid example of resistance to change as we look at the technological advances of the nineteenth and twentieth centuries. The manufacture of blackpowder as a seriously mechanized industry did not begin in Europe much before the sixteenth century, but thereafter it grew dramatically, with 60 to 70 known sites in the United Kingdom alone over these 300 years.

The Chinese consider that the properties of gunpowder were discovered during the period 475 to 221 BC with the rise of interest in alchemy and the urge to perfect the making of both gold and the elixir of life. During the Han Dynasty (206 BC), alchemy became increasingly popular, but precise details of the origins of gunpowder are not clear because this was not what the alchemists primarily were looking for.

By the fourth century AD the search for the elusive elixir was still active. Saltpeter was still the main ingredient, but as the experiments could be somewhat explosive, aqueous methods were also in vogue to produce the elixir utilizing saltpeter and realgar. The fusion method used mixtures such as saltpeter, realgar, resin, and hog intestines; it was found that the mixture could be quite explosive under some conditions.

The first serious formulations for gunpowder as we know it seem to stem from writings about AD 808, when sulfur and charcoal clearly entered the trials, although the object was not necessarily to produce explosions as much as to subdue them. It was reported that mixtures of saltpeter, sulfur, realgar, and honey created fires that burned hands, faces, and houses. The addition of charcoal to hot mixtures of saltpeter and sulfur certainly would have been inflammatory!

The documents of AD 808 establish that the first formulation for gunpowder worked out as 46% saltpeter, 46% sulfur, and 8% charcoal. It seems that the charcoal was from Birthwort (*aristlochia debilis*).

It has always been stated that the origins of blackpowder are Chinese, and much of the evidence has been found in books printed in the West. We are happily now in the position to have more authentic evidence from a document forwarded by Professor Changgen Feng of the Beijing Institute of Technology. The authors of the paper "The Origin and Development of Ancient Gunpowder" were Yang Shuo and Ding Jing (42).

Blackpowder is described as "a bright pearl in the ancient Chinese civilization" with its three components being known from the earliest of times. Charcoal was already used in the Bronze Age in China and was recognized as a fuel for the reduction of metal ores. Sulfur was recorded in the sixth century BC, and in other documents it was regarded as remarkable for its ability to dissolve metals such as gold, silver, copper, and iron. It also is featured in medicine.

There is considerable evidence of the existence and knowledge of saltpeter. Once again, its reactivity with metals and its use in medicine in the elixir of life recipes established its importance in these times. In later times, the Arabs called saltpeter "Chinese snow" and the Persians called it "Chinese salt." There is no doubt that this protogunpowder can burn well and produce quite a quantity of gas.

While there have been several writings concerning the use of gunpowder and weapons in India and in Muslim countries, Professor Feng Jia Sheng, an authority on the study of gunpowder, is of the opinion that the origins were in China. It is possible that by 1225 gunpowder techniques were transmitted from China via India to the Muslim countries. Later on, the Arabs then transmitted gunpowder to Spain. By the middle of the thirteenth century, it was transmitted to Europe. It is suggested that Roger Bacon discovered gunpowder through translations of Arab books.

Early Chinese formulations recorded in 1044 work out at approximately 60% saltpeter, 30% sulfur, and 10% charcoal, but formulations include sesame fiber, bamboo fiber, arsenic, lead salts, oils, and resins. The assumption is that the organic materials would need to be carbonized. The addition of lead and arsenic compounds may suggest a poisonous smoke of military nature.

Table 3.1 indicates that the proportions are getting nearer to the present.

Inevitably, the biggest problem in some parts of the world, not the least in England, was the availability of potassium nitrate. In *The British Explosive Industry* (24), Brayley Hodgetts tells how vegetable and animal refuse and the sweepings of slaughterhouses were mixed

Table 3.1 Ming Dynasty (1368–1644) Tables

	Saltpeter	Sulfur	Charcoal
Night gunpowder	45.5	9.0	45.5
Flame-throwing gunpowder	74.3	3.9	21.8
Cannon gunpowder	78.7	7.9	13.4
Small cannon gunpowder	75.9	10.4	13.6

with limestone, old mortar, earth, and ashes. Piled into heaps, this material was sheltered from the rain but was kept moist with the runnings from stables and urine. The collection, fermentation, and treatment of this filth frequently led to litigation. It also is on record that "the saltpeter men abused their privileges and were hated and abhorred by the rest of the community." The conditions are very important for the production of nitrates from the decaying organic material, and, needless to say, it worked much better in the Middle East and India.

By 1626, The East India Company was importing saltpeter into the United Kingdom, and this seems to have settled the problem. Importations were also made from Spain, Germany, and Poland. In a paper to the Surrey Industrial Group, B. J. Buchanan (25) sheds extra light on the saltpeter. Records reveal that saltpeter was imported into Bristol from Danzig in the eighteenth century. It also seems to be clear from the documents relating to the Danzig-owned ships that the material was produced there and not transshipped from India.

Commercial gunpowder costs many times the cost of the basic ingredients, and, as such, it is an expensive product to purchase. However, it was expensive even in the sixteenth century, presumably because of the price of saltpeter. Again quoting Hodgetts, in 1788 George Napier said that he considered Russian saltpeter to be the best, but did not think that it was advisable to refine it more than four times "as it was probable that on repeated evaporation part of the elastic and expansive fluid contained in the nitre might be liberated!"

Dr. Buchanan pointed out that Bristol became an important center of the U.K. gunpowder trade because of its strategic position on the west coast. Gunpowder was required in the colonies for blasting and for weapons and became a very important part of the Bristol triangle. Criminals and workers were sent out to work in the colonies. Black West Africans were bartered and purchased with a variety of manufactured goods, including gunpowder.

It is assumed that most of the sulfur that was used in earlier times came from Sicily or Marseille. Sulfur-bearing rock was ignited on a hillside so that molten sulfur would run down into wooden boxes. This crude material was distilled in Marseille, where it appeared in two forms in the cooling chamber. Liquid sulfur was cooled into sticks to form "roll sulfur," and the powder that collected on the walls was sold as "Flowers of Sulfur." These "Flowers of Sulfur" tend to be less dense and can be slightly acidic. Firework makers prefer ground rock or roll sulfur or "Flour of Sulfur," and in the 1990s feel a little uneasy about the abundant, but pure "refinery" sulfur, which often smells too much of hydrogen sulfide for peace of mind. Sulfur was imported into Bristol in earlier times from Ancona, Italy.

Earlier writers seemed to suggest that the type of charcoal used to make gunpowder was of no great significance, but history reveals that this was not true and that in many ways it was the most significant problem once the nitrate issue had been solved. Every firework maker knows that many different effects can be obtained with available charcoals, but ultimately it often is a question of what is readily available (see "Charcoal" in Chapter 4).

In the earliest times, charcoal came from the forest where wood was burned in piles covered with earth and, needless to say, there must have been great variations in the ash content. The resulting powders were extremely variable, and in the course of time the charcoal was made by destructive distillation in an iron cylinder.

The wood was cut and selected before it was packed into long iron cylinders, which then were sealed in some way, but with a number of holes to allow the gas and moisture to escape. Alder and willow charcoal have always been favored for the manufacture of gunpowder, but it also has been said that dogwood made the cleanest-burning powder and was the easiest to ignite. After the distillation process, great care was taken to avoid air coming into contact with the charcoal in order to avoid spontaneous combustion.

In the Mediterranean countries, charcoal made from grapevine cuttings is greatly prized, and there is no doubt that it produces a good result. Dr. Shimizu gives details of the types of wood used in Japan in Chapter 19.

In a delightful scholarly book called *Cornish Explosives* (26), Bryan Earl described the growth and decline of the industry in great detail. As a general rule, it must describe the industry over this period of history, and he has made a splendid effort to document a part of it. It must be said that this short chapter owes much to Mr. Earl.

In the early days, much of the powder was ground or stamped by hand, but in the course of time, the familiar incorporating mills came

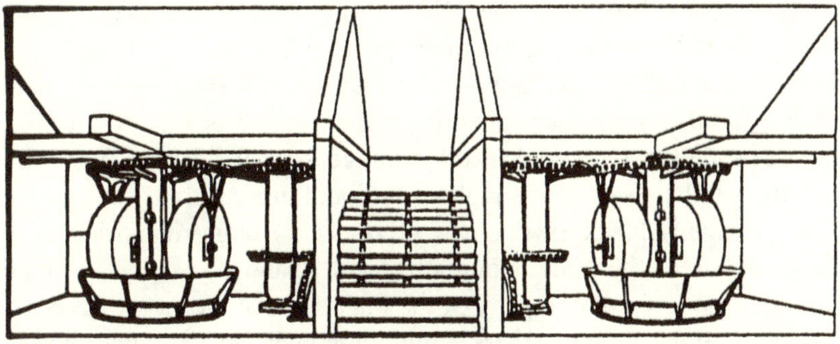

Figure 3.1 A pair of water-powered incorporating mills based on a drawing from the Royal Gunpowder Factory, Waltham Abbey, 1830 (Courtesy the Gunpowder Mill Study Group)

into being. These mills inevitably required power, so it is not surprising that powder mills grew up in places where water mills could be utilized (Fig. 3.1). Many of these factories were situated by water courses in ravines where sufficient heads of water could be used at various stages of the process. However, this was not the only factor; some mills grew up near places where it was easy to import saltpeter and sulfur and where there were suitable woodlands for making charcoal.

The first stage of the process was to weigh the three components and roughly mix them together. This rough "green" mix often was milled in a wooden barrel turned by a waterwheel.

The incorporating mills were a great advance because they incorporated the "green" mix into a very fine powder, bringing each ingredient into close contact with each other to produce a faster, cleaner-burning powder. However, there are limits to this milling, and it is said that beyond a certain point, further incorporation actually decreased the burning speed, even though it continued to improve the cleanness of the burning residue.

In some parts of the world, stamp mills have been used to work the powder. Using a mechanical version of a pestle and mortar, the powder was worked in a marble or stone mortar set in the ground and stamped with a wooden stamper operated with a belt drive and cam shaft arrangement to lift and drop the "pestle." The author has seen these in Spain and the Far East after World War II.

Gunpowder has been manufactured in Europe for hundreds of years. It was made in Spandau near Berlin in 1431, and possibly earlier. Over many of these years, its manufacture has been something of a craft, and it was not to reach its zenith until the nineteenth century, when the study became a little more scientific. It might be argued that it saw its most powerful form as the prismatic powder that came in 1868.

In England, the Royal Gunpowder Factory at Waltham Abbey dated to the middle of the seventeenth century and was one of Europe's largest producers (Fig. 3.2). Production ceased after World War II and now the site has been reopened as a national museum.

In Germany, blackpowder was produced in large quantities in the Hannover area, the Harz mountains, and Spandau, as well as many other places. A small company of the beginning of the twentieth century eventually became the Werk Kunigunde near Goslar, where powder is still produced today under the trade name Wano (25).

In recent times, the author has seen a set of old stamp mills in the factory of Sachsenfeuerwerk in Freiberg. Such mills were used in the United Kingdom in earlier times but were banned as long ago as 1772.

Although the words *gunpowder* and *blackpowder* are synonymous, the two words also exist in German. It has been suggested that the word *blackpowder* was adopted at the close of the nineteenth century to distinguish it from the newer smokeless nitro-compounds.

The manufacture of gunpowder in recent times was a monopoly of ICI, but manufacture ceased in Ardeer in Scotland in 1976. Batches of

Figure 3.2 Early blackpowder manufacturing sites in the nineteenth century

Figure 3.3 "The Master Gunner supervising the pounding of the powder," from Firework Book [I.34], South German, mid-fifteenth century (Reproduction courtesy The Board of Trustees of the Armouries, Great Britain)

200 pounds (90 kilos) were milled for 2.5 hours. The charcoal was beech and ground initially to pass a 40 mesh screen. Powder presently is imported from Germany or Spain.

Meal A is a reasonably dense powder of about 40 mesh, and this density is quite important in manufacture. Some powders are so light and fluffy that they are both dusty to handle and difficult to charge into tubes. An examination of some recent imports revealed four powders of low density that all produced different burning speeds in the traditional column test; a given column height pressed in similar increments at the same dead load produced the following burn times:

Powder	A	5.6 seconds
	B	6.4 seconds
	C	8.4 seconds
	D	9.5 seconds

Grain powders vary in size and are still made to the original ICI specification by Wano of Germany. NPXF is a soft grain powder that is useful for pressing, and it runs easily in machine filling with cam shaft rammers—a method formerly widely used in the United Kingdom. Shells and candles tend to use a grain called 4FA, and the smaller 5FA was used in "bangers" and for ignition.

There also are blasting powders on the order of 2 mm to 8 mm used as shell fillers. A grade called NPCG is used to burst the old cylindrical comet shells where a slow burst is required along with heavy string binding.

All recent writers have revealed the sad story of the shrinking use and production of gunpowder. It is extremely difficult now to find ships and harbors willing to handle the material, and ever-increasing health and safety officers anticipate dangers that seldom existed in the past.

Edge runner mills generally have been preferred in the West. Again quoting Earl, "These consisted of two large fine-grained limestone, or marble discs rolled on their edge like two wheels on a common horizontal axel, the powder charge being spread over a limestone bed on which the wheels ran." By the mid-nineteenth century, iron shod stone or cast iron "runners" working onto an iron bed had been adopted by some powdermakers, weighing more than seven tons each. Iron shod or solid iron runners were run on iron beds. "Stone on stone" or "iron on iron" was necessary to avoid producing sparks.

Some of the stones were massive, up to 1.5 meters in diameter and 0.4 meter thick, and it appears that starting up was quite dangerous because of the intense strain and pressure on the cake at that point, resulting in ignition. To assist in this dangerous stage, leather sheets were placed under the wheels as they came to a halt, and these were pulled off when the machine restarted.

Consistently produced material can be much more commercially reliable for firework manufacture than the sieved fine ingredients, which tend to be more variable in burning speed and, of course, less dense. Modern chemical mills are very efficient, and newly milled

Figure 3.4 Gunpowder incorporating mill (Courtesy John E. Dolan)

potassium nitrate along with charcoal and sulfur available at 50μ can produce a fast powder, but it tends to be rather light and fluffy and must be mixed with water and, if possible, consolidated.

This was clearly recognized in earlier times, for the incorporating mills produced a mixture of fine powder and hard lumps that did not ignite evenly. This led to the process of "corning" or granulating. At first water was added at the end of the incorporation to make a thick paste. It was then forced through sieves and dried to make hard pellets.

In the course of time, this part of the process was improved by pressing the mill cake in a damp state. The mill cake was moved with wooden shovels into a wooden box. Powder and copper sheets 18 mm thick were loaded alternately into the box, and this was then subjected to powerful pressure. It appears that there were many accidents at this stage, but it was concluded that these layers of mill cake and the copper sheets made up a voltaic pile and that sparks could be drawn from this.

Slabs of powder made in this way were broken down into small pieces. At first this was done by hand with wooden mallets, but eventually they were replaced with toothed roller drums.

Earl makes the point that the powders were dried in drying rooms that were heated with iron flues from furnaces. Accidents still happen in the twenty-first century, but it does make us think today when we realize that the old factories had open fires, oil lamps, and quite a lot of ignitions. Such conditions lasted until 1850.

Gunpowder was transported in wooden casks, and there were many problems caused by the shrinkage of the wood and consequent spillage. However, the main problem was the absorption of water, particularly when it was transported by ship or stored on floating magazines. Experiments at the beginning of the nineteenth century found that storage in copper barrels was a vast improvement in preserving the quality of the powder (25).

The transportation of gunpowder had a number of obvious problems, with the result that the Port of London published a set of 12 rules (43), which make interesting reading. However, as a fascinating comment on the times, the following quotation reveals how they came about.

> Nevertheless, in such carriages, and under the charge of such people, are thousands of barrels of Gunpowder moved from the Mills at Hounslow, through the different areas into London, by Piccadilly, along the Strand, Fleet Street, Past St. Paul's, the Bank, the India House, the Tower, and last of all by the London Docks, to Union Stairs, Wapping, where the carmen, with iron nails in their shoes, get into the wagons and carts, in order to unload them, which, by the time they arrive, are often strewed with Gunpowder from one end of the floor to the other, jolted out of the barrels through the saltpeter bags. That neighborhood has hitherto escaped an explosion, which certainly would be destructive to all around; there these thoughtless carmen, in the midst of drunken sailors of all nations, and the most wretched and abandoned women, drinking and smoking, unload their carriages subject to accident from fire, and the wickedness of ill-disposed people, and carry the Powder to the waterside through a passage, into which a door of a public house opens, surrounded by the lowest orders of society and watermen smoking and priming their boats, there are boats appointed for the purpose receive it; and the saltpeter bags are, contrary to the Act of Parliament, taken off the barrels, and returned to the wagons. The Act directs also, that the boats should proceed immediately down the river, which is rarely completed with; and it is a fact, that these small boats are often long on their passage to the sloop magazine or ship, and are very fre-

quently and improperly managed by boys, who lay them most imprudently alongside houses, ships, the entrance of the docks, & c. & c. On the 27th January last, the sloop "Success," lying above Blackwall, was with two boys, very nearly blown to atoms, in consequence of her being left by the master, with a fire on board, (in direct opposition to the Act of Parliament), by which means two boys, (who had got possession of one pound cartridge of Powder) in making squibs, exploded it and were nearly killed, and nine barrels of Powder, each containing 100 lbs had nearly exploded, the force whereof, from the improper situation in which the sloop was, might have done irreparable mischief.

The last part of the corning process consisted of sifting the grains in a revolving drum to remove the dust, and after that the grains were revolved in a drum for the "glossing" or polishing process. In some cases, graphite was added at this stage to produce the characteristic shining powders that are still produced. A lively comment made by a writer in 1859 mentioned that some varieties of gunpowder, especially those manufactured for the African market, were made to shine with blacklead (graphite). He said, ". . . the negroes seemingly think that gunpowder which approaches their own complexions most surely is the best."

Glazing certainly helped the flow of these large grains, particularly the slower-burning and larger-grained blasting powders. However, because of the heavy pressure, the grains are not porous, and it has been suggested that the sulfur has colloidal properties that fill the space between the particles of the other ingredients.

Tenny L. Davis (7) includes quite a lot of information on the issue of graining. Little did they realize then how important a matter like this would become in modern rocketry, in relation to the shape of burning surfaces with newer materials.

Gunpowder grains burn from the surface inward, and, as a consequence, the production of gas decreases as burning continues. It is clear that larger pieces required more time to burn, and Napoleon's army apparently used cubic grains about 8 mm thick in the smaller guns as well as in larger ones. Later on grains were made in the shape of a prism with a central hole, the idea being that a constant amount of gas would be produced since the area of the inner surface would increase as the area of the outer surfaces decreased.

Davis makes the point that cocoa powder was the most successful form of blackpowder that was developed for long-range rifles. In prism shape, it looked like milk chocolate and gained its color from partly burned charcoal made from burning rye straw. There was little sulfur in these powders, which made graining easier, but it was much more sensitive to friction than ordinary blackpowder. English cocoa powder was made with:

	(%)
Potassium nitrate	79
Brown charcoal	18
Sulfur	3

In the United Kingdom, the Acts of Parliament of 1875 made many changes, and it often has been remarked that the work done at that time was so excellent that it still stands us in good stead today. Every age has its inexperienced young bureaucrats who have splendid ideas, but we were well served by the 1875 Acts. Limitations were placed on quantities, safety distances, and a host of other factors. Workers wore woolen suits without pockets and leather boots without nails. Wet mats were situated at the building entrances to remove grit from shoes. Many of the buildings were lined with wood and frequently gloss-painted to make them much easier to keep clean and to prevent dust and grit from falling into the operations. The somewhat dangerous press boxes were eliminated.

The 1875 Acts also required the publication of many notices. However, as Earl points out, many of the workers could not read, so the foreman periodically was required to read the notices to the workers. However, although people can read nowadays, there is no guarantee that they do read the notices.

The past 100 years have seen some developments and refinements, but fundamentally the processes are very similar. It is interesting to note that the price of blackpowder in Cornwall in 1895 was about £33 per ton. In 1990, blackpowder manufactured in Germany and sold in the United Kingdom cost at least 120 times that price. It was noted earlier in this chapter that commercial blackpowder costs about eight to fourteen times the cost of the ingredients. It is clear that production has dropped dramatically in the past 50 years. Part of the problem is that small family businesses gradually have been taken over by large corporate bodies, and in a small specialized business it is important to have small groups of dedicated people who have blackpowder in their blood and a feel for the business.

For their part, the large corporate bodies have only one motive for their modus operandi, and they tend to get rid of anything that does not bring financial return. In the United Kingdom, for example, a number of small manufacturers of both blackpowder and fireworks gradually were bought out and then closed down. Gunpowder has not been made in the United Kingdom for many years, and there cannot be many people left who have the experience to restart operations, even though some plants have been carefully kept "in mothballs," we hope.

The high cost of manufactured blackpowder is a serious problem to the firework manufacturer, so it is no surprise that it is still made by firework makers themselves in some parts of the world. This quite frequently is still done with steel or wooden ball mills. In some cases, copper-lined barrels are charged with the "green" charge, and then a number of bronze balls are added. The author has even seen cement mixers used in some places.

The mixtures are not always milled dry, but the addition of water makes it difficult to mill the material properly. The best and probably the safest way is to mill the nitrate with some of the sulfur in one mill, and then the sulfur and the rest of the charcoal in another mill. Finally, the two powders must be incorporated in still another mill. The resulting powder can then be damped down, possibly stamped, and finally granulated by forcing it through a sieve. Clearly, it is a dirty and hazardous business, but it may be necessary in the future if the firework industry is to survive.

Most manufacturers make their own "rough" powders with the basic ingredients. One granulated shell filler is often called "polverone," with manufacturers having their own convenient way of doing it. The Kimbolton method is to grind the sulfur and the nitrate together and then mix it with 300 mesh charcoal. A small amount of gum arabic is added, and then the powder is moistened in the normal way for making stars. The powder is then pressed into 45-mm comets under a heavy load in the presses. The comets are finally forced while damp through a 4-mm screen to produce a granulated powder, which is then dried.

An interesting book, *Blackpowder Manufacturing, Testing and Optimizing* by Ivan von Maltitz (77), has gathered a wealth of material and practical experience (for which there is no substitute). He rightly explains the problems and snags, but in this context he explains the wet method of making gunpowder—the so-called CIA method. Clearly, only for emergencies because of the cost, the principle is to make a saturated solution of the potassium nitrate at boiling point with the minimum amount of water and then add this to a heated mixture of the fine sulfur and charcoal. The mixture is poured quickly into a large quantity of alcohol (which absorbs the water). The gunpowder is filtered off and dried.

The speed of the activity, the high temperature, and the rapid cooling lead to the production of the smallest potassium nitrate crystals in the charcoal. Some nitrate is lost in the process, and allowance has been made for this as well as for the cost of the alcohol. Some interesting experiments are described in the book.

Others have reported that the commonly held view that potassium nitrate is milled into the porous structure of the charcoal may not really be the case and that electron micrographs do not support this. It has rather been suggested that the 4% water added to the milling process is absorbed onto the surface of the charcoal particles and that this water is saturated with potassium nitrate.

It is also interesting to note in Hardt (78, p. 263) that attempts to find substitutes for gunpowder have not been very successful for a number of reasons, but this text makes the point that it seems that static electric discharges are the most likely causes of blackpowder accidents. This may be the reason for various unexplained explosions in blackpowder plants and why high levels of humidity are acceptable. A minimum ignition energy of resting blackpowder of 6mJ is rather worrisome.

Table 3.2 shows the grades that are used in fireworks.

Table 3.2 Blackpowder Types and Grades

Type	Grade	Microns
ENGLISH		
Meal A	B	400–Dust
NPXF	B	780–140
FO/A	B	780–250
4FA	C	1,580–940
5FA	C	780–440
Sulfurless Meal		105–Dust
Sulfurless Grain		Various
GOEX		
FA		5,660–4,000
2FA		4,760–1,680
3FA		2,000–1,190
4FA		1,680–840
5FA		840–297
7FA		420–149
Meal D		420–Dust

(continued)

Table 3.2 *(continued)*

Type	Grade	Microns
Fine Meal		149–Dust
Extra-Fine Meal		105–Dust
ELEPHANT BRAND		
FA		8,000–4,000
2FA		4,760–1,680
3FA		2,000–1,190
4FA		1,680–840
5FA		840–297
6FA		590–297
7FA		420–149
Flour		420–Dust
Fine Flour		149–Dust
Extra-Fine Flour		105–Dust

Note: English powder is 75:15:10 mixture using beech charcoal and contains a maximum of 1.2% moisture and has a minimum density of 1.70

Sulfurless powder consists of 70.5% potassium nitrate and 29.5% charcoal and is manufactured in much the same way. It is used mainly in priming and igniter compositions, particularly where they come into contact with chlorates or magnesium. The sulfur in the ordinary black-powder, coupled with the moisture in the charcoal, tends to accelerate the decomposition of magnesium.

One of Her Majesty's Inspectors of Explosives produced a poem summarizing the progression to detonation, and we, too, end with these words:

Initiators fire the chain
Acceleration boards the train
Fierce and fast reactions zip
Ingredients, self-sufficient, whip
The pace beyond "*combustion*"
Past the point of "*Deflagration*"
Atoms fly with mounting pressure
"*Explosion*" then becomes the measure
Ah! but for some that's not enough
For they are made of rougher stuff.
And still the pace goes up and up

Until it reaches ceiling, top
The pace by now extremely hot
No more acceleration can be got
Energy loosed in *shocking wave*
Atoms agitated so behave
With truly violent reputation
The label then is "*Detonation*"

4

FIREWORK MATERIALS

The materials used in firework manufacture can be divided into the following categories:

1. Oxidizing agents
2. Fuels
3. Color-producing agents
4. Substances that improve particular effects (e.g., color, light)
5. Substances that produce smoke
6. Binding agents
7. Phlegmatizers, which reduce the sensitivity of mixtures
8. Stabilizers, which help to prevent chemical reactions
9. Substances that accelerate or retard combustion
10. Aids in production, such as solvents and lubricants

The following list of materials would be more useful if it could be arranged in accordance with the aforementioned categories, but because there is a tendency for functions to overlap, the list has been arranged alphabetically. The list in no way represents a full list of materials used over the wide pyrotechnic range and represents only materials commonly used for firework manufacture.

Aluminum

Over the past 70 or 80 years aluminum has added tremendously to the brilliance of fireworks, and yet the great variety in production techniques has caused problems in the production of uniform effects. Ellern has covered the various grades of aluminum very well in his two works (5), but there is a little more to be said to supplement the specialized nature of this book. The powders are prepared in hammer mills, in ball mills, or by atomization. The first two techniques produce the so-called "flake" powders and hitherto have been the techniques most frequently used for fireworks. Powders made by this method are stamped or milled with stearic acid or other lubricants so that they form tiny plates of irregular shape and large surface area. Foil is used as a starting point and can end up in particles as small as 2μ and finer. This type of aluminum also is used for making paint and is known as aluminum bronze, although paint aluminums can contain as much as 3% or 4% grease. The resulting "bright" or "brilliant" silver powders are used in the slower-burning silver effects and usually are 120 to 200 mesh. Three samples of this powder from English sources were in accordance with the following:

Retained on 120	10	10	3
120–200	47	43	32
200–325	26	30	30
Pass 325	17	17	35
Grease content	0.03	0.25	0.4

Flake aluminum in larger mesh sizes is known as "flitter," and there has been a tendency for manufacturers to sell this to the firework trade under the three categories of "fine," "middle," and "coarse." It must be admitted that, as a general rule, a wide variation in mesh size is permissible with flitter, but it makes a critical difference in some compositions. Table 4.1 illustrates the unfortunate variation from one company to another.

Naturally, with such wide variations in mesh size, there inevitably will be considerable fluctuation in aluminum samples because fine and coarse particles may tend to separate in the drums. As a general rule, the following mesh sizes would appear to be the best for flitter:

Coarse	10–30 mesh
Middle	30–80 mesh
Fine	80–120 mesh

According to Ellern, stamped powders are less common in the United States and have been replaced with those made in ball mills. This was not the case in England, where the firework manufacturers preferred the denser stamped powders that are slightly easier to handle and more consistent in quality.

Table 4.1 Flitter Variations

	English						German Sample			Swedish Sample	
	Sample A			Sample B							
	Fine	Middle	Coarse	Middle	Coarse		Fine	Middle	Coarse	Fine	Coarse
5–10	—	—	5	—	10	5–10	—	2	22	—	100
10–20	2	—	54	45	55	10–20	—	45	66	59	—
20–40	32	7	41	30	35	20–40	68	53	12	28	—
40–80	30	78	—	20	—	40–80	30	—	—	13	—
80–120	22	17	—	5	—	80–120	2	—	—	—	—
Per 120	14	—	—	—	—	Per 120	—	—	—	—	—

The so-called dark pyro aluminum consists of a very fine flake powder that is produced in varying shades of dark gray, and although it has a nominal mesh size of about 200, it contains particles as fine as 2μ.

It has been said that in the past this powder was made from burned paperbacked foil or foil that had been sprayed with varnish that subsequently was charred. The author has not seen this practice, but manufacturers incorporate a small percentage of carbon black in their pyropowder. Large quantities of firework aluminum of good quality are produced by the Germans, who also use the terms *bronze* and *flitter*; they also use the terms *schliff* for flakes and metal inks and *pyroschliff* for the finest firework grades.

Atomized aluminum is being used increasingly in fireworks, but up to the present time only limited use has been found for this material. The problem stems mainly from the fact that it is more difficult to light a spherical or spindle-shaped particle than it is to ignite a flake. One English production method seen by the author consisted of the dispersion of molten metal that was poured into a stream of compressed air. The fine metal particles were then carried along into a large collecting chamber. Powder produced in this way certainly varies in shape and is not as regular as some of the spherical grade described by Ellern. This also may explain why some grades have better burning characteristics than others.

After extensive trials with atomized material, the author has not been able to use any such material coarser than 120 mesh in firework compositions of the normal type (i.e., excluding thermites and incendiaries). Beautiful effects can be obtained with powders such as 120-Dust or 300-Dust and a grade that passes 120 and is retained on 200 mesh, but these are extremely limited and tend to be useful only in gunpowder-type mixes. Perhaps time will remedy the situation and make life a little easier for everyone by reducing the dirt and sensitivity that characterize the flake aluminums.

Ammonium Salts

With the exception of perchlorate, ammonium salts do not find any place in modern firework manufacture. The possibility of the formation of highly unstable and explosive ammonium chloride by an ion exchange in the presence of water precludes their use. In the past, white smokes have been made with potassium chlorate and ammonium chloride, and it must be admitted that the mixture appears to be reasonably stable, although at first sight theoretically this should not be so (see Chapter 5).

Ammonium Perchlorate, NH_4ClO_4

Ammonium perchlorate has been used extensively in recent years, not only in fireworks for the production of rich blue and red colors, but also in the manufacture of propellents. It can be safely mixed with pure potassium perchlorate, but it must not be used with chlorates, and it is unwise to mix ammonium perchlorate stars with other chlorate stars in the same shell or rocket. This material usually is imported into England as a white crystalline powder of about 120 mesh.

Anthracene, $C_{14}H_{10}$

The pure form occurs as fine blue fluorescent crystals that melt at about 213°C. It is insoluble in water and rather sparingly soluble in most organic solvents. The main source of supply is from the distillation of coal tar, and the impure commercial grades are frequently greenish yellow because of the presence of other hydrocarbons.

Anthracene is used mainly in combination with potassium perchlorate to produce black smokes; an oxygen-negative mixture is required for this purpose. During World War II the Germans used anthracene (in combination with hexachloroethane, magnesium, and naphthalene) for smoke production. In addition, the shortage of shellac and accaroid resin (which come from India and Australia) caused experiments to be undertaken with a view to replacing these substances with anthracene and naphthalene. A German patent (No. 677532) was concerned with this exercise. Anthracene is becoming increasingly difficult to obtain.

Antimony, Sb

This powdered metal, also known as antimony regulus, comes to the trade as a dark gray powder fine enough to pass 200 and often 300 mesh. It usually is prepared by heating the native sulfide (stibnite) with scrap iron or with poorer quality ores by burning off sulfur in a reverberatory furnace to produce the oxide, which is then reduced with carbon. The metal melts at 630°C.

Antimony is used mainly in combination with potassium nitrate and sulfur to produce white fires or in combination with aluminum to aid ignition. Antimony also is responsible in part for the well-known glitter effect, which basically is a combination of gunpowder, antimony (or the sulfide), and aluminum; glitters also can be made without antimony. Although many antimony compounds are toxic, this material, with its low solubility, is not unduly toxic (79).

Antimony Trisulfide, Sb_2S_3

The black powder used for firework manufacture usually is the ground native ore, stibnite, which is mined in Bolivia, China, Hungary, and South Africa. The Chinese powder used in commerce is very fine, usually passing 200 mesh, and has a melting point of about 546°C. As a fuel, its uses are much the same as the metal powder, although it ignites more easily. It also has the disadvantage that it is dirtier to handle than the metal powder.

Synthetically produced material usually is not used in fireworks, and it can be difficult to get good glitter effects from it. The red precipitated form of this sulfide can be used for some firework mixtures, but this is not common practice. Although many antimony compounds are toxic, the material, with its low solubility, is not unduly toxic (79).

Arsenic Disulfide, As_2S_2 Realgar

The native ore, realgar, is sometimes ground to a fine powder and used to make white fires. It also has been used for making smokes. Realgar also is produced by sublimation when arsenopyrite is roasted. It comes from the United States, Canada, France, Sweden, and Britain, and consists of a red powder, soluble in acids and alkalis, with a melting point of 307°C. It is no longer used in fireworks and is virtually unobtainable.

Arsenic Trisulfide, As_2S_3 Orpiment

This sulfide also occurs in a mineral form. The commercial powder has two forms, one yellow and one red. The yellow form changes to red on heating to 170°C. It is insoluble in water and hydrochloric acid, but it dissolves in alkaline sulfide solutions and nitric acid.

The red form often is used to make white stars, which have the advantage of being easy to ignite when moving at very fast speeds. Apart from the occasional use in smokes, orpiment is used in combination with carbon black for making Flower Pots with their characteristic golden spur fire.

Arsenic is safe to handle, provided, of course, that precautions are taken to keep it out of the nose and mouth. Perhaps the name itself conjures up images of other uses to which some arsenic compounds have been placed, but it should be remembered that soluble barium, for example, is equally toxic. Arsenical compounds have been phased out of fireworks.

Asphalt, Gilsonite

Asphalt is a blackish brown solid imported from Syria, Egypt, and Trinidad. It possibly is formed by chemical changes and the oxidation of mineral oils that have a high boiling point. Asphalt with a melting point of about 100°C normally is used since those with lower melting points are more difficult to keep in powder form. It is insoluble in water and alcohol, but it dissolves in coal tar naphtha, turpentine, and petroleum.

Asphalt is seldom used in English firework manufacture, but it seems to have found considerable use in American formulations. Its possible sulfur content could make it hazardous with chlorates, and there is the additional disadvantage that it is very dirty to handle and tends to leave sticky deposits on tools.

Barium Carbonate, $BaCO_3$

Barium occurs naturally in England as witherite (barium carbonate) or barytes (barium sulfate). The barium carbonate of commerce is made either by precipitation or by the conversion of the natural sulfate into the sulfide and then its interaction with sodium carbonate.

Barium carbonate is of no use as a coloring agent in low-temperature flames, but it is often used to reduce acid formation in mixtures or to slow the speed of some compositions.

Barium Chlorate, $Ba(ClO_3)2H_2O$

The fine white powder of commerce usually is of 99.5% minimum purity and is prepared by the electrolysis of barium chloride. It has a melting point of 414°C and is imported into England.

Barium chlorate is one of the most sensitive chemicals used in firework manufacture, but it is difficult to manage without it when deep green colors are required. It is wise to use this substance as little as possible and to use it in combination with other substances, which will tend to reduce its sensitivity.

Barium Nitrate, $Ba(NO_3)_2$

Barium nitrate is perhaps one of the most useful and stable of the nitrates, but it is somewhat limited in use. It is manufactured by dissolving the native carbonate in dilute nitric acid or by mixing solutions of sodium nitrate and barium chloride. It is marketed as a fine white powder of about 200 mesh and has a melting point of 575°C. Low-temperature green colors are not very strong when they are made with barium nitrate, although the salt frequently features in compositions made with barium chlorate as the main coloring agent. On the other hand, recent years have seen extensive use of barium nitrate to make high-temperature green flames using magnesium or magnalium.

More than anything else, this substance is used in combination with aluminum powder for the production of silver effects. Silver stars, flares, and waterfalls invariably use barium nitrate, and the aluminum combined with it frequently is mistaken for magnesium by the uninitiated. Below 1000°C aluminum burns with a silvery gold effect, and this is characteristic of mixtures of gunpowder and aluminum. Above this temperature silver effects can be obtained with the aid of barium nitrate. Barium nitrate greens occasionally are used in situations in which there is a danger of ignition from friction (as in filling green lances with a funnel and wire).

Barium Oxalate, $BaC_2O_4 \cdot H_2O$

Occurring as a fine white powder made by precipitation, barium oxalate is insoluble in water, but it is soluble in dilute hydrochloric and nitric acids. The price and the weak flame coloration of barium oxalate make it unsuitable for ordinary firework manufacture, but it sometimes is used in more specialized items in combination with magnesium.

Barium Peroxide, BaO_2

Although barium peroxide is used in pyrotechnics, it is not suitable for use in fireworks because of its very reactive nature. It decomposes in water and also at 800°C, and mixtures containing aluminum are likely to heat up in the presence of water. It is prepared by heating the monoxide in a stream of oxygen and has a melting point of 450°C.

Beta Naphthol, $C_{10}H_7OH$

Manufactured mainly for the dye industry, this substance has found an occasional use as a fuel in colored stars, mainly because of its carbon content. It melts at 122°C. Unfortunately, this substance irritates the mucous membranes, is not pleasant to handle, and consequently is no longer used. It may be carcinogenic.

Boric Acid, H_3BO_3

This very weak acid sometimes is used in firework compositions to prevent the decomposition of mixtures containing aluminum. Since the decomposition of aluminum is an alkaline reaction, 1% or 2% of this acid will help to prevent the acceleration of the decomposition. Heat sometimes accompanies this reaction, but the boric acid acts as a buffer, preventing an alteration in the pH. The wet slurry of barium nitrate and aluminum that is used for making sparklers would be quite likely to react if allowed to stand for some time unless there were some buffering.

In dry mixtures, zinc oxide sometimes is added for the same reasons; in this case, the oxide is not a buffer, but, being amphoteric, it may slowly react with any acidity or alkalinity present. Boric acid is a good lubricant in dry mixtures.

Boron, B

Boron is a relatively rare nonmetallic element that does not occur free in nature, but always occurs in combination with oxygen and other elements, notably as sodium and calcium borates in the ores borax and colemanite.

Amorphous boron is a very finely divided dark brown to black powder of a particle size of usually less than 2μ. It normally is prepared by the Moissan process, in which boric oxide is reduced by magnesium at high temperatures and the product, after acid washing, is boron of about 90% purity.

Boron has found wide application in military pyrotechnics over the past 40 years, and its properties and performance in these compositions have been the subject of much scientific study. As a pyrotechnic fuel, boron has a high heat output and reacts readily with a large number of oxidants and other fuels to provide compositions having a range of properties.

Mixtures of boron with oxidants, such as oxides, peroxides, chromates, nitrates, and sulfates, burn in different ways, the combustion ranging from rapid burning accompanied by long flames and showers of sparks to very slow combustion and the evolution of little or no gas.

Mixtures of boron and potassium nitrate produce compositions that consolidate well when pressed and are easy to ignite because they are sensitive to flash. These are used as priming or first fire compositions because of their ability to transfer heat to other compositions. The addition of silicon enhances this effect.

Boron mixed with oxidants, such as bismuth oxide, barium chromate, or potassium dichromate, form a range of gasless delay compositions that can provide a range of burning rates from 0.3 to 40 seconds per 25 mm. These delays perform with high reliability and great accuracy at the extremes of environmental temperature and pressure.

The majority of compositions containing boron are very sensitive to friction and shock. The more reactive mixtures also are very sensitive to ignition by electrostatic spark, and these require special process rooms fitted with electrically conducting equipment.

Calcium Carbonate, $CaCO_3$

The precipitated form of this compound finds occasional use as a neutralizer in some mixtures, Armstrong's mixture, matches, and snakes made with nitrated pitch. It also is used to make rockets. Additionally, it is used in high-temperature flames for orange color.

Calcium Oxalate, CaC_2O_4

A fine white powder made by precipitation, calcium oxalate is not commonly used in ordinary firework manufacture. It has been employed mainly in signals to give depth of color and as a burning moderant to mixtures of sodium nitrate and magnesium and in other signaling devices. It is insoluble in water, but it dissolves in dilute hydrochloric and nitric acids.

Calcium Orthophosphate—Tricalcium Phosphate, $Ca_3(PO_4)_2$

This well-known "flow agent" is used in firework manufacture. About 1% can be added to many mixtures to enable them to move freely in a funnel and wire apparatus or in automatic machines and presses. The white amorphous powder is made by precipitation. It is insoluble in water, but it dissolves in acids and has a melting point of 1670°C.

Calcium Silicide, $CaSi_2$

This material is a gray-black powder that is insoluble in cold water but soluble in acids and alkalis. It is decomposed by hot water. It finds use mainly as a fuel for self-heating cans of soup and often is an important component in smoke compositions. It has been used quite effectively in star compositions, but silane may be produced, even though it seems not to be a hazard.

Carbon Black—Lampblack

These blacks are essentially carbon in a very fine state of division and are prepared by burning oils in such a way that incomplete combustion takes place, so a large volume of smoke is produced; this smoke consists of unburned particles of carbon and is collected in a specially constructed system of flues.

The quality of blacks vary, but they usually contain approximately 98% carbon, have a pH of about 8, and usually pass a 350 mesh screen. It is important to make sure that lampblack does not contain traces of unburned oil, which would be dangerous. Carbon blacks, vegetable blacks, and gas blacks are all closely related, being made similarly by burning oil, gas, or vegetable matter. As the quality varies so much, the firework maker must experiment to find the material that suits his purpose.

Carbon black is used to make Flower Pots, the unusual little golden fountains with their own special type of gold spark. Golden streamer stars also employ carbon black for the best effects. It is a pity that the material is so filthy to handle; otherwise it might find more extensive use. When compositions are made wet, it is advisable to add a proportion of alcohol to reduce surface tension and make it easier to wet the mixture.

Castor Oil

Little use has been made of castor oil in English manufacture, but it appears to have found considerable use in the United States in the past. It is used mainly as a protection for magnesium, but it also acts as a binder or lubricant in that it reduces the friction of the powder against the walls of the container into which it is pressed.

Charcoal

The different kinds of charcoal that might exist are a constant talking point among the firework fraternity. The plain fact, however, is not so much what one would like, as what one can get locally in reasonable amounts. Consistency in quality is a problem, and the last U.K. manufacturer went around his woodyard with a truck collecting different types of wood so that a constant mixture went into the still to produce a constant blend.

When wood is heated in the absence of air, the volatile products distill off and the charcoal remains behind. In large commercial stills, the burning products are fed back into the still as fuel.

In earlier times, the wood was stacked in the forest like a wigwam, covered with inverted turf, and then set on fire. Dr. Jeacock (43) quotes a Forest of Dean charcoal burner who was asked by a radio interviewer how he knew what was going on inside,

The reply rather shocked the establishment when he said, "I spits on the outside and watches it sizzle." The combustion process had to be watched 24 hours a day and turf had to be replaced as necessary. This could be a very hazardous business, and men have been burned to death after walking on the stack. In a private communication, Dr. Jeacock makes the point that the operator also cut a suitable piece of wood, stuck it in the ground, and then sat on it to watch the process. Perhaps this was the origin of the nitroglycerine one-legged stool that operators had to sit on to prevent them from falling asleep as they watched the nitration process.

That a particular type of charcoal has vastly different burning characteristics from another, whether it be for the production of sparks or of gas, is well known. The author, however, got one of the surprises of his life when testing a high-quality Japanese hemp charcoal alongside a commercial U.K. grade. Using the traditional column test, the U.K. charcoal burned for 27 seconds, and the Japanese burned for only 7 seconds with the normal gunpowder mix.

Much traditional experience in charcoal production has been lost, but it is generally known that the temperature of the distillation is crucial. Most people recommend that it should be below 500°C. A Swiss source states that black alder is used for blasting powder and that a lot of 1,495 pounds would yield about 340 pounds of charcoal when carbonized at 500°C for six hours. Sporting powder was carbonized at 300°C to 320°C. Undoubtedly, there has been a preference over the years for willow, dogwood, and alder, but even these woods vary depending on the time of the year they were felled.

One unusual grade of charcoal made from rye straw was used to make the so-called "cocoa powder" because of its color. It was said to have been invented by Castner and had the proportions of 78:3:19.

All charcoals contain percentages of ash and moisture. The lower the ash content, the better; about 2% or 3% would be ideal, which seems to be the case with willow and dogwood. Some charcoals can have an ash content as high as 15%.

However, carbon and ash are not the only two components of charcoal, for the volatiles also are a vital and variable part of it. The volatile liquids and gases start to come off the wood at something like 270°C and are highly flammable. In some cases, they are directed back into the kiln with some oxygen to sustain the reaction, which is exothermic.

The percentage of volatile hydrocarbons is somewhere between 20% and 40%, but this is related to the temperature of carbonization. In practice, this means that the higher the temperature, the less volatiles and an increased ignition temperature of any gunpowder made with it. Conversely, the ease of ignition of "cocoa" powder may be due to the charcoal (77).

It often has been a pleasant thought that one could produce different kinds of charcoal and then employ them in the various firework mixes as a means of comparison. Happily, such a series of tests have been carried out by Roger O'Neill and colleagues and published in

Pyrotechnica XVII (29). The results make interesting reading and conclude that handmade charcoals can give superior performances.

For charcoal-tailed comets, hemp, peach, apple, and grape seem to be superior. For blackpowder, plum and paulownia are superior, with oak, willow, maple, peach, and apple giving excellent results.

Once again, it must be stressed that the commercial manufacturer must use what he can get, and in practice two main types have been used. One is a very fine powder—as fine as possible—for making gunpowder or fast-burning effects; in Europe, black poplar and lime are sometimes used. The other is a softwood blend that the manufacturer may well put through this mechanical sieve to give grades of 40 to 80 mesh and material retained on 40 mesh.

More precision is required for military and other special purposes. A paper (47) gives details of work carried out on charcoals for the Ministry of Defence Procurement Executive.

The three types of charcoal used were identified as hardwood charcoals. They were alder buckthorn (*Frangula alnus*), known earlier as "dogwood" or "black dogwood," alder (*Alnus gultinosa*), and beech (*Fagus sylvatica*).

Alder buckthorn is used in gunpowder for making evenly burning fuses. Alder is used in many commercial gunpowders, and beech is used where precise burning is not required. It also was found that alder buckthorn had the lowest spontaneous ignition temperature and the highest porosity; it possibly was the best charcoal to use for ease of ignition.

Air float is a U.S. term for finely ground charcoal, and it is possible that the wood is not specified.

Chlorinated Rubber, Alloprene, Parlon, Pergut

This is a white, odorless, granular powder that sometimes is used as a color intensifier because of its chlorine content. Parlon is an American product and is not the same as Parlon P, which is chlorinated polypropylene, and which is not used as a color intensifier. Parlon contains 66% to 67% chlorine and is soluble in acetone.

Clay

Clay is an important part of firework manufacture because it is used to block up the ends of tubes or to provide a washer through which fire can be forced to produce pressure. Almost any kind of clay can be used, but it must be thoroughly dried and sifted to remove stones. When the dried clay is struck a few blows with a mallet and drift or pressed, it forms a very solid mass that does not crumble easily. White fireclay is the type most commonly used, but it does not bind well and easily absorbs water. Deep blue clay used for house bricks is excellent and can be strengthened with a small amount of linseed oil.

Copper Powder

Use has sometimes been made of the bronze and electrolytic copper powders for the production of the blue colors, or as intensifiers for green colors, but this is not very common, for the same effects can be achieved by more efficient means.

Copper Acetoarsenite $(CuO)_3As_2O_3Cu(C_2H_3O_2)_2$—Paris Green

This substance has been known by a variety of names (e.g., brilliant green, imperial green, schweinfurtergrün), and it frequently has been wrongly named in the firework literature. Copper arsenite (Scheele's green) and copper arsenate are different substances and are less useful. Paris green is prepared by the interaction of sodium arsenite, copper sulfate, and acetic acid. It has an intense green color, is insoluble in water and alcohol, but is soluble in acids. Needless to say, it is toxic and sometimes caused nosebleeds and skin rashes.

Apart from the compositions employing ammonium perchlorate, Paris green still provides the best blue colors, but it is difficult to obtain because it no longer has any commercial uses.

Basic Copper Carbonate

Occurring in native form as malachite $CuCO_3 \cdot Cu(OH)_2$ and azurite $2CuCO_3 \cdot Cu(OH)_2$, the compound used for fireworks usually is made by precipitation. It is the best copper compound to use in combination with ammonium perchlorate for the production of blue colors, and quite deep shades are obtained in this way. An adequate blue star can be made with potassium chlorate and copper carbonate by adding PVC and cool burning fuels.

Copper Oxides

Black copper (II) oxide was used many years ago (along with the fused sulfide) in the production of blue colors, but calomel also had to be used as an intensifier, and it fell into disuse. Both copper (I) and copper (II) oxides are now used for ignition and starter compositions in conjunction with silicon and lead (IV) oxide. However, the availability of good halogen donors and the disadvantages of Paris green have resulted in a revival of the use of copper oxide for making blue colors.

Copper Oxychloride

This basic chloride appears to have a variable composition and is possibly $3CuO \cdot CuCl_2 \cdot 3H_2O$. It is formed when copper (I) chloride is exposed to air. It is soluble in acids and ammonium hydroxide, but not in water. Blue colors frequently have been made with this chloride, and it still finds some use because it is less expensive than some of the other copper compounds and the blue color is good.

Cryolite-Greenland Spar, Na_3AlF_6

The production of yellow color fireworks is something of a problem because of the liability of sodium salts to absorb water or to react with other substances and incidentally absorb water. Cryolite is one of the few sodium compounds that is completely insoluble and unreactive. The natural material appears to be better than the synthetic form in many compositions.

Dextrin $(C_6H_{10}O_5)n$

Dextrin is the generic name for products of the partial hydrolysis of starch. Because it is soluble in water and has good adhesive properties, it is used extensively as a binder in fireworks. It is fairly usual to add a small percentage of the dry powder to a star composition during the mixing operation and then add water before star formation. Although dextrin is very useful for these purposes, it also tends to be somewhat hygroscopic, so it is unwise to use more than about 5%, and preferably less. Dextrin is insoluble in alcohol; if alcohol is added to the water to promote faster drying during star manufacture, no more than 30% should be added or the adhesive quality of the dextrin will be inhibited.

Dyestuffs

The quality of the materials used for smoke production is important, particularly the particle size of the dyes and their freedom from inorganic salts. The sublimation temperature should be as low as possible. The field is very specialized, but there is an adequate technical literature.

Flour

In the past, wheat flour was extensively used to manufacture paste, but most manufacturers now purchase their adhesives. It also was used in some compositions to retard burning speed.

Gallic Acid 3,4,5-Trihydroxybenzoic Acid

The acid crystallizes as almost colorless silky needles of the composition $C_7H_6O_5 \cdot H_2O$. It is soluble in water, alcohol, and ether, and decomposes at 220°C. The fine commercial powder sometimes is used for the manufacture of firework whistles and is the most dangerous of all the whistle compositions. Being highly sensitive to both impact and friction when combined with potassium chlorate, it frequently has been known to catch fire or explode when charging operations are taking place. It cannot be used with potassium perchlorate in whistle compositions.

Glue

Hide glues invariably are used in match heads and certain types of sparklers where a wire or stick is dipped in a wet slurry. Hide glue also is employed in some kinds of inexpensive smokes to cool the burning process.

Hide and fish glues have always been traditionally used for sticking various firework parts together, but these gradually are being replaced by the more modern PVA emulsion adhesives. These emulsions have the advantages of toughness and a certain amount of elasticity and do not require any heat for application.

Glass Powder

Glass powder is not often used in fireworks, but occasionally it is used in match head and striker compositions.

Graphite

The fine black powder of commerce is not used as firework fuel, although it frequently is added to compositions that are to be pressed in molds to ease their ejection. It also is added to magnesium flash powders.

Guanidine Nitrate, $NH{:}C(NH_2)_2{\cdot}HNO_3$

This organic compound is a white powder that melts at about 208°C and decomposes at 235°C to generate ammonia gas. Unlike nitroguanidine, which can be explosive when detonated, guanidine nitrate is quite stable and is not affected by either shock or friction.

The material is not hygroscopic, and it can be quite useful in strobe compositions where it is decomposed by various substances such as copper or chromium oxides, which act as catalysts. It also is used in some smoke formulations and was used as a fuel in JETEX, which was a toy rocket fuel.

GUM RESINS

Accroides Resin, Red Gum

Information in previous editions was provided by the U.K. importer of this material from Australia, but this has been considerably updated through the interest of Australian pyrotechnists, particularly Dr. Mike Stanbridge (29).

Clearly, there has been some confusion about Accroides resin, which is derived from the Liliaceae family, *Xanthorrhoea preissii* and *X.*

reflexa. As previously stated, the resin exudes from the trunk and during the passage of bushfires the glasslike boules fall to the ground. However, at this stage the chemical composition of the material has been changed by the heat from the bushfire (pyrolysis).

According to Dr. Stanbridge, it is more than likely that the material reaching Europe has been gum yacca from Kangaroo Island off South Australia. Apparently, the Xanthorrhoea trees, known as grasstrees or blackboys, depending on the area, were felled and whacked on canvas sheets to gain small diamond-shaped fragments of resin. Grit and vegetable debris must be primitively removed, which may account for the vegetable debris in the coarse powders available in the United Kingdom. Dr. Stanbridge speculates on the etymology because the aboriginal word *yacca* refers to hard work. He also suggests that the destruction of trees, which may be 600 years old, will necessitate the need for alternatives.

True red gum apparently comes from eucalyptus trees, particularly the red gum, marri, or *Eucalyptus calophylla*. Calo-resin is produced from the mill fines of this wood and may have promising commercial possibilities.

The chemistry of these resins is beyond the scope of this rather practical work, but the benzoin-like odor is very characteristic, and chemists say that the variable composition contains benzoresinotannols, benzoresorcinol, and traces of other materials such as cinnamic acid.

The brown powder varies in mesh size and can be used as a control for the burning rate. In slow-burning illuminations, a powder passing 30 mesh and retained on 80 mesh would be useful. For fast-burning compositions such as stars, a powder passing 120 mesh is best, particularly with the more modern high-temperature flames where smaller percentages of the gum are used. The gum replaced shellac long before World War II because it was less expensive and was one of the substances called "shellac substitute." However, in many respects it is superior to shellac in colored flames. It also is surprising to learn that it has not been extensively used in Japan for making fireworks. The gum is soluble in alcohols and alkalis and has a melting point below 100°C.

Further research can be found in a *Journal of Pyrotechnics* article (77), Issue 15, page 77.

Gum Arabic

Also known as acacia gum, this material is obtained mainly from the various species of acacia trees. Gum arabic comes to the English market from Aden and Bombay and is mainly of use because it is more or less soluble in water. It occurs in roundish or ovoid pieces of various sizes and varies in color from colorless or pale yellow to brown. The best qualities take up about 1.5 times their own weight of water for solution, but the gum is insoluble in alcohol. Alcohol precipitates the gum from aqueous solution.

The solution tends to be acidic, particularly when it has been allowed to stand and ferment; it is best avoided as a firework adhesive, particularly in chlorate mixtures. It quite frequently is used for making quickmatch and in other blackpowder mixtures.

Gum Copal

The East Indian Copals, as their name implies, generally are obtained from Malaysia, particularly from the Malay Peninsula, Borneo, Java, Celebes, Papua, Sumatra, and the Philippines. Pontianac is the hardest material readily available, originating from Borneo, whereas the softer variety is known as Manila, Macassar, or Papua. These latter copals are won from *Agathis alba* by tapping and subsequent collection some three months later. Finally, they are sorted and graded according to purity, color, and hardness.

The material has a softening point of 80°C to 88°C and a melting point of 110°C to 125°C. The powder used in fireworks is light gray in color and has a characteristic smell with a particle size of approximately 60 to 200 mesh. It is soluble in alcohols, esters, and alkalis, but insoluble in water.

Shellac

Shellac is the name given to the refined form of lac, and the word, derived from shellac, specifically refers to refined lac in thin flakes in the form in which it is most commonly marketed.

Lac is the resinous secretion of the lac insect (*Laccifer lacca*), which is parasitic on certain trees principally in India, Myanmar, and Thailand. Large numbers of the tiny red larvae, about 0.5 mm long, come out of each mother cell and settle on tender portions of fresh twigs of certain trees, called lac hosts. They feed on their sap and exude the resin from glands under the skin. The insect completes two life cycles in a year, yielding two lac crops. When the material is collected, it is known as sticklac and usually contains about 60% to 70% of the lac resin and 6% to 7% of lac wax, which also is secreted by insects.

The first part of the refining process consists of grinding and washing to remove coloring matter, sand, insect debris, and pieces of wood. The lac resin then is finally produced by melting sticklac in cotton bags and pressing out the melted resin, or by dissolving out the resin with alcohol, filtering, and subsequently recovering the resin and alcohol by distillation. The exact nature of the monomeric lac complex, whose molecular weight lies between 1,000 and 1,100, is still not known. The present state of knowledge indicates the constitution to be as follows:

1. Dyes. Two distinct coloring matters are present to the extent of 1/2%
   (a) Laccaic acid (b) Erythrolaccin
2. Wax. Lac wax exists as about 4.5% by weight of whole lac and consist of even-number primary alcohols from C_{26} to C_{34} and esters from C_{30} to C_{36}.
3. Resin. 90% of the total material is resin and consists in the main of hydroxy fatty acids together with their lactones and lactides.
   Treatment with ether produces two fractions:
   (a) Soluble, known as soft Lac Resin, having an approximate molecular weight of 520, and whose principle component is aleuritic acid, 9, 10, 16-trihydroxypalmitic acid of melting point 100°C to 101°C.
   (b) Insoluble, known as hard Lac Resin, of approximate molecular weight of 1,900, and consisting of about 70% of the lac resin. Its components are derivatives of aleuritic acid.

Shellac used for firework manufacture is an orange-brown powder, softening at 50°C and melting at about 75°C. Powder as fine as 200 mesh sometimes is used, but the normal powder of commerce has a wide particle size range and is approximately 30 to 200 mesh or 60 to 200 mesh. Material made by the original machine-made/hot solvent process frequently is used. It is insoluble in water, but it dissolves readily in alcohols, organic acids, and aqueous solutions of alkalis. Shellac, accaroid resin, and copal gum are all used as fuels, mainly for the production of color. Shellac substitutes have been used in the past mainly because shellac is expensive. One is a similar-looking resin obtained from *Araucaria augustifolia*. It is less waterproof than shellac.

Gum Tragacanth

This expensive gum comes from several species of *Astragalus* growing in Lebanon and Syria. The gum is exported chiefly from Baghdad and

Basra. It is soluble in water, but not in alcohol. Very occasionally it is used as an adhesive in fireworks.

Wood Resin—Pine Resin—Colophony Resin—Rosin

Resin is the residue of the distillation of turpentine oil from the crude oleoresin obtained from the pine trees, principally from *Pinus palustris* and *Pinus caribea*. The resin consists mainly of acids of the abietic and pimaric types with the general formula $C_{19}H_{29}COOH$. Colophony is soluble in alcohol, ether, and benzene, and it has a melting point between 100°C and 150°C. Vinsol resin (Hercules Co.) is a form of rosin used in fireworks.

The resin is used in blue colors and occasionally in smokes, but it is difficult to powder because, although it is brittle and straightforward, it tends to agglomerate on standing. A small amount of starch sometimes is added as an aid to pulverization.

The resinates, which are products of the reaction between colophony and hydrated oxides or salts of certain metals, appear to be more useful. For example, calcium resinate is made by fusing rosin with slaked lime. Resinates have the advantage of softening at higher temperatures. This resin may aggravate asthma in some people.

Hexachlorobenzene, C_6C_{16}

Extensive use of this material as a chlorine donor in color mixtures appears to have been made in the United States in the past. According to Ellern, its use has been superseded. Little use of the material has been made in Europe since PVC has been preferred. Hexachlorobenzene occurs as white needles that melt at 229°C. It must not be confused with Lindane, gamma-benzene hexachloride ($C_6H_6Cl_6$), which consists of white crystals with a slightly musty odor and a melting point of 112.5°C. This substance is a well-known insecticide.

Hexachloroethane, C_2C_{16}

This is a white crystalline powder with a slight camphor-like odor. It melts and sublimes at 185°C. The main use of this substance is in the manufacture of smokes, and for this purpose the chemical is mixed with substances such as aluminum, zinc oxide, calcium silicide, and so on. Unfortunately, it is very volatile at room temperature, and therefore the chemical itself and the smoke composition made with it can be kept only in sealed tins.

Hexamine, $(CH_2)_6N_4$

Hexamethylenetetramine is formed when six molecules of formaldehyde condense with four molecules of ammonia. The white crystals burn with a steady yellow flame and have been used for small indoor fireworks in combination with magnesium and lithium salts. Metaldehyde has been used for the same purpose but has the disadvantage of giving off poisonous formaldehyde during burning. Hexamine would be quite useful as a fuel, but it does not appear to have found much application, possibly because of its cost.

Iron, Fe

There can be little doubt that iron has been used in fireworks from the earliest times. Apart from the steel dust used for the manufacture of sparklers, the best iron for fireworks is the ordinary cast iron borings that have been broken down to a rough powder and which will pass 20 mesh. The long needle-like fragments give the best effects.

Unfortunately, it is only too well known that iron corrodes in firework mixtures and it therefore must be treated. The oldest method, which employs linseed oil, is still the best. About 16 pounds of iron are mixed with 1 pound of linseed oil, and the mass is slowly roasted in an iron pan until most of the oil has burned off. Another method is to coat the iron in oil and then allow it to dry slowly at room temperature, spread out in thin layers on trays. If preferred, the iron also may be coated with paraffin wax. In some cases in which it is very necessary to be sure that corrosion cannot take place easily, the iron is pretreated with a solution of dibasic ammonium phosphate. Polyurethane also has been used for coating iron, but it does not burn as well as linseed oil.

Iron Oxides

In recent times, the two oxides have been increasingly used in pyrotechnics. The black magnetic form, Fe_3O_4, is used in thermite and incendiary compositions, and the brown form, Fe_2O_3, has been used in first fires and ignition compositions where high temperatures are required.

Lead Oxides

The red form, Pb_3O_4, and the chocolate-colored dioxide, PbO_2, have been used in first fires and ignitions for military purposes and also are potentially useful for fireworks. Combined with magnalium, it is used for crackling microstars.

The checking of lead in blood levels is now mandatory in Europe.

Linseed Oil

For many years, boiled linseed oil was used to coat magnesium powder to protect it from corrosion. Magnesium powder and oil are mixed together and allowed to stand in a warm place in shallow trays for about 48 hours before the other chemicals are added. In practice, lithographic varnish is preferred to ordinary linseed oil. Lately there has been a tendency to replace these oils with polyesters or to use no coating at all, but there can be no doubt that linseed oil renders good protection, and its period of usefulness is not yet over. Stores made with magnesium coated with lithographic varnish are good for several years, which is not the case with uncoated magnesium. Layers of curing composition should have a maximum depth of 24 mm.

Lithium Carbonate, Li_2CO_3

The red coloration of lithium has no distinct advantage over strontium in firework manufacture, so the high cost of lithium salts preclude their use. An occasional use has been found for lithium carbonate in certain types of indoor firework, some of which use nitrocellulose as fuel and oxidizer.

Magnesium, Mg

Because of its high cost and somewhat reactive nature, magnesium is not used very much in commercial items. However, it is indispensable in situations in which it is essential to gain high candlepower in signal flares. The readiness of magnesium to react with other materials and decompose has been mentioned elsewhere, thus making it essential to choose the other materials carefully and to coat the magnesium before mixing. Linseed oil and drying oils have been extensively used in the past for these protective purposes, but in recent years there have been attempts to replace the oil with polyester resin. Although these are very good, they are slightly messier to handle, and pressing must take place soon after mixing. Experiments have been made with potassium bichromate as a means of protection, and it is now common practice. The modern practice of pressing magnesium compositions dry in aluminum or paper tubes is a great timesaver, but the finished product cannot be guaranteed for a long period.

The metal used for pyrotechnics usually is pure magnesium, but small amounts of manganese sometimes are added. Magnesium-aluminum alloys also are employed in varying proportions and are rather more stable.

A rough sieving of the English grades 0 to 5 produced the following result:

	0	**3**	**4**	**5**
40/80 mesh	51	22.5	—	—
80/120	43	50	—	—
120/200	6	32.5	47.5	5
200/300	—	4	42.5	25
+ 300	—	1	10	70

Comminution takes place by a process of atomization when the molten metal is dispersed by gas in an inert atmosphere or by a process of cutting from large pieces of metal. Of course, the density varies with the method of cutting. Grades 0 and 3 are turnings, or raspings, made by a process using a rotating cylinder rather like a very large food grater bearing on a cylindrical billet of metal. The fine raspings are then milled and separated by sieving. Grades 2 and 4 are produced by a process of abrasion. A rotating cylinder 12 inches in diameter has continuous strips of wire brushes wrapped around its periphery. Against this rotating cylinder is pressed a slab of magnesium about 2 inches thick. The product from this process is called "cut" powder and is again separated by sifting.

Magnalium, Mg/Al

The reactivity of magnesium somewhat precludes its use in fireworks. However, the need to produce brilliance to colors has led to the extensive use of magnalium, and it is quite probable that we have the Chinese to thank for precipitating the move in this direction.

There is the additional advantage that the use of magnalium can mean that smaller percentages of oxidizer can be used. The reactivity of the alloy tends to be somewhere between the properties of each of the metals alone, and while the normal alloy is a 50/50 mixture, other alloys have been made and used.

The metal is rather brittle and consequently produces quite sharp splintery particles that can increase the sensitivity of mixtures. It also can affect the eyes. The somewhat new use of this material to make crackling granules with lead oxide has led to some surprisingly sensitive mixtures.

Magnalium is used in fireworks for a variety of purposes:

1. In coarse powder of about 30 to 50 mesh the metal produces a sharp crackling noise in colored flame fountains. The presence of cryolite seems to enhance this effect.
2. Dust-free coarse powders produce pleasant sizzling effects in some stars.
3. The finer powders from 120 mesh onward are used to enhance colored flames, produce some silver effects, and produce the newer blinker/strobe effects. Strobe effects come from magnalium burning with the nitrates of barium or strontium where the vibrational burning can be varied with catalysts.

Magnalium is attacked by weak acids (e.g., boric acid) and sodium oxalate and, to a lesser extent, by other firework chemicals.

It is, however, protected by potassium bichromate, which is often used for this purpose (see Chapter 5).

Magnesium Carbonate, $MgCO_3$

It appears that this material is useful for making potassium chlorate or perchlorate free-flowing. American smoke compositions employ 3% of the total weight of the chlorate itself. One percent of the light carbonate was not as successful in a lance composition as 1% tricalcium phosphate for promoting free flow in the funnel and wire apparatus, although it obviously is potentially useful as a neutralizer.

Since calcium chlorate and magnesium chlorate are very hazardous, it probably is unwise to use magnesium carbonate as a neutralizer with chlorate mixtures.

Mercury Chloride, Hg_2Cl_2, Calomel

Mercury chloride is a white, tasteless, insoluble, extremely fine powder that formerly was used as a chlorine donor in color mixtures. The fantastically high cost of mercury salts precludes their use in fireworks, but there are good substitutes. Lead chloride has been used as a substitute, but it is not used now. These substances are no longer used and are only of historic interest.

Naphthalene, $C_{10}H_8$

Naphthalene is produced commercially from coal tar from the fraction boiling between 180°C and 250°C. It is a white solid crystallizing in plates with a characteristic tarlike odor. It melts at 80°C and is almost insoluble in water, but it readily dissolves in benzene. Its main use in pyrotechnics is for the production of black smoke. See Anthracene.

Paraffin Oil and Wax

Paraffin oil finds some use in fireworks and frequently performs more than one function at a time. When added to a colored fire mixture, for instance, it not only helps to reduce the influx of moisture and reduce the sensitivity, but also makes the mixture easier to press. Normally about 1% is used for these purposes.

Paraffin wax usually is used to coat metal powders or to waterproof finished fireworks, which are merely dipped in the molten wax.

Phthalic Acid Compounds

It has been known for some time that the modern whistling compounds from China contained phthalic acid compounds. In particular, potassium hydrogen phthalate has been identified. The author tried the normal laboratory grade without success. Some light could have been shed on this problem as a result of an article by Richard Dilg in *American Firework News* (November 1987). The article suggested that the material used could be made from the para isomer rather than the ortho isomer. The whistles therefore might be made from a terephthalic acid salt (1,4 benzenedicarboxylic acid). Potassium hydrogen terephthalate is now the subject of a patent. Phenolphthalein can be used as a substitute for charcoal, albeit at a price!

Phosphorus, P

Only the red variety finds any application in the manufacture of fireworks, and only then in the production of amorces and match striker surfaces. The production of amorces is a very dangerous and somewhat specialized business, and it is quite essential to ensure that dry amorphous phosphorus and dry potassium chlorate never get near each other or they will explode violently when they are subjected to even the slightest friction. Over time, red phosphorus slowly oxidizes in contact with air to phosphoric acid. It must be freshly prepared and its acidity carefully monitored as toxic phosphine is evolved.

Weingart describes liquid fire rockets, which require yellow phosphorus to be placed in the tops of rockets. There obviously is no commercial future in this, and it is not particularly recommended.

Pitch

> *And the streams thereof shall be turned into pitch, and the dust thereof into brimstone, and the land thereof shall become burning pitch.*
>
> Isaiah 34:9

Hard pitch, the residue from the distillation of coal tar, sometimes is used in powdered form in the production of cheap smokes used for making drain testers, but liquid tar absorbed on sawdust often gives better results. The powder also has found some use in the production of colors; good red and green lances, for example, have been made with pitch as a fuel. However, the possible impurities could be hazardous with pitch as a fuel. On the other hand, the possible impurities could be hazardous with chlorates, and pitch is not much used. German formulations sometimes have employed beechwood pitch—*buchenholzpech*—and the Japanese have shown a partiality for pine root pitch, a by-product of the distillation of oil of turpentine.

Polyvinyl Chloride, PVC. $(CH_2{=}{=}CHCl)_n$

The Germans appear to have been the first to use this material extensively as a chloride donor during the late 1930s. It is a white powder or granular substance with a softening point of about 80°C. It is soluble in dichloroethane and cyclohexanone. At about 160°C it begins to decompose with the liberation of HCl.

PVC contains 56% chlorine and normally is used in magnesium compositions. It is marketed under various trade names such as Vestolith, Corvic, Igelith, and Vipla. There also is a double-chlorinated PVC marketed as Rhenoflex. It is very superior to ordinary PVC for the production of barium nitrate–magnesium green stars, but it also is much more expensive.

Closely related to PVC is Saran, a copolymer of vinylidene chloride $CH_2 = CCl_2$ (85%) and vinyl chloride $CH_2 = CHCl$ (15%), which can be used in similar ways to PVC. It is insoluble in most liquids but will dissolve in cyclohexanone.

Notes on PVC

In recent times, greater use has been made of chlorine donors with a larger percentage of chlorine. Cereclor 70, for example, is a chlorinated hydrocarbon that contains 70% chlorine, and there is a chlorinated PVC called Lucalors.

These chlorine donors come in powdered form or as fine granules. The granules are apt to become charged with electricity and adhere, but they are useful for producing denser mixtures (e.g., for ball stars).

Some processes use the PVC in the composition as an adhesive for making stars, but this also has the additional problem of using organic solvents. Cyclohexanone could be a solvent of choice, possibly with the addition of octyl phthalate, but it is not pleasant to handle industrially.

Potassium Benzoate, $C_6H_5.COOK$

Whistling compounds normally use this substance in combination with potassium perchlorate. The powder must be at least 120 mesh and must be kept very dry because it has a tendency to absorb water from the atmosphere, and the mixture will burn but not whistle.

Potassium Bichromate, $K_2Cr_2O_7$

Potassium bichromate is an orange red crystalline powder that is soluble in water and decomposes at about 500°C. It is poisonous and corrosive, so care must be taken to avoid the material coming in contact with the mucous membranes.

Although it is not very useful as an oxidizing agent, it is of great value for preventing the corrosion of magnesium and magnalium. In solution, potassium bichromate forms a reaction at the surface of the metal, thus preventing further corrosion (see Chapter 5).

Potassium Chlorate, $KClO_3$

One of the most important chemicals used in the firework industry, this material is prepared by the electrolysis of potassium chloride solution and frequently is imported into England from Spain and Switzerland. It is a fine white powder passing 170 mesh and normally is of 99.5% minimum purity. The melting point is 360°C. The sensitive nature of potassium chlorate is a problem for the firework manufacturer, particularly in the presence of sulfur, ammonium salts, and phosphorus, and none of these materials should be used with it. Wherever possible, potassium perchlorate should be used in place of the chlorate, but this is not always possible. There also is a tendency to use it more in commerce because it is less expensive than the perchlorate.

Potassium Nitrate, KNO_3

The modern method of manufacturing potassium nitrate is to mix hot saturated solutions of potassium chloride and sodium nitrate. Sodium chloride crystallizes out first, followed by the potassium nitrate. Purification then takes place by recrystallization. It has a melting point of 334°C and usually is imported into England. Two grades are used in fireworks, both 99.8% purity, but vary only in particle size. The fine powder passes about 180 mesh, but some use is made of a coarser crystalline powder of about 60 mesh. Unfortunately, the fine powder tends to cake and invariably must be reground before use.

Potassium Perchlorate, $KClO_4$

Potassium perchlorate is obtained by the electrolysis of cold concentrated sodium chlorate solution, followed by precipitation with potassium chloride. It can be freed from potassium chlorate by digesting it with concentrated hydrochloric acid, with which the perchlorate has no reaction. Material of at least 99.5% purity is used, and this is imported. The fine white powder usually passes 170 mesh. It decomposes at 400°C. Wherever possible, the perchlorate is used in preference to the chlorate, but it still requires careful treatment. Mixtures of sulfur and potassium perchlorate are slightly less sensitive than those mixtures with potassium chlorate. In view of this, people frequently add sulfur to stars made with the perchlorate to improve the somewhat difficult ignition, but this is not recommended. Charcoal usually will help the ignition of perchlorate stars without the addition of sulfur.

Potassium Picrate, $C_6H_2(NO_2)_3OK$

This uncommon salt normally is prepared only in small quantities for firework manufacturers. It forms yellow crystals that are slightly soluble in water and decompose on heating to about 300°C. As with picric acid and other picrates, it is sensitive to shock, but it is reasonably safe to handle if care is taken. The salt is used for making whistles and can be pressed alone or mixed with potassium nitrate, stearine, or asphalt. Asphalt is the least suitable because it is rather unpleasant to handle in fine powder form. It is worth mentioning that because lead picrate is dangerous, lead ramming blocks should not be used. Picrate whistles have been superseded by other types.

Silicon, Si

The dark gray powder of commerce is known as "fuzed" silicon. It usually passes 240 mesh and contains about 94% to 95% silicon. The material is used in fireworks only as an igniter in certain types of fireworks that require a hot slag to initiate the reaction. For this purpose, it frequently is mixed with potassium nitrate and gunpowder. Silicon melts at 1420°C.

Silicon Dioxide, SiO_2

Recent years have seen the appearance of very pure forms of silicon dioxide obtained by flame hydrolysis. Products such as Cab-o-sil or Aerosil have an average particle size of about 16μ and are used as grinding, sieving, and "free-flowing" aids for hygroscopic and other products. They also help to prevent the electrostatic charge of powdered substances and have many other uses such as thickening and thixotropic agents for liquid systems, pastes, and emulsions.

Pyrogenic silicas are either hydrophilic or hydrophobic. Hydrophobic aerosil R 972 adsorbs only 0.8 millimol water even at 80% humidity. Hydrophilic aerosil, however, adsorbs 20 times this quantity.

The material is of extremely low density and flies about the workshop, but there is no evidence that it caused silicosis. Some firework manufacturers are said to use the material for coating iron and for adding to flash powders in the proportion of about 0.1%.

Sodium Azide, NaN_3

Most azides are explosive and demonstrate high thermal and mechanical sensitivity. Exceptions are the alkali metal azides, which decompose on heating but do not explode. Sodium azide is formed as follows:

$$NH_2Na + N_2O = NaN_3 + H_2O \ (T = 200°C)$$

Sodium azide is used in pyrotechnics as a nitrogen generator, but it has a toxicity almost akin to that of cyanide.

Sodium Nitrate, $NaNO_3$

Chile saltpeter, originating in the caliche deposits, comes to Europe for final refinement, but sodium nitrate also is manufactured from synthetic nitric acid by neutralization. The crystals are readily soluble in water and melt at 308°C. Sodium nitrate normally is not used in fireworks because of its hygroscopic nature, although the degree of water absorption is related to the purity of the salt. In combination with magnesium, it is useful for illuminating flares, but the stores must be sealed so that they do not come into contact with the air.

Sodium Oxalate, $Na_2C_2O_4$

Prepared commercially by neutralization of oxalic acid, sodium oxalate forms white crystals that are soluble in water but insoluble in alcohol. It melts at about 250°C and normally is bought as a fine powder of 120/200 mesh.

There are two main uses for this substance. The first is for the production of yellow colors in combination with potassium perchlorate and suitable fuels. Potassium chlorate is unsuitable because it is likely that some double decomposition will take place with formation of deliquescent sodium chlorate. It also is advisable to use spirit only when making stars with this material because water will tend to accelerate any reactions that may take place. The second use is the production of yellow glitter effects, with gunpowder, aluminum, and antimony.

Sodium Salicylate, HOC_6H_4COONa

This substance is made by heating sodium phenate in an autoclave with carbon dioxide. It forms white, lustrous, pearly scales, which are soluble in water and alcohol. The powder used for making whistles must pass 120 mesh and is used in combination with potassium perchlorate. As with potassium benzoate, the whistles are effective but are liable to deteriorate in storage because of the gradual absorption of moisture. The absorption of water also seems to be accelerated in the presence of natural resins.

Starch, $(C_6H_{10}O_5)n$

Although starch is almost insoluble in cold water, it dissolves in hot water to form a thick solution that can be used as an adhesive. It frequently is used for making quickmatch, and some forms of it are used for making stars in Japan. Maize starch sometimes is added to compositions to cut down the burning speed. Wheat flour also can be used for this purpose. Japanese soluble glutinous rice starch is a white or slightly brown powder that dissolves easily in water to form a very adhesive paste. It is manufactured by soaking glutinous rice in cold water, steaming, and subsequently pounding the material into a rice cake. The rice cake is then roasted and finally pulverized.

Stearine-Stearic Acid

This material probably is a mixture of stearic and palmitic acids obtained by hydrolysis from their glyceryl esters. The powder used for firework manufacture usually passes 80 mesh. The main use for this material is for adding it to some compositions that are somewhat sensitive to friction. It also can be used in those fireworks where it is desirable to have a long flame.

Strontium Carbonate, $SrCO_3$

Strontium occurs naturally in England and elsewhere as strontianite, $SrCO_3$, or celestine, $SrSO_4$. The natural carbonate is frequently used in firework manufacture and usually is a pale pink, fine powder of about 120/200 mesh. Purer forms are made by boiling celestine and ammonium carbonate or fusing celestine and sodium carbonate. Strontium carbonate is used more than any other salt for making red flames and stars, but it also finds occasional use as a retardant in some gunpowder mixtures. It is insoluble.

Strontium Oxalate, $SrC_2O_4{\cdot}H_2O$

This material is manufactured by precipitation. It is a fine, white insoluble powder, losing its water at about 150°C. It often is used to obtain red color as with carbonate, but it is considerably more expensive. The water content also is something of a drawback. As with the carbonate, it also finds occasional use in the manufacture of glitter stars with gunpowder and aluminum.

Strontium Nitrate, $Sr\,(NO_3)_2{\cdot}4H_2O$

Needless to say, only the anhydrous salt is used for firework manufacture. It melts at 570°C and decomposes to the oxide at 1100°C. In many ways, this is the best of the strontium salts because it gives a very rich flame coloration, but unfortunately the technical grades very quickly become damp when mixed with potassium chlorate. The inconvenience of having to dry the salt and use the fireworks made with it fairly quickly somewhat limits its use. It is most commonly used in combination with magnesium and PVC for the manufacture of red stars and flares that are pressed dry. It also is used in the manufacture of bengal illuminations and road flares.

Sugars

Lactose, $C_{12}H_{22}O_{11} \cdot H_2O$, is sometimes used as a fuel in firework manufacture. The water can be removed only at about 125°C, and lactose melts with decomposition at 200°C. Used in compositions that are required to react at low temperatures, it is of use in the manufacture of some blue colors. It also replaces sucrose, which is more sensitive with chlorates. Perhaps its most extensive use is in the manufacture of smokes using organic dyes.

Sucrose. beet or cane sugar, $C_{12}H_{22}O_{11}$, is used only in the production of colored smokes along with potassium chlorate and an organic dye or mixture of dyes. The finely powdered confectioners' sugar is used for this purpose, and it appears to be fairly safe in combination with the chlorate and an excess of dye. It melts at 160°C. Sorbitol, $C_6H_{14}O_6$, with a melting point of 98°C, and mannitol, $CH_2OH \cdot (CHOH)_4$, CH_2OH, with a melting point of 166°C, have also found some use in special pyrotechnic products, the former in smokes and the latter as a fuel in some signal stars.

Sulfur

> *. . . and he shall be tormented with fire and brimstone in the presence of the holy angels.*
>
> Rev. 14:10

One of the most important fuels of the firework industry, sulfur is a pale yellow, fairly dense powder, all of which should pass 120 mesh. Some use also is made of a rather coarse powder in the manufacture of pinwheels. During the distillation process and presumably the melting, which takes place in the Frasch process in the United States, the sulfur solidifies to a solid mass of rhombic crystals. This mass, which is soluble in carbon disulfide, is powdered and forms the powdered sulfur of commerce. It must not be confused with Flowers of Sulfur, which is produced during the distillation process. The powder produced by cooling sulfur vapor always contains some free acid, maybe as much as 0.25%, and therefore is quite dangerous to use.

Sulfur melts at 113°C and ignites at about 260°C. It is in plentiful supply, and it is nonpoisonous, inexpensive, and safe to handle. Sulfur is produced extensively by the Middle Eastern oil industry.

Synthetic Resins

Recent years have seen the introduction of urea-formaldehyde, phenol-formaldehyde (Bakelite), and resorcinol-formaldehyde resins. They generally are used in the form of syrups that can be mixed with accelerators and oxidizing agents, and the resulting mass sets to a hard solid under various conditions of temperature and pressure. Powdered resins also feature as fuels in color compositions, but they may irritate the nose and lungs.

Thio-Urea, $(NH_2)_2CS$

Thio-urea is a white crystalline solid that melts at about 180°C. It finds some use in the production of white smoke in combination with potassium chlorate and so on. An explosion in a grinder used for thio-urea and potassium chlorate led to an investigation, which indicated that such mixtures are very reactive and sensitive to mechanical forces.

Titanium, Ti

Although titanium has found use in military compositions for some time, it probably was not until the 1972 publication of the first edition of *Fireworks: Principles and Practice* that the metal became widely used in the firework industry. Trials were carried out in about 1960 in the United Kingdom with material that consisted of swarf from the airplane industry.

The first material to be used consisted of sharp acicular fines of varying mesh size, and there also was some other rubbish and oil, as one would expect to find in a machine shop. The material is very hard and, although it is not corroded by water—or dirty to handle—it is possible that some fires were caused by friction. There certainly were enough to give cause for concern, but the waste material was also inexpensive.

More recently there has been a move to using the sponge material, and it appears that there have been fewer accidents as a result. Although no one can be sure, it generally is agreed that sponge material is the safest to use.

The metal is easily ignited to produce brilliant silver sparks, can be used the same way as iron powder, and has the advantage of not needing to be coated. Like all new things, it may have been overused, but no one regrets the absence of the dirtier old aluminum powders.

Ferro-titanium also is very useful for making fireworks. Very much depends on the percentages in the alloy and the temperature of the flames. High percentages of titanium tend to be silver or silvery gold, but some beautiful gold sparks can be produced at levels of only 30% to 40% titanium. The alloy does not corrode easily.

Two accidents in the United Kingdom during the processing of blackpowder-titanium mixtures prompted the Government Health and Safety Laboratory to conduct experiments to determine the effects of particle size and proportion of titanium on the mechanical sensitivity of such mixtures. The titanium in the study was 10 to 20 mesh, as used at the time by parts of the firework industry to manufacture display gerbs; the titanium flake was the same material as had been involved in an explosion during pressing. The conclusions of the work have been published (48) and, indicate that,

1. The particle size of the metal is not critical. Tests used sizes of 355 to 500 μm and 710 to 1000 μm.
2. The most sensitive mixture to mechanical forces (such as can be encountered in pressing) contains 20% to 30% titanium by weight.
3. Significant safety gains should result from working with formulations containing less than about 8% titanium.

Whether it is relevant or not, there is a view in the industry that the sharp acicular turnings that are available from both titanium and its alloys are less acceptable for use than sponge titanium.

In a personal communication, Professor Yurii Frolov spoke of a novel type of titanium supplied by the Institute of Chemical Physics at the Russian Academy of Science in Moscow.

Powders of high surface area of titanium are made by the high temperature reduction of TiO_2 using magnesium as a reducing vapor. Commercial grade TiO_2 and fine magnesium powder (99.7%, 44 μm) were put in a reactor under argon (1 bar) and heated locally for ignition.

The heat of the reaction

$$TiO_2 + 2\ Mg = Ti + 2MgO\ (H = -330\ kJ/mole)$$

is sufficient to sustain the conversion reaction once ignited. In the reaction front of the combustion wave, the measured maximum temperatures were in the range of 2100°C to 2500°C.

The solid product, which consists of a mixture of Ti, unreacted Mg, MgO, and unreacted TiO_2 was leached with 10% hydrochloric acid, washed with distilled water, dried, milled, and sieved. Analysis revealed the following impurities by weight: Mg 0.8%, MgO 0.5%, TiO_2 2.0%, and Cl 0.01%.

The industrial Kroll method for the production of titanium sponge reduces titanium chloride in molten magnesium and yields metal powders that have a surface area on the order of 0.1 m/g after milling. However, the combustion wave process has a higher figure (up to 20 m/g). Powder made in this way is more reactive than aluminum, magnesium, and commercial titanium powder and is more stable for use in pyrotechnics.

Titanium Dioxide, TiO_2

This oxide is not much used by the firework industry, but it occasionally features in some smoke compositions and sometimes is added to waterproof paints.

Vaseline—Petroleum Jelly

The jelly has sometimes been incorporated into illuminating compositions, where it waterproofs, binds, and helps to desensitize, but little use is now found for it.

Wood Flour

Softwood powder of about 60 mesh is used to dilute or slow down some slow-burning items. It also is used in some molesmokes. Hardwood flour is said to cause respiratory disease.

Zinc, Zn

Certain recipes in the old firework books included zinc, but little use is made of the finely powdered variety except in the manufacture of certain types of smokes. Zinc smokes are very efficient and very good, but they are somewhat sensitive to moisture and have been known to react and ignite themselves. Most schoolboys know that a mixture of zinc, ammonium nitrate, and ammonium chloride will catch fire when a small amount of water is added to the powder.

Zinc Oxide, ZnO

The principal use for this material also is in the manufacture of smokes. Many of the smokes used in warfare basically are mixtures of zinc oxide, hexachloroethane, and calcium silicide. Zinc oxide also has been used by the Germans in mixtures of barium nitrate and aluminum in the manufacture of Very stars. It apparently helps to melt the aluminum tube.

5

GENERAL PYROTECHNIC PRINCIPLES

The burning speed of a firework mixture is governed by a variety of factors, the three most important being particle size, temperature, and pressure. In commercial practice, particle size produces the greatest problems, mainly because any material can vary from a coarse powder to a powder that is fine and impalpable. Naturally, variations exist between these two extremes, and chemical batches can vary from one manufacturer to another. Indeed, it is even possible to obtain variations between different batches from a single manufacturer. In practice, materials are bought to a specified mesh size; potassium nitrate, for example, might be bought as 150 mesh powder.

A 150-mesh sieve originally had 150 wires to the inch, but so much depended on the diameter of the wire itself that this inevitably led to standardization. In more recent times, greater accuracy has come about by measurement of the opening between the wires. In this system, 1μ represents 1/1,000 mm, as shown in Table 5.1.

British and American sieves are much the same and are still used in commerce.

It is well known that a firework composition usually will burn much faster with the increasing fineness of the chemicals used. On some occasions, the firework maker will use this principle to make fast-burning stars or explosives, but he also may do just the opposite and employ slow-burning compositions in portfires. The burning speed of a pyrotechnic composition also is affected by the surrounding temperature, so it will be found that as the initial temperature increases, so will the combustion rate. Large quantities of gases usually are produced in these burning processes, and although these gases normally escape into the air, the story is rather different if the burning takes place in con-

Table 5.1 Standard Sieves

American Sieves	British Sieves	Sieve Opening in Micron
12	10	1,680
18	16	1,000
20	18	840
30	25	590
35	30	500
40	36	420
60	60	250
70	72	210
80	85	177
100	100	149
120	120	125
140	150	105
170	170	88
200	200	74
230	240	62
270	300	53

finement. In an enclosed space, the combustion rate rises quickly and may become explosive.

Turning to the important matter of sensitivity, the firework maker is concerned primarily with sensitivity to shock and friction. To some extent, he also is concerned about ignition temperature, but sensitivity to shock and friction is of greater practical importance, mainly because firework mixtures usually are more sensitive than ordinance compositions. In general, firework mixtures that are more sensitive to friction also are more sensitive to shock, but this is not necessarily the case.

Shock sensitivity usually is determined with a drop hammer. A small quantity of composition is placed on the anvil of the apparatus and a known weight is allowed to fall on the anvil. The weight is dropped from various measured heights until a minimum distance can be recorded at which the material explodes. Several trials are necessary for any degree of accuracy, but the method provides a useful comparison between various mixtures. Sensitivity ratings can be produced in this way; one method, for instance, uses picric acid as a standard with a figure of 100.

Friction sensitivity sometimes is vaguely determined by striking the mixture a light glancing blow with a polished hardwood mallet, but the other method uses two weighted porcelain plates. A small amount of the mixture is placed between the plates, which rub against each other. A comparative set of sensitivity figures can be obtained by weighting the plates and determining when the mixture explodes or ignites. The Germans use a mixture of potassium chlorate and gallic acid as a standard with a sensitivity figure of 1. On this system, a chlorate color star would have a figure between 2 and 8, a perchlorate and aluminum mix would be about 6, and gunpowder would be approximately 36 or more.

The use of these specialized methods of testing can be both time-consuming and expensive, but they are essential, particularly for military processing. The average manufacturer is unlikely to be able to purchase such costly apparatus and is more likely to resort to more primitive tests with a wooden mallet or a hammer on an anvil, a method still used in some military manuals.

A useful set of results has been published by BAM in Germany with a mixture of potassium chlorate and lactose. At the extreme, with 10% of potassium chlorate and 90% lactose, the mixture is not sensitive to impact and friction. However, as the potassium chlorate percentage increases, so does the sensitivity rating for both the friction and drop hammer tests, right up to 80% of potassium chlorate. For example, on the 90% lactose mixture, the friction number was as low as 36 and the drop hammer 2.5 kpm. With a mixture of 20% lactose, the figures were 4 and 0.5 kpm. Most of these mixtures ignited at about 200°C.

From this follows the obvious question of how the sensitivity would compare with the use of potassium perchlorate instead of potassium chlorate. Some 20 years ago Dr. Pfeil of BAM very kindly indicated that although the ignition temperatures were somewhat higher at 360°C, the sensitivity is a little bit less. As might be expected, mixtures of 50/50 or 60/40 were more sensitive to impact.

In turn, this then led to the question whether potassium perchlorate and sulfur was less sensitive than potassium chlorate and sulfur. Dr. Pfeil's opinion was that such mixtures were a little bit less sensitive than chlorate and lactose, but not enough to get enthusiastic about. On the BAM friction scale of 1 to 36, perchlorate/lactose 60/40 gave a figure of 36, with no reaction; perchlorate/sulfur 60/40 was 8; and perchlorate/lactose/sulfur was 16.

These figures provide only a useful guide, but we are grateful to those who have done the work. Just as an indication, one or two other items are worth inclusion.

Still on the friction scale of 1 to 36, the following are typical:

Barium chlorate star	2–8
Red chlorate star	8
Perchlorate/pyro aluminum 70/30	6
Nitrocellulose	24
Silver star nitrate base	26
Nitrate/sulfur/charcoal mix	36

In an experiment, a friend, Dr. Krone, using 200 kilos of red and silver stars in a drying room, found that ignition caused very rapid burning in which the stars were thrown out of the building, which was made of substantial concrete. It was not a 1.1 mass explosion, but the behavior of stars when ignited in bulk can be highly dependent on the degree of confinement.

It is worth emphasizing that these figures are bound to vary because of the wide variety of materials used. A barium chlorate star with just accaroid resin, charcoal, and dextrin will give a different rating if potassium chlorate is part of the mixture.

Studies of sulfur-chlorate mixtures at the Health and Safety Executive laboratories in the United Kingdom investigated the effect of particle size and additives on the thermal stability and sensitivity of such mixtures. The studies showed that stoichiometric sulfur-chlorate mixtures ignite at 119°C, around the melting point of sulfur. They also showed that the addition of up to 70% of other materials can still produce mixtures capable of igniting at temperatures as low as 119°C. These mixtures also are friction- sensitive and have values well below the UN transport criteria (82).

It is interesting to note that metal powder especially increases the sensitivity of these mixtures. An accident took place with braided Chinese fuse that contained both chlorate and iron powder. It was fastened to a piece of black match mechanically with an aluminum hog ring, catching fire immediately.

The sensitivity of flash powder mixtures is of special concern in situations in which large quantities are used to make salutes. We are indebted to friends who conducted tests in the United Kingdom using the Rotter Impact Test to produce figures of insensitivity.

The scale is:

0–30	very sensitive
30–90	sensitive
90+	comparatively insensitive

With a figure of about 70

Potassium perchlorate	37%
Magnesium fine	57%
Graphite	6%

is a mixture that has a good record industrially allowing for the normal precautions.

The mixture of potassium perchlorate 66% with dark pyro aluminum 34% gives a figure of 58. The question arose therefore whether the addition of graphite would improve this figure; it does. Potassium perchlorate 60%, pyro aluminum 38%, graphite 2% produces a better figure of 78.

Interestingly, barium nitrate as the oxidizer may not be an advantage because the mixture of barium nitrate 68%, pyro aluminum 23%, and sulfur 9% is sensitive with a figure of 66. Devastatingly, the Chinese mixture of potassium perchlorate 45%, sulfur 15%, and magnalium 40% could be as low as 20 as a result of the hardness of the magnalium particles. It may well be that the presence of sulfur sensitizes the mixture.

It can be seen from these figures that the chlorate mixtures are the most sensitive, and there is no doubt that barium chlorate presents the greatest hazard in that compositions made with it are almost as sensitive as mixtures of potassium chlorate and gallic acid. Part of the problem seems to lie in the fact that potassium chlorate, for example, begins to decompose at a fairly low temperature, about 350°C, and that this decomposition also is accompanied by the evolution of a small amount of heat. Perchlorates and nitrates are much safer in this respect.

The extreme danger of mixing potassium chlorate and phosphorus is almost too well known to comment on. Similarly, chlorates and sulfur present a serious hazard when placed together and normally are not used. Indeed, such mixtures were banned by law in England as long ago as 1894. The discovery of potassium chlorate by Berthollet at the end of the eighteenth century was a major step forward for the firework industry. By 1830, the industry was able to produce colors that had never been seen before.

During the early years of the nineteenth century, fireworks were very much a cottage industry, with historical works like Brock describing factories in city suburbs and people drying stars in front of open fires. In the United Kingdom, the Gunpowder Act of 1860 and the Explosive Acts of 1875 did much to regulate the manufacture of explosives, but they did not prevent accidents, which occurred with monotonous regularity, many of them appearing to be caused by the use of a chlorate.

The annual report of Her Majesty's Inspectorate of Explosives in 1894 cites a number of interesting accidents. One of these concerns a Mr. Hicks, who had manufactured signal lights that could be attached to a kite and fired by electricity. Mr. Hicks had seven of these items hanging on a wall outside his central London premises, and after about two years, exposed to all weathers, two of them suddenly exploded. The signals were made of barium chlorate and shellac, and they were primed with a sulfur composition. They had been wet, and they exploded during a period of hot weather.

In the report issued in 1890, there is an account of an explosion of a magazine at the factory of James Pain. This took place on July 6 when approximately 7 tons of fireworks exploded just before noon on a Sunday when nobody was at work. As quite a number of possible factors could be eliminated in the circumstances, it was assumed that the explosion was caused by spontaneous combustion. In the same report, a number of accidents all involved the destruction of drying rooms, thus raising the suspicion that some chemical instability was causing these problems.

Chemical investigations instituted by Colonel Majendie, the Chief Inspector, came to a number of conclusions about the chemicals in use at that time. These words are stressed because there may be some significance in this. Lampblack was found to contain free sulfuric acid. Chertier's copper, made by treating copper sulfate and potassium chlorate with ammonium hydroxide, was alkaline when freshly made but "acidic when it lost ammonia." They do not mention the possibility of ammonium chlorate. Paste for damping pillbox stars was made with alum and found to be acidic.

Colonel Majendie issued a memorandum to the industry on January 1, 1882, instructing the industry that care must be taken. Two further accidents established that sulfuric acid had been found in stearine; that linseed oil had been found to cause spontaneous heating, even when it was partly polymerized by boiling; and that fountains containing chips of colored stars had been known to cause spontaneous combustion. Warnings also were issued about the need to perfectly wash Flowers of Sulfur, the subject being discussed after a spontaneous accident from colored fire aboard a pleasure steamer in Norwich. The report mentions as many as 12 accidents between 1873 and 1880.

Of particular interest is the manner in which lances were made in the factory of James Pain. The lances were charged with color compositions containing either potassium or barium chlorate, and then they were topped with a white fire igniter that contained potassium nitrate, Flowers of Sulfur, and antimony sulfide. The report made the point that either of these mixtures is quite safe when separated, but dangerous when brought into contact. In fairness to Pain, the investigator makes the point that careful charging had been done, but he goes on to suggest that if a suitable intermixture took place at the junction, then this would be quite dangerous.

Alas, exactly 10 days after this report had been made, fire broke out in one of Pain's filling sheds. This fire was caused by the observed spontaneous ignition of green lances that were lying on a shelf in the shed. Only one workperson was in the shed at the time, and she was able to put out the fire promptly with a bucket of water that was there for the purpose. Shortly afterward, Pain informed the Inspectorate as follows: "I have received advices from America to the effect that some of our green lances caused an explosion there some days ago."

A number of accidents occurred as a result of the charging of Roman candles with chlorate stars, and the report goes on to say,

> . . . by themselves these stars would not, I believe, give rise to danger, but closely packed into a mixture containing sulfur they are not, I think, quite free from danger of spontaneous ignition. On the one hand, no doubt the hard mass of the stars offers less chance of action than the contact of two but slightly compressed mixtures. On the other hand, however, the surfaces in contact are far greater, assuming no actual mixture to have taken place, and in view of the present accident it certainly seems in the highest degree advisable never to imbed a star containing a chlorate in a mixture containing sulfur.

Roman candles have been made with chlorate stars in the West until quite recently, and in some cases they were embedded in compositions that contained as much as 28% sulfur, although few people would use Flowers of Sulfur in these times. Good-quality ground rock sulfur contains very little free acid. A candle composition that contained no sulfur but did contain a percentage of soft grain commercial gunpowder would be well within guidelines referred to later.

Work carried out by Colonel Majendie produced the Order in Council of 1894, which finally made mixtures of chlorates and sulfur illegal. This was little more than 100 years ago, and the problem still exists today. The research that had taken place also involved the detonation of star shells, which had caused some very serious accidents. Indeed, Brock had been involved in these exercises, and it was widely believed that many detonations had been caused by soft chlorate stars that had crumbled to some extent. It also was stated that stars made with either sodium oxalate or copper acetoarsenite were particularly liable to create this problem. As a result, Brock stated that large shells were always charged with these colored stars mixed with lampblack stars to lessen the likelihood of detonation.

This information has been placed in its historical context to show how well these events have served those of us in the Western Hemisphere. Changes of compositions had to come about in 1894, and the absence of sulfur in colored mixtures was not a very great problem, and, after all, the prohibition was only for clear mixtures of chlorate and sulfur. However, some effects such as stars did not ignite very well, and it was accepted that a coating with a sulfur-containing blackpowder-type mixture did not constitute an admixture, and usually the migration of sulfur into the chlorate mix was not very great. This was a very important and necessary guideline, and it will be seen that it influenced the thinking when designing a suitable Roman candle fuse. It could contain manufactured grain gunpowder but not sulfur.

Some years ago the British government once again looked at the matter, particularly with the increase in importation from the Far East, and raised a number of interesting issues. The first of these raised the question of how pure the chlorates were in the previous century, when these accidents were so common.

A part of the problem in this case is due to the tendency for sulfur to be acidic and therefore to accelerate decomposition. An article on the instability of chlorate-sulfur mixtures (21) postulates a trigger reaction caused by the formation of polythionic acids on the sulfur. It is possible that air oxidation produces sulfurous acid on the surface of the sulfur and that this immediately reacts with the sulfur again to form polythionic acids. Evaporation of aqueous polythionic acid or a rise in temperature causes the partial decomposition of the polythionic acid as follows:

$$H_2S_nO_6 = H_2SO_4 + SO_2 + (n-2)S$$

Sulfuric acid is likely to react with the chlorate to form chloric acid and is thought to be a cause of instability, but sulfur dioxide also reacts with the moist chlorate to form the highly explosive chlorine dioxide.

$$SO_2 + 2KClO_3 = 2ClO_2 + K_2SO_4$$

$$2ClO_2 + 4S = 2SO_2 + S_2Cl_2$$

The chain reaction is influenced by friction, impact, sunshine, and heat. Experiments seem to suggest that natural sulfur is less reactive than sulfur formed by cooling sulfur vapor, and that the formation of sulfur dioxide contributes more to instability than the possible formation of chloric acid.

The difference between the chlorate and the perchlorate ions undoubtedly is related to their structure. It could be that the chlorate ion is flat but that the perchlorate ion is tetrahedral, with the possibility that elemental sulfur can approach the chlorine atom in one case but not the other. With regard to stability, the fact that perchlorate involves a chlorine atom sharing its four pairs of electrons with four oxygen atoms must be an advantage over chlorate, with three oxygen atoms and an unshared pair of electrons. Whatever it may be, it is agreed that a true mixture of chlorate and sulfur is hazardous and that a "contact" is subcritical.

It must be remembered that potassium chlorate made by electrolysis is much purer than it used to be. The old-fashioned method of manufacture from chloride of lime must have had calcium chlorate impurities, and this substance is highly unstable. It is common these days to make sure that chlorate compositions are damped down with pure water, particularly if the local supply is "hard" (i.e., contains calcium and magnesium salts). A wet process and a rise in temperature in sulfur-containing mixtures may lead to a deterioration of the sulfur and a risk of spontaneous combustion.

Chlorate stars are coated with blackpowder-type mixtures, and it must be admitted that in present times, this does not seem to be a major problem. Sulfurless gunpowder also can be used, but its price militates against this because manufacturers have to dedicate special machinery for the purpose.

The fact that it seems reasonably safe, therefore, to coat chlorate stars with blackpowder mixtures means that some degree of compromise is needed in those countries that have legislated against chlorate-sulfur mixtures.

As a result of various investigations, it has been unofficially agreed (by some) that:

1. The chlorate could be present as the ingredient of a consolidated star.
2. The sulfur could be present only as an ingredient of gunpowder or a similarly formulated composition used as an igniting or bursting charge. The charcoal:sulfur ratio should always be at least 3:2 (i.e., with charcoal in excess).

3. The stars may be primed with gunpowder provided that there is no migration of the sulfur into the star composition.

Chlorate-sulfur mixtures have not existed in many areas of the West for almost a century. However, the economic situation means that huge volumes of fireworks are being imported from the Far East and the growth continues. While importation was in the hands of manufacturers who understood the issues at stake, all was well. Now, however, the situation is out of control.

Importers now bring in fireworks in much the same way as they might import bananas, and there is much ignorance. Where laws still exist, the onus is on the importer to test extensively, and this is very expensive and time-consuming.

More recent studies by the U.K. government conclude that there is enough evidence from sensitivity and reactivity data to retain a ban on sulfur-chlorate mixtures.

Some firework makers add sulfur to perchlorate mixtures, and while it must be admitted that such mixtures are less sensitive than chlorate mixtures, they are best avoided. Tests seem to indicate that mixtures of perchlorate and sulfur are only marginally less sensitive to impact and friction. The main point really is that they are chemically more stable.

Mixtures of chlorates with magnesium, aluminum, metallic arsenic, the sulfides of arsenic, antimony, and phosphorus are all highly sensitive to both shock and friction.

It is clear that nitrates are by far the safest oxidizers, but it is obvious that even they can be dangerous in some situations (e.g., when mixed with fine metal powders).

Many factors play a part in the ignition of firework compounds, but the following general points contribute to an increase of sensitivity:

1. Mixtures with fine particles
2. Intimate mixtures of the chemicals
3. When the components are in stoichiometric proportions, an excess of fuel often decreases sensitivity
4. Increase in temperature
5. Loose powder is more sensitive than compacted masses
6. The presence of grit and impurities
7. Increase in pressure

An increase of outside pressure will lead to an increase in the burning rate, and this could lead to explosion, with the destruction of both the charge and the equipment. Under high pressure, the combustion products—the hot gases—can easily penetrate the pores of the charge and thus preheat the mixture. This leads to a higher rate of combustion, and what started as a stable process converts to convective combustion and possibly detonation.

Needless to say, granulation of explosive composition increases the burning speed. A pile of colored star composition probably will flare up when ignited, causing much heat and flame. If, however, the composition had been made into stars, their ignition could lead to detonation, with considerable destructive force. Loose powder containing metal powder also would detonate if the quantity were appropriate.

Phlegmatizers frequently are added to firework compositions so that friction sensitivity can be reduced; stearine, vaseline, and paraffin oil can be particularly useful in this respect. In some cases, inert substances can be added and, although they do not take any part in combustion, they effectively reduce sensitivity. Carbonates can be useful for this purpose and also are useful for reducing any tendency toward the formation of acidity. It is advisable, for example, to add a small percentage of barium carbonate to green stars that are made with barium chlorate. Strontium carbonate is particularly useful in red fire compositions because it serves the double purpose of providing both color and phlegmatization.

The decomposition of chlorate mixtures requires little heat for their initiation and therefore are unlike either the perchlorate mixtures, which do require a little, or the nitrate mixtures, which require a large quantity of heat before decomposition can take place. In addition, the presence of fine metal powders provides an extra liability to decomposition; mixtures of chlorates and magnesium are particularly hazardous. Potassium perchlorate and aluminum is much safer than the chlorate, as one would expect, but barium or potassium nitrate and aluminum would be the better choice if the required effect could be obtained with these substances, although this is not always possible. Nevertheless, barium nitrate and aluminum mixtures present problems and have been known to produce very serious explosions when handled in large quantities. The high ignition temperature of aluminum can make these mixtures difficult to ignite, particularly if atomized aluminum is used, but with the fine flake powders, the mixture can be quite explosive. Naturally, the addition of sulfur to a mixture of barium nitrate and pyro aluminum will make the mixture easier to ignite, but it also is considered that this increases the sensitivity and it is best avoided if possible. Wet mixtures of these substances are apt to heat up from time to time, and because the decomposition of aluminum is an alkaline reaction, it is advisable to add something to counteract this; 1% or 2% of boric acid can be added for this purpose.

Moisture is always a problem to the firework maker, and because fireworks must be stored for periods of time in varying conditions—and invariably in unheated buildings—they are bound to absorb some moisture. When large amounts of water are absorbed, physical changes are likely to occur, making the fireworks useless or even dangerous to handle. Chemicals vary in their degree of water absorption, but it is essential to avoid the use of those materials that have a tendency to absorb moisture. Consideration also must be given to the possibility that, in the presence of water, electrolytes will exchange their partners to form a partial double decomposition. The chemistry of this reaction is well known, but it clearly would be unwise, for example, to use potassium chlorate and sodium oxalate together since sodium chlorate is so deliquescent. Sodium oxalate can be used much more satisfactorily with potassium perchlorate, of course, but all oxalates occasionally present their problems in the presence of water and impure chemicals. This does not seem to be the case, however, in gunpowder-type mixtures.

It is possible to produce some degree of protection from moisture by the addition of oils, resins, and the like, but it is essential to be careful in the choice of the most stable materials. Magnesium and aluminum are particularly reactive in the presence of moisture. Fortunately, aluminum tends to oxidize only on the surface, with the formation of a protective film of aluminum oxide. Magnesium, on the other hand, does not form such a protective oxide and the corrosion continues. As a rule, the magnesium is coated with a protective layer of linseed oil, lithographic varnish, paraffin wax, or one of the natural or synthetic resins.

Titanium is now becoming increasingly useful and easier to obtain, and it has the great advantage of being both unaffected by water and quite safe, provided that its particles are not sub–sieve size. Titanium filings seem safer and more stable than most other metals used, although it is very hard and friction sensitivity must be considered. It is worth recording that several fires have been caused by people charging titanium compositions with a mallet and drift; pressing is safer (see "Titanium" in Chapter 4).

The use of magnesium or aluminum with nitrate oxidizers has become increasingly popular since World War II, and although this has certain advantages during processing, it has been established that such mixtures are more likely to decompose in the presence of water, with the formation of ammonia.

$$8\ Mg + 16H_2O = 8\ Mg(OH)_2 + 8H_2$$

$$Ba(NO_3)_2 + 8H_2 = Ba(OH)_2 + 4H_2O + 2NH_3$$

Since moisture is so critical in these mixtures, it is clear that barium nitrate is the most useful oxidizer, and for much the same reason, aluminum is used much more than magnesium for firework manufacture.

The addition of sulfur to mixtures of barium nitrate and aluminum does not particularly affect stability, but it is unwise to add sulfur to magnesium mixtures because they are liable to react in the presence of moisture. During World War II, German igniter compositions for magnesium stars contained sulfur or blackpowder or a combination of these. Problems were encountered with the influx of moisture, and an attempt was made to overcome the problem by using a mixture of potassium nitrate, aluminum, and gelbmehl (tetranitrocarbazole) (12). According to Schidlovsky (4), powdered alloys of magnesium and aluminum are much more stable chemically because they are coated with a film of aluminum oxide.

From time to time, barium peroxide appears in pyrotechnic formulas, but it is rather too dangerous for use in fireworks and, when mixed with aluminum, it is liable to heat up in the presence of water. Ammonium salts must be used with great caution and certainly must not be mixed with materials that can cause reactions in the presence of water. Potassium chlorate is hazardous with ammonium salts, particularly the nitrate, because the highly explosive ammonium chlorate can be produced by double decomposition. Magnesium is particularly liable to heat up in the presence of ammonium salts, the reaction proceeding at great speed with the evolution of hydrogen. (The author once had a "close shave" when ammonium perchlorate instead of the corresponding potassium salt was accidentally added to a magnesium mix.) Certain old-fashioned commercial smoke formulas were compounded by mixing ammonium chloride with potassium chlorate and are somewhat unusual because they appear to be relatively stable. In general, ammonium salts are not now used, with the exception of the perchlorate.

Dr. Shimizu (34) has conducted many experiments on the stability of pyrotechnic mixtures. Although most pyrotechnic mixtures are mixtures of solids, the presence of limited amounts of moisture may allow the movement of some ions. It is therefore suggested that in planning combinations one should think of the mixture in terms of a concentrated solution.

Shimizu sets out many examples to show that the constituents of a double decomposition reaction will proceed to the left or the right, depending on the solubility of the materials. The reader must look this up for himself or herself, but examples cited are the abovementioned dangerous mixture:

$KClO_3$ +	NH_4ClO_4 →	$KClO_4$↓ +	NH_4ClO_3		A
10.1	29.9	2.6	53.4	Solubility gm	

However, the comparatively safe mixture is:

$KClO_3$↓ +	NH_4Cl ←	KCl +	NH_4ClO_3	B
10.1	41.4	37.2	53.4	

In A, the reaction proceeds to the right because the $KClO_4$ has the minimum solubility; it is dangerous and could lead to spontaneous combustion.

In B, there is no danger because $KClO_3$ has the minimum solubility.

Shimizu also points out that mixtures of ammonium perchlorate with nitrates have a similar explanation. A mixture of ammonium perchlorate and potassium nitrate will proceed to the right, with the formation of hygroscopic ammonium nitrate. However, when ammonium perchlorate is mixed with barium nitrate, no moisture is formed, and the reaction stays to the left because of the minimum solubility of barium nitrate (11.4) compared with ammonium nitrate (238) and barium perchlorate (320).

Related to these matters is the treatment of metal powders with potassium dichromate. The technique is well known for passifying the surface of metals. For many years powders were treated with hot solutions of potassium dichromate, quickly dried, and then further treated with other materials such as magnesium stearate. Shimizu again has done much research on this important matter, and more precise details are given in Chapter 23.

It has been pointed out that ammonium perchlorate is not the best material to use with magnesium, and reactions occur during storage. The use of potassium dichromate also has its risks in such combinations, but because of the preceding theory, Shimizu makes the point that it would be better to use ammonium dichromate in this special instance.

The use of potassium dichromate can lead to the formation of small amounts of acid, and it would be wise to add small amounts of the appropriate carbonate. For example:

$$K_2Cr_2O_7 + Ba(NO_3)_2 + H_2O \rightarrow K_2CrO_4 + BaCrO_4 + 2HNO_3$$

$$K_2Cr_2O_7 + SrCO_3 + H_2O \leftarrow K_2CrO_4 + SrCrO_4 + H_2CO_3$$

$$K_2Cr_2O_7 + SrSO_4 + H_2O \leftarrow K_2CrO_4 + SrCRO_4 + H_2SO_4$$

As a percentage of the total weight, the amounts of potassium dichromate are small.

Further studies by Shimizu (44) have concluded that the coating of magnesium is greatly enhanced by adding a sulfate (other than ammonium sulfate) and some guanidine nitrate to the dichromate solution. Experiments over many years including equally long storage times have concluded that these three together create a black film of CrO_2 on the surface of the magnesium. It appears that the additional chemicals make the film stick well on the surface of the magnesium grains.

Shimizu proposed the following technique:

200 g	Hydrated magnesium sulfate
90 g	Ammonium dichromate
20 g	Guanidine nitrate

The chemicals are dissolved in 500 mL of water.

The solution is stirred into 1,000 g of magnesium powder in a suitable aluminum bowl and stirred until the coating is complete and uniformly brown.

The powder is dried well on paper and finally sieved through 30 mesh.

Care must be taken to avoid breathing the dichromate containing dust, and protective gloves should be worn.

It has been well known for many years that magnesium and magnalium will burn intermittently with the nitrates of barium and strontium. Sulfur also was a part of these reactions. However, the chemistry was not well understood, and there was an inherent fear of using such mixtures for long-term storage.

Recent years have seen a great fascination in the development of these strobe or blinking lights. The ideal of every manufacturer would be to have the ability to produce sharp colors that would flash on and off at something between 1 and 3 hertz. This can be achieved, but the manufacturing techniques are not particularly straightforward and there are problems with regard to shelf life and the influx of moisture. These comments come at this point because the protection of the metal involves the use of a dichromate, and some strobes are made with ammonium perchlorate (*Pyrotechnica* X1, 29).

Magnalium has always been used in fireworks, but usually in a coarse form, where it produces a pleasant crackle with colored flame-based fountains. There is no doubt, however, that in the post–World War II period, the advent of Chinese fireworks in the West has encouraged a greater interest in the use of magnalium powder for strobe and color effects.

The earlier Chinese mixtures were based on mixtures such as barium nitrate, magnalium, and phenolic resins with or without sulfur. There have been many imitations of these stars, but quality control is not easy and the stars require careful ignition. An excellent summary of the state of the art was made by Robert Cardwell, editor of *Pyrotechnica*, in *Pyrotechnica* No. 5, 1979 (29). He quite rightly reports at length the studies made by Uwe Krone of Nico Pyrotechnik Hamburg. These studies clearly were a landmark in this field (45). Krone gives a number of leads in composition trials, most of them based on nitrates of barium or strontium, magnalium, a nitro-compound, small amounts of perchlorate, and the important catalyst. The variations in the composition are quite critical, and this certainly applies to the very small percentages of copper-chrome oxide or other catalysts. The flares formulated by Krone were pressed dry at heavy loads.

Good strobing bengals can be made in this way, but as with so many delicate devices, the result will depend on the particle size of the ingredients and the catalyst in particular. The metal powder must be treated if a long shelf life is desired, and even the tube is critical. It probably will be discovered that wide-bore tubes are not suitable for bengals with loosely filled composition, and the wall thickness of the tube can be quite critical in affecting the strobe period, leading to continuous burning in some cases.

Further trials by Shimizu (*Pyrotechnica* VIII, 29) came down heavily in favor of strobes made with magnesium, ammonium perchlorate, a sulfate, and potassium dichromate. Undoubtedly, the use of magnesium gives a much better color and flash precision, but problems arise with the priming of stars. A nitrocellulose lacquer must be used as a binder, and a specially formulated igniter must use a similar binder.

As these stars usually are coated with blackpowder-type mixtures for ignition in shells, great care must be taken to ensure that the central nitrocellulose-based core is quite separate from the water-based igniter layer. Tiny bridges between the layers would allow a chemical reaction to take place, and this in turn would lead to the cracking of the stars and deterioration.

Shimizu concluded that strobe lights are produced from a composition that has two groups of components: a dark reaction group and a flash light reaction group. The temperature of the dark reaction is lower than that of the flash light group at first, but it increases with the accumulation of heat until it reaches the ignition point of the flash light group. A flash is then produced. At the same time, a small part of the dark reaction is preparing for the next flash ignition.

Materials that provide a suitable dark reaction are

1. Sulfur with magnalium or magnesium
2. Ammonium perchlorate with magnesium or magnalium
3. Potassium perchlorate, guanidine nitrate, and PVC

The components of the flash reaction might be

1. An ordinary oxidizer with magnalium or magnesium
2. A sulfate and magnesium or magnalium

The reader is referred to Chapter 6 and *Pyrotechnica* VIII (29).

The best blue colors can be made with ammonium perchlorate and one of the usual copper salts. A good green also can be made with barium nitrate and ammonium perchlorate, but these mixtures tend to be slower-burning and have been reported to explode occasionally because of the amount of gas they produce. Perhaps we need to exercise some caution in using this oxidizer, although the fear that some firework manufacturers show perhaps is based on lack of experience and may be unjustified. Needless to say, it would be foolish to mix stars made with ammonium perchlorate with other stars made with a chlorate, and ammonium perchlorate requires a shed of its own in firework plants. Basic copper carbonate is the best copper salt to use with ammonium perchlorate for blue colors, but care must be taken not to add either magnesium or aluminum to such a mixture (or indeed to any blue mixture) because the corrosion of aluminum powder is accelerated in the presence of copper or mercury.

The most common oxidizers used in firework manufacture are the nitrates of potassium, barium, strontium, and occasionally sodium; the chlorates of potassium and barium; and the perchlorate of potassium and ammonium. It is essential that nonhygroscopic salts be used, and it is not usual to employ sodium or strontium nitrate unless they are very pure. Lead nitrate occasionally was used in former times, but it has fallen into disuse. Potassium bichromate sometimes is used in match heads, but it has little use in fireworks because it does not yield enough oxygen.

Table 5.2 shows how Ellern (5), quoting Schidlovsky, provided useful information to the firework maker on the relative position of oxidizers.

A percentage figure lower than 92.5 is hygroscopic, thus tending to eliminate sodium nitrate. Strontium nitrate is useful when it is mixed with materials that do not encourage ionization. A good red star, for example, can be made with strontium nitrate, magnesium, and PVC (12). Provided that a good igniter is used and that moisture is excluded, this star keeps well for a limited period. However, it is unwise to use potassium chlorate and strontium nitrate together.

Table 5.2 Relative Position of Oxidizer

Salt	% RH over Saturated Solution, 20°C
$KClO_4$	99
$Ba(NO_3)_2$	99
$KClO_3$	97
NH_4ClO_4	96
$Ba(ClO_3)_2$	94
$Pb(NO_3)_2$	94
KNO_3	92.5
$Sr(NO_3)_2$	86
$NaNO_3$	77

In general, firework compositions contain one or more oxidizing agents and a fuel or mixture of fuels. Charcoal and sulfur are the most common fuels, but extensive use also is made of natural gums and resins such as accroides, shellac, and copal. The list is quite long, including materials such as lactose, sawdust or woodmeal, starch and dextrin, and so forth. (The reader is referred to Chapter 4.) It should be mentioned that charcoal and woodmeal can absorb large quantities of water and must be carefully watched. Similarly, dextrin will absorb water, but it is used extensively in small percentages as a binder. Aluminum and magnesium are fuels, although they also provide special spark or illuminating effects. Iron, antimony, copper, and titanium also are quite useful, especially titanium, which produces the most beautiful silver sparks and requires no treatment of the metal. Zinc powder sometimes is used in smoke compositions, but it is little used in firework manufacture, mainly because it is so reactive with water and various electrolytes.

Colored light is produced by the vaporization of the compounds of certain elements. Red, green, and blue colors are produced by the elements strontium, barium, and copper, and although the halides of these metals would be the best compounds to employ, they are unsuitable in practice. The chlorates, chlorides, and perchlorates of these three elements are either hygroscopic or deliquescent (with the exception of barium chlorate) and are quite useless in practice. Strontium monochloride, barium monochloride, and copper (I) chloride are the three compounds required for the color production, and the excess of chlorine must be present to ensure their formation. Polyvinyl chloride, Parlon and Chlorowax, chlorinated rubber (Alloprene), polyvinylidene chloride (Saran), hexachlorobenzene, and mercury chloride (calomel) have all been used. Calomel is not used now because of its extraordinarily high price and toxic nature.

For ordinary firework purposes, red colors usually are made with potassium perchlorate and strontium carbonate, while greens are made with barium chlorate or barium nitrate. Blue is more of a problem, and care must be taken to keep flame temperature as low as possible and reduce the oxygen balance. Paris green was often used in blue stars, but in recent times copper (II) oxide is more common. Yellows normally are produced by using potassium perchlorate and sodium oxalate or cryolite. The latter is particularly useful because it is not affected by water. Sodium nitrate would be very useful if it were not so susceptible to attack from moisture.

The ignition of the majority of the firework compositions is a simple matter because they invariably ignite at temperatures less than 500°C, and a gunpowder prime or paper impregnated with potassium nitrate may be all that is required. The matter is not quite as straightforward when the ignition temperature is higher, as, for example, with a mixture of barium nitrate and atomized aluminum. Similarly, it may be difficult to ignite a highly compacted surface, such as a magnesium star that has been pressed at several tons. In these circumstances, it is usual to employ a composition that produces a very hot slag, which will ensure the transfer of sufficient heat to cause ignition. Mixtures used for these purposes usually contain potassium nitrate and silicon or lead dioxide, copper (II) oxide, and silicon.

Time fuses and delays for ordinary firework purposes usually are made with compressed gunpowder, which burns at the rate of about 2.5. seconds per inch. The so-called Bickford fuses usually are trains of modified gunpowder that are covered with flexible and waterproof covers. Ellern, Shimizu, Hardt, McLain, and Schidlovsky have written excellent textbooks in recent years, and the reader is referred to these works if he or she wants to know more of these principles in detail or to learn of the more complex or exotic materials used in pyrotechnics (4,5,30,34).

The following summary may be useful:

- Never mix chlorates with: sulfur or sulfides
ammonium salts
phosphorus
pitch or asphalt
picric acid or picrates
fine metal powders
- Avoid chlorate and oxalates (moisture problems)

Chlorates and gallic acid are very hazardous. Avoid friction or impact with chlorate mixtures, and do not charge them with a mallet.

- Avoid using potassium perchlorate with: sulfur and sulfides
phosphorus
picric acid and
picrates

Potassium perchlorate is hazardous when mixed with fine metal powders. The same points concerning friction and impact also apply. Treat barium nitrate and pyro aluminum with great respect. Never use fulminates of silver or mercury.

TESTING FIREWORKS

Since a number of countries have started to ban certain mixtures in fireworks, it has become necessary to conduct chemical tests at the point of import. This may be a requirement banning admixtures of chlorate and sulfur, for example, or the prohibition of certain toxic materials.

Since some degree of chemical knowledge is required and as chemical testing becomes ever more sophisticated, there is little scope in this book for anything more than simple guidelines. There is little point in describing the usual ions and acid radicals, which can be found in a textbook. The following, however, could be quite important.

Chlorate

A drop of aniline hydrochloride solution will turn a blue-green color in the presence of chlorate. This can be done as an aqueous solution of the material in distilled water or as a spot test. The color depends on the amount present, but a drop of the test solution added to a crystal of chlorate produces a strong violet-blue that will stain even a plastic spot test plate. Perchlorate produces no reaction.

The test solution is made by slowly stirring 18 mL of colorless aniline into 370 mL of concentrated hydrochloric acid in a large beaker. Slowly add 107 mL of distilled water to the aniline hydrochloride, and store it in an amber bottle.

Perchlorate

Simple spot tests are not straightforward. A 0.05% solution of methylene blue in distilled water will turn violet in the presence of perchlorate on a white spot test plate, but not with chlorate. It may not be quite as obvious in aqueous solutions. Furthermore, the test using a saturated solution of cadmium nitrate in concentrated ammonium hydroxide will give a positive result with potassium perchlorate and potassium chlorate.

Having first eliminated chlorate, it may be necessary to reduce any perchlorate with sodium nitrite and concentrated nitric acid. The chlorate produced by this reaction can then be identified. However, care must be exercised here lest the chlorate be further reduced to chloride and so missed. The presence of chloride only (having checked its absence to start with) will be an indicator.

Zwikker's reagent reacts with an aqueous solution of potassium perchlorate to form minute blue crystals in the already deep blue solution. The crystals do not form with chlorate. The solution is made with 4 mL of a 10% aqueous solution of copper (II) sulfate, 1 mL of pyridine, and 5 mL of distilled water.

Sulfur

Piperidine placed on a spot of the material on a spot test plate will produce a yellow-brown color. This is not foolproof, however, because red gum also produces a yellow color with piperidine.

As a check, it is better to place about 0.3 g of benzoin in a small test tube and add a few milligrams of the material to be tested. The tube is gently heated over a flame with a piece of filter paper moistened with lead diacetate in the mouth of the tube. If sulfur is present, the filter paper turns black and there is a smell of hydrogen sulfide.

Ensure that the two materials are well mixed.

Ammonium Compounds

A simple physical test is to warm the mixture with a hydroxide, when the characteristic smell of ammonia will be apparent. The standard test is to use Nessler's reagent, which produces a brown precipitate.

Lactose and Sucrose

These materials occur in many chlorate combinations and can be sensitive to impact and friction. The red color obtained when mixed with Fehlings solution and heated will indicate the presence of lactose. Sucrose is more commonly used in smokes, but it does not react positively to Fehlings unless it first has been heated with hydrochloric acid.

6

CHEMISTRY OF FIREWORK COMPOSITIONS

Takeo Shimizu

DEGENERATION

Firework compositions generally are a mixture of selected materials for a particular purpose. Each substance is in a powdered solid state, but some of them are liquid or a semiliquid state. The different substances are mixed and consolidated into solid masses and, as such, are not natural but artificial.

In the course of time, pyrotechnic articles have a tendency to revert from this artificial state to the more natural state, striving to reach a more stable condition. The compositions, therefore, may crack, shrink, or change in quality, particularly when they absorb moisture. Flares, for example, can burn excessively quickly and burst out of their tube even though they had been safely tested a short time beforehand. In this case, the loaded mixture shrunk in the paper tube, creating a gap between the mixture and the paper wall. A composition containing shellac as a fuel and its relatively low softening point (ca. 50°C–80°C) could cause the composition to shrink.

Blackpowder rockets quite often burst, especially when they are large. This could come about by the bridging of the potassium nitrate particles and crystallization. In the Chichibu district of Japan, an old rocket-making technique is preserved as a traditional art form. Large bamboo tubes are loaded with blackpowder to which a small amount of water has been added. The rockets quite often burst, possibly as a result of the bridging of the potassium nitrate crystals.

To avoid such degenerations, the selection of the materials and the binder is quite important so that mutual reactions can be avoided. However, the manufacturing process also is important. As a general rule, the materials must be mixed uniformly. This is quite important where a small quantity of an ingredient is mixed with a large volume of premixed materials. The former small amount should be mixed well with a small part of the large mass and, in turn, should be scattered into the rest of the mass. Inadequate mixing time often causes depression after loading, particularly when the mixture contains some liquid.

There are exceptions. The late Bill Withrow wrote to the author to point out that too much mixing is not good for the so-called glitter (sparkling) effect.

To avoid chemical or physical changes after loading, it may be better to age the material beforehand. Materials may be coated or mixed and then matured for many hours at elevated temperatures. As an example, magnesium may be coated with boiled linseed oil and then polymerized in heat to avoid reactions after pressing.

BURNING REACTION

A firework composition generally consists of several materials: an oxidizer, a fuel, special items to obtain desired effects, and a binder. These mixtures use the oxidizing effect of the oxidizer (e.g., potassium nitrate, potassium perchlorate, potassium chlorate, etc.). On the contrary, however, there are mixtures that use reduction effects to obtain a fire (e.g., magnesium can be used as the reducing agent).

The author calls the former "positive explosives" and the latter "negative explosives," where inert materials that contain oxygen become the reactants (e.g., H_2O, $SrCO_3$, $SrSO_4$, $MgSO_4$, etc.). In general, when negative explosives are used (e.g., as a propellant) for lifting a shell in a mortar, they create silane-type substances in the ash left in the mortar. This material ignites when the mortar is cleaned with water. The following composition may be a representative example of a negative explosive that does not create such a phenomenon:

Magnesium passing 80 mesh	60%
Anhydrous magnesium sulfate	40%

This material is granulated with nitrocellulose solution in acetone for use as a lifting charge or a noisemaker. The effect is much milder

Table 6.1 Colored Flames of Negative Explosives

	Red	Orange	Yellow	Green	White
Magnesium passing 80 mesh	50	50	50	50	50
Magnesium sulfate hydrated $MgSO_4.7H_2O$	25	25	50	25	50
Strontium sulfate	25	—	—	—	—
Calcium sulfate	—	25	—	—	—
Glass powder	—	—	10	—	—
Barium sulfate	—	—	—	25	—
Chlorinated isoprene rubber	10	10	—	10	—
Nitrocellulose solution in acetone as needed					

than that of blackpowder or a chlorate mixture. From experiments, the force of explosives is about 12% as large as that of blackpowder.

Table 6.1 shows how it is possible also to produce colored flames.

BLACKPOWDER REACTIONS

Many curious burning reactions occur with mixtures of nitrates, sulfur, and charcoal—with or without some additives. They produce a large amount of ash when they burn, which plays an important role in the effect. The ash is liquid at a high temperature. When it is dispersed into the air, the temperature of the particles rises because of the reaction with the oxygen in the air. A second reaction takes place at this stage, creating sparks explosively.

The following are examples of blackpowder reactions:

Senko Hanabi

This produces a most curious effect that is not clearly understood at present.

	Weight (%)
Potassium nitrate	60
Sulfur	25
Anthracene soot low temperature–type	4
Hemp charcoal	8
Anthracene	3

The sulfur content is larger than that of blackpowder, and while the soot is very difficult to burn, the hemp charcoal burns very easily. The good spark effect may come from the mixture of two carbons with different burning characteristics. The anthracene helps to stop the burning fireball from falling after the mixture burns in the paper twist.

Firefly

This effect occurs in a narrow range of composition: between 45% and 55% of potassium nitrate, 0% and 10% sulfur, and 35% and 65% charcoal. The sulfur content is very small and the opposite to Senko Hanabi.

The charcoal is vital to get the best effect, and it is tested by heating it on an iron plate with some form of heat underneath. The charcoal is gently stirred to gain a certain amount of ash, which must not be a powder but granular. The author has been able to find this charcoal only as a hemp charcoal produced in Tochigi-ken in Japan. In the absence of such a charcoal, additional material is necessary as follows:

Firefly *(using pine charcoal)*	Weight (%)
Potassium nitrate	47.5
Pine charcoal	47.5
Sulfur	5
Flake aluminum (18–30 mesh)	4.6
Barium sulfate	7 (additional)
Glutinous rice starch	5 (additional)

The aluminum particles flicker as if they were a group of fireflies. However, the flickering lights are not visible over long distances.

Micro Stars

Micro stars consist only of a nitrate and sulfur with a metal powder—magnalium or magnesium. They can produce an explosive flash when the mixture is formed into small stars.

		Weight (%)			Weight (%)
(A)	Barium nitrate	40	(B)	Strontium nitrate	30
	Sulfur	20		Sulfur	30
	Magnalium (50/50)	40		Magnalium (50/50)	40

One hundred grams of the mixture is kneaded with 20 grams of a 10% nitrocellulose solution in acetone to form a homogeneous mass. It is rolled out on a star plate of about 8 to 10 mm thick and cut into a few pieces. These are dried at room temperature for one day and then further dried in an oven at 50°C for five hours. They are then crushed into angular grains with a wooden mallet and an aluminum roller. Finally, the grains are separated through a sieve to obtain small stars of 1.5 mm to 3.0 mm.

The small stars do not ignite on their own, but when they are mixed with a delay-type composition and loosely loaded into a paper tube to form a small fountain. When the fountain is ignited, the stars are ejected from the tube and burn with a flash of light after a short delay.

Micro Star Delay Composition	Weight (%)
Potassium nitrate	50.0
Pine charcoal	37.5
Sulfur	12.5

The delay should be mixed with the small stars in the ratio stars/ delay: 3:7 by weight.

Glitter Effects

Potassium nitrate, charcoal, sulfur, metal powders (atomized aluminum or magnalium), and some additives (Fe_2O_3, $CaCO_3$, or $Na_2C_2O_4$) cause the glitter effect when they are mixed and made into stars or sparklers.

There is some variation in interpretation of what is meant by glitter and some forms differ (e.g., when speaking of stars or sparklers). The two compositions shown in Table 6.2 are similar in effect.

The corrosion of the aluminum or magnalium caused by the other components is quite critical, especially the potassium nitrate. The use of boric acid is not very good with the two mixtures shown in Table 6.2 because it is likely to interfere with the glitter effect.

To make stars, about 17% water is added, but to make a sparkler slurry, about 33% water will be needed. As boric acid is not used, the batches must be kept small to avoid rises in temperature. The stars should be dried at room temperature—between 5°C and 32°C.

Dextrin should be used as a binder for larger stars so that they can dry more quickly. The stars should be kept in a cool, dry store.

Table 6.2 Aluminum and Magnalium Glitter Mixtures

	Aluminum Glitter Weight (%)	Magnalium Glitter Weight (%)
Potassium nitrate	50.0	50.0
Pine charcoal	6.9	8.3
Sulfur	20.0	25.0
Atomized aluminum	17.3	—
Magnalium (50/50) 60 mesh	—	12.5
Ferric oxide Fe_2O_3	5.8	4.2

NITRATE AND SULFUR REACTIONS

Nitrates mixed with sulfur do not burn easily, but there are some additives such as realgar or blackpowder that promote smooth burning, even in small amounts. When a mixture of potassium nitrate and sulfur is in the ratio of 50:50, it will burn to produce a white smoke. However, if the ratio of potassium nitrate increases, the burning temperature increases and will even melt the iron nozzle of a rocket. The temperature increase may be due to a very large amount of molten ash (K_2SO_4, $SrSO_4$,or $BaSO_4$) and a small amount of gas (SO_2). The weight ratio of the gas and the ash in solid or liquid state is about 35:65 to 18:72. The value for the specific heat of the gas at high temperature may be much smaller than that of the gas. Thus, the heat capacity of the total products should be very low in this case to create a high temperature.

An experiment showed that the flame melted a platinum wire, which indicates a temperature exceeding 1769°C; that is sufficient to obtain a flame color. This effect is used for long-burning smokeless road and rail signals, where the large quantity of high temperature ash and SO_2 gas is no problem [H. Ellern, *Modern Pyrotechnics*, 1961, p. 275, (5)]. Using the same principle, Dr. C. Jennings-White has written about nitrate colors [*Pyrotechnica* XV, p. 23, 1993 (29)].

Spark Composition Without Nitrate and Sulfur

Christoph Bosshard in Switzerland suggested the following composition (1989).

Perchlorate Sparks	Weight (%)
Potassium perchlorate	30
Red gum	20
Atomized aluminum	10
Iron (III) oxide Fe_2O_3	40

Further studies may be developed with this type of effect.

STROBE REACTIONS

The strobe reaction is a solid mixture consisting of several different materials. When the mixture is ignited, a part of the mixture that reacts easily does so to produce a quantity of heat.

This heat then promotes the other substances to rise in temperature in order that they may reach ignition point. At this stage, they ignite with a flash. Many mixtures could behave in this way, but, clearly, those compositions that produce strobe lights are few in number.

The mechanism may be most clearly explained in the following composition guide.

Strobe Light	Weight (%)
Magnesium (coated with $K_2Cr_2O_7$)	20–30
Ammonium perchlorate	50–60
Metal sulfate	10–20
Potassium dichromate (stabilizer)	5

This example is a mixture of a positive explosive (magnesium + ammonium perchlorate) and a negative explosive (magnalium + metal sulfate)

The whistle may come from a strobe-type reaction because it also is a repetition of burning and detonation:

burning — detonation — burning — detonation ------

To clarify the mechanism of the whistle, the author conducted the following experiments.

In Figure 6.1, the whistling zone is surrounded by the burning zone and the detonation zone. When using other materials, such as benzoic acid or anthracene in place of gallic acid or phthalic acid, the effect is

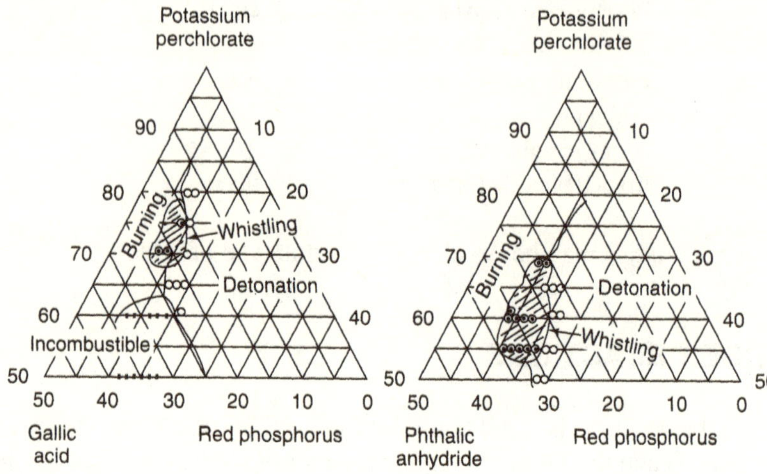

Figure 6.1 Strobe reactions. Whistling mechanism

the same. From these effects, the whistling must be a repetition of burning and detonation.

The experiment was carried out on a small scale: 2 g of composition were loaded at a pressure of 1,200 kg per cm into a tube of 12 mm inside diameter and 76 mm long. The wall thickness was 1.2 mm. The author does not recommend the experiments because they are quite dangerous. Such mixtures can be highly explosive under pressure.

CHARACTERISTICS OF WATER-SOLUBLE BINDERS

Water-soluble binders have a great influence in the manufacturing techniques for fireworks. In the manufacture of cut stars, for example, dextrin is quite good because the stars dry rapidly and no cavities are created in the drying process. However, when they are then primed with a water-based slurry composition, the stars absorb the water and collapse. Similarly, when a star is dropped into a beaker of water, the star rapidly goes out of shape.

On the contrary, when glutinous starch is used as the binder, this does not occur. However, the drying time must be much longer. When the stars are thick, it is easy for cavities to form. Nevertheless, there is one significant bonus: Stars made with glutinous rice starch become waterproof. This is a great advantage for making color-changing stars.

The charts in Figures 6.2, 6.3, and 6.4 may be useful for the selection of binders.

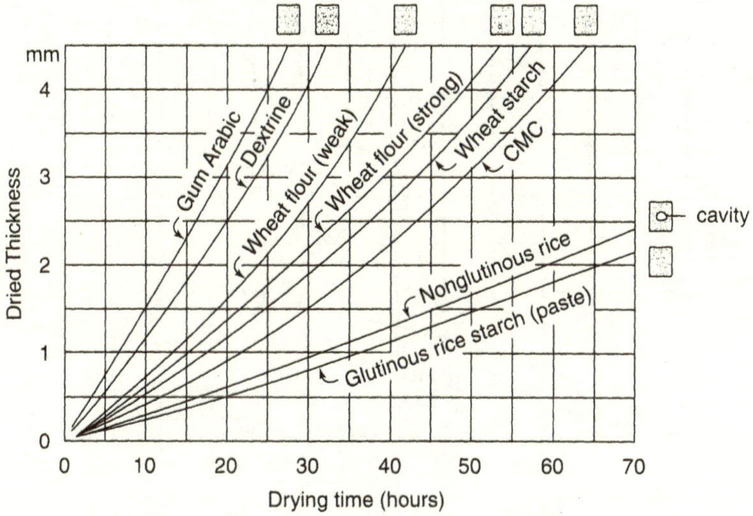

Figure 6.2 Drying time of 9-mm cubic red stars with 6% binder at room temperature

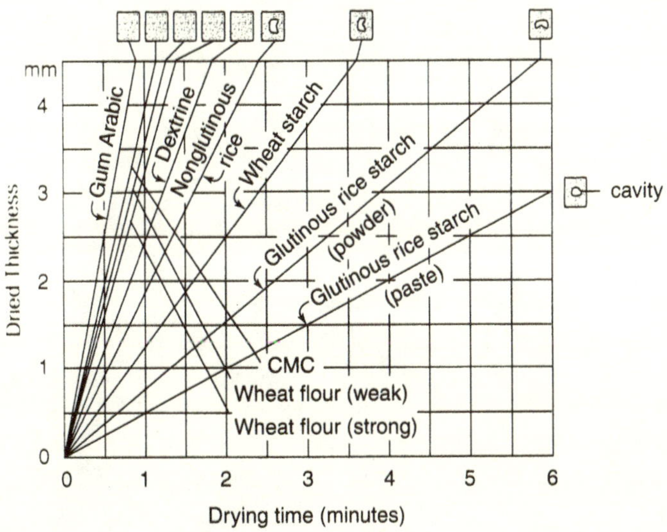

Figure 6.3 Drying time of 9-mm cubic red stars with 6% binder at 51°C in a drier with warm air

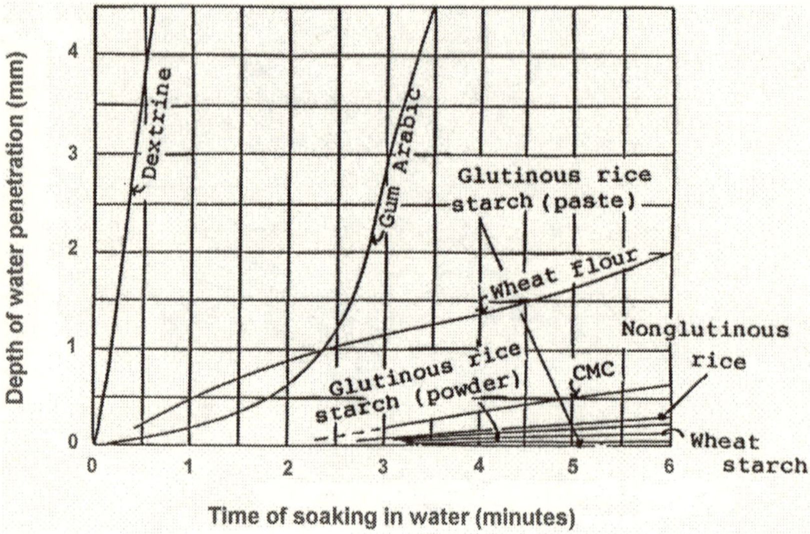

Figure 6.4 The waterproof characteristics of 9-mm cubic red stars with 6% binder when dipped into water

7

THE LEGISLATIVE FRAMEWORK OF FIREWORK CONTROL

T. A. K. Smith

The explosives industry is said to be the most highly regulated industry after the nuclear industry, and indeed it often seems the case that it is overregulated. However, the need for regulation is clear. First, explosives in general undoubtedly are potentially dangerous. Second, the public perception of the explosives and chemical industries is such that a high level of legislative control is demanded by them. Recent events such as those at Toulouse (involving ammonium nitrate) and Enschede (involving fireworks) have highlighted that when things go wrong, they can go spectacularly and tragically wrong. Such events, in which the site, operation, or legislative controls in place failed to adequately predict and manage the hazard on-site and off-site are, thankfully, rare, but nevertheless serve only to make a concerned public even more concerned, and the perceived threat of terrorism only makes this situation worse.

This chapter necessarily focuses on the legislative situation in the United Kingdom, where most regulations are administered by the Health and Safety Executive (HSE, see http://www.hse.gov.uk), but where possible draws parallels with that in other countries, especially the United States. The natural emphasis is on fireworks, but references to more general topics are made when necessary.

THE UNITED NATIONS RECOMMENDATIONS FOR THE TRANSPORT OF DANGEROUS GOODS

Much current legislation is based on concepts derived from the United Nations recommendations for the Transport of Dangerous Goods (often referred to as the "Orange Book"). Although these recommendations are not mandatory, they are widely accepted and fairly widely understood. The fundamental problem with using the recommendations for general legislation is that they address only the hazard of materials in their state for transport. For fireworks this usually means packaged and does not (and indeed cannot and should not) address their storage, use, or manufacture. It is in these areas that legislation based on the UN scheme of classification of dangerous goods breaks down.

The UN recommendations classify all explosives in one of the following groups:

- 1.1—those packages that pose a mass explosion hazard
- 1.2—those packages that pose a projectile hazard
- 1.3—those packages that pose a hazard from fiery projections
- 1.4—those packages that pose only a limited hazard
- 1.5—extremely insensitive substances that pose a mass explosion hazard
- 1.6—articles containing extremely insensitive substances that do not have a mass explosion hazard

Furthermore, the UN assigns a compatibility group to packaged goods and defines which compatibility groups may be packaged or transported together. The result of testing assigns any item to a particular hazard category and a compatibility group. The item as presented is then assigned a four-digit UN number, which identifies both the hazard of a particular packaged item and its correct shipping name. Thus all fireworks are assigned one of the following UN numbers:

- 0333 (1.1G) Fireworks that pose a mass explosion risk
- 0334 (1.2G) Fireworks that pose a projectile risk
- 0335 (1.3G) Fireworks that pose a fiery projectile risk
- 0336 (1.4G) Fireworks that pose a low risk
- 0337 (1.4S) Fireworks that pose a very limited risk

It is essential to understand that the assignment of a particular firework package into one of these five groups is critically dependent on the packing of that item.

For instance, a hypothetical example shows how packaged 75-mm maroon (salute) shells could be assigned to any of the five UN numbers if packaged in particular ways.

- If a large number are packaged together, there may be a mass explosion hazard and the resulting package is correctly classified as 1.1G, or
- If a small number are packed in a metal box, there may be no mass explosion hazard—but metallic debris from the box may present a 1.2G hazard, or
- If a small number are packed individually in a fiberboard box, there may be a "fiery projection" hazard resulting in a 1.3G assignment, or
- A single maroon in a fiberboard box may pose a 1.4G hazard, and finally,
- A single maroon in a very large box may have effects confined within the box leading to a 1.4S hazard.

Until recently most fireworks were classified as either 1.3G (0335) or 1.4G (0336). However, results of trials conducted to support the development of a UN "default" classification regime have meant that many larger fireworks, or those containing flash powder, are now classified as 1.1G (0333).

It is pointless to try to define which fireworks are suitable for consumer use by using hazard classifications. Although it undoubtedly is true that more powerful fireworks are likely to have greater classifications (e.g., 1.3G rather than 1.4G), careful packaging can reduce the hazard. Indeed, there are several ongoing trials that attempt to legitimately classify even large shells as 1.4G by using wire mesh–lined fiberboard boxes.

In addition, the UN recommendations define, in the most general terms, the types of packaging that are suitable for the carriage of all dangerous goods, and explosives and fireworks in particular. This leads to the assignment of a UN Mark (not to be confused with the UN hazard code) to a particular package that has been tested to be suitable for transport of a particular item or combination of items.

This may appear to be a classic "chicken and egg" paradox. You need to have a suitable package, that is certified to hold a particular hazard type, in which to test the item to assign it that hazard type. Fortunately, most regulatory authorities have agreed upon methods by which this paradox may be overcome.

The basis of the UN recommendations for fireworks is that if they are packaged in a suitable way, the hazards arising from accidental ignition of a package are reasonably well defined and understood. This is particularly important for the emergency services in case of an accident. They must be able to assess the risk of attempting to control the incident. However, there are anomalies in the UN classification scheme, and there is a danger that the emergency services will not respond appropriately to an event unless there are changes to the UN scheme.

This is particularly true in the case of explosives presenting a "fiery projection" hazard—1.3. Strictly applied, any "fiery projection" emanating from a UN series 6c test that passes 15 m must mean that a 1.3 hazard is assigned (presuming that the material is not assigned to 1.1 or 1.2). However, there is a great difference between the occasional Roman candle star being shot beyond this distance over an extended period of time and the short-duration fireball produced by the rapid ignition of 1.3C propellant. There is a danger that a firefighter who has witnessed the former will treat the latter completely inappropriately and be severely injured or killed as a result. There is a case for a new compatibility group for pyrotechnics that function over an extended period (we suggest 1.3P) or even a completely new class (1.7) for such items, but so far such suggestions have fallen on deaf UN ears!

To determine the classification of a package of fireworks for transport, it is theoretically necessary to undergo full burn and drop tests as defined in the UN recommendations. However, many countries have adopted simplified means of assigning a firework package to a UN class by analogy or by default classification. In the United Kingdom, the firework industry and government have agreed upon a "default" classification regime by which fireworks may be assigned a UN classification in the absence of any further information. Naturally, if the applicant for classification can demonstrate a lower hazard (it would be unusual to request a higher hazard classification!), by special packaging or otherwise, this will override the default classification.

In this way, the process of classification should become a relatively simple exercise. The default classification regime is not perfect, and there are many arguments about the assignments made, but it has worked well for the U.K. firework industry for many years.

The UN working groups have been charged with developing a similar default regime for fireworks to be appended to the next version of the "Orange Book" and which will be applied in all UN signatory countries. This approach would have many benefits and simplify transport across national borders considerably—but unfortunately, at the time of writing, the approach is destined to fail because of various entrenched positions taken by the participants—each of whom sees no reason to deviate from the systems already in place in their own countries!

There has been agreement at the United Nations on some items, and in many cases these have been adopted into national regulations so that even if the UN proposals fail, there will be greater consistency than before. The following abbreviated table outlines the position at the time of writing.

Type	Description	Default
Shells	All report shells	1.1G
	Color shell ≥ 200 mm longest dimension	1.1G
	Color shell $<$ 200 mm and $>$ 50 mm longest dimension	1.3G
	Color shell $<$ 50 mm longest dimension	1.4G
Shell of shells	$>$ 120 mm diameter with report effects	1.1G
	$>$ 300 mm without report effects	1.1G
	≤ 120 mm with report effects	1.3G
	≤ 300 mm without report effects	1.3G
Roman candles	≥ 50 mm inner diameter containing flash composition	1.1G
	≥ 50 mm inner diameter not containing flash composition	1.2G
	≥ 30 mm inner diameter	1.3G
	$<$ 30 mm inner diameter	1.4G
Rocket	Containing only flash composition effects	1.1G
	Flash composition $>$ 25% of overall composition	1.1G
	Pyrotechnic composition $>$ 20g	1.3G
	Pyrotechnic composition ≤ 20g	1.4G
Mines	$>$25% flash composition as loose powder or effects	1.1G
	≥ 200 mm longest dimension	1.1G
	$<$ 200 mm longest dimension	1.3G
	$<$ 150g composition	1.4G
Fountain	≥ 1 kg pyrotechnic composition	1.3G
	$<$ 1 kg pyrotechnic composition	1.4G

It is suggested that readers consult the UN web site (http://www.unece.org) to see the latest version of the default classification table.

The default classification scheme relies on the use of UN-approved packaging for the fireworks and follows the usual UN rules that in mixed packaging, or combination items, the highest classification takes precedence (i.e., $1.1 > 1.2 > 1.3 > 1.4$).

The default scheme does not remove the legal requirement to have all items classified in the United Kingdom; it merely aids the process.

Applications for classification and authorization of all fireworks manufactured in, or imported into, the United Kingdom still need to be made to the HSE.

It also has been agreed that full compliance with all aspects of firework design (e.g., fusing times, labeling, etc.) will be ignored when it comes to assigning a UN classification to a firework. The presence or absence of the correct label will not affect the hazard in transport packaging.

Combination items (e.g., a frame of 5 mm × 30 mm Roman candles) must have a separate classification and authorization from the individual components. This can cause problems, not least because of the almost infinite number of combinations that are possible.

In the United Kingdom, the List of Classified Explosives and Fireworks (LOCEF) is now available as an online database at http://www.locef.co.uk.

HAZARD AND RISK[1]

There is much confusion between the use of the terms *hazard* and *risk*, even within dictionary definitions. In formulating and interpreting legislation, however, the distinction is clear and essential.

Hazard—the possible consequences of an event
Risk—the combination of the hazard and the likelihood of an event

It is right that the UN recommendations should confine themselves to the assessment of hazard; however, the UN tests themselves are relatively artificial and neither relate to actual transport situations nor reflect the likelihood of events occurring. It is up to the users of fireworks to determine and minimize the risks of handling potentially dangerous items.

However, it also is important that the regulatory authorities do not allow themselves to overregulate industry on the basis of high hazard when the likelihood of that hazard occurring is extremely small, and thus the overall risk is small. Accidents involving the manufacture, storage, and transportation of fireworks are mercifully rare in the West, Enschede notwithstanding, although there have been a number of spectacular incidents in the East in recent years. The failures at Enschede

1 See also *Journal of Pyrotechnics*, No 18 (Winter 2003), pp. 32–42.

revolved around the incorrect assessment of the hazards posed by the fireworks stored and poor management and enforcement. In the United Kingdom, there has not been an "off-site" fatality arising from an incident at an explosives facility for over 50 years, but this is no excuse for complacency!

TYPES OF FIREWORK LEGISLATION

In general, there are two main types of firework legislation: those pertinent to the producer (which is taken to include the manufacturer, importer, and professional user) and those pertinent to the consumer (i.e., of domestic fireworks).

In the United Kingdom, most firework safety legislation is administered by the HSE and occasionally by the recently renamed Department for Transport (DFT—see http://www.dft.gov.uk). In the United States, legislation is controlled mainly by the Department of Transportation (DOT).

These regulations cover the time from the importation of a firework into the country, as well as all the manufacturing operations that take place either for home-produced or modified fireworks. In most countries, there also are regulations for the use of fireworks by professionals.

In the United Kingdom, consumer legislation is administered by the Department of Trade and Industry (DTI—see http://www.dti.gov.uk), whereas in the United States, this role is mainly administered by the U.S. Consumer Product Safety Commission (CPSC).

These regulations broadly cover the fireworks available for retail sale and may include references to an individual company's established standards for consumer fireworks. The availability of fireworks (e.g., within the United States) may further be controlled on a state-by-state basis, and in some cases the sale of fireworks may be prohibited altogether.

FIREWORK LEGISLATION IN THE UNITED KINGDOM

Much of the early history of fireworks and firework legislation in the United Kingdom has been covered earlier in this book. However, a very basic summary of early legislation is presented here. A comprehensive list of U.K. explosives legislation pertinent to the fireworks industry has been published by the Confederation of British Industry (CBI) in the United Kingdom and is available from the EIG web site (http://www.eig.org.uk).

The current basis for most explosives legislation in the United Kingdom is still the venerable Explosives Acts of 1875, the guide to which is known as the "Green Book." This remarkable piece of legislation, essentially written by one man—Sir Vivian Majendie—was a response to the poor safety record in the U.K. explosives industry at that time.

On April 26, 2005 the U.K. Explosives Act was superseded by modern, risk-based law—the Manufacture and Storage of Explosives Regulations (MSER). The development of these new regulations, started in 1983, has taken a fundamentally new approach to assessing the risks from, particularly, higher hazard explosives as a result of the realization that fragment throw from buildings containing 1.1 explosives often poses a greater risk than the blast wave from such events.

The MSER recognize that the hazards posed in storage are not the same as those posed and quantified in transport using the UN regime. The MSER introduce the concept of "Hazard Type," related to the "Hazard Classification" of the UN regime—but allowing for changes (normally increases) in hazard where bulk storage introduces confinement of the explosives. In extreme cases, such as that which occurred at Enschede, inappropriate storage in highly confined situations could lead to a mass explosion (HT1) hazard.

Hazard Type 1—explosives that present a mass explosion hazard in storage
Hazard Type 2—explosives that present a major debris hazard in storage
Hazard Type 3—explosives that present a major thermal effect in storage
Hazard Type 4—explosives that present only a minor hazard in storage

The 1991 Control of Explosives Regulations (COER) control the security of explosives, but fireworks generally are exempt from these regulations.

The remainder of modern U.K. explosive legislation is centered on the adoption of the UN Recommendations by the United Kingdom and their enactment in the Classification and Labeling of Explosive Regulations (CLER) in 1983.

In the summer of 2004 all of the existing U.K. regulations concerning the transport of dangerous goods in general, and explosives in particular, will be replaced by the Carriage of Dangerous Goods and Use of Transportable Pressure Equipment Regulations (CDGUTPER), which finally will adopt the European Agreements on the transport of Dangerous Goods (ADR) into U.K. law.

Major Accident Hazards—COMAH

The Control of Major Accident Hazard Regulations (COMAH), which enact the European Seveso Directive in the United Kingdom, require operators, among other duties, to prepare on-site and off-site emergency plans to cope with any foreseeable accident, including accidents to the environment. As originally drafted, the impact on the fireworks sector was limited only to the most major establishments, but, following Enschede, the qualifying quantities for regulation under COMAH have been reduced significantly.

COMAH has two levels of control—so-called "upper tier" and "lower tier"—and the amount of information required to be provided by the operator is related to the amount of dangerous goods held. The qualifying quantities for all explosives except those classified as 1.4 are 10 tons and 50 tons for COMAH lower and upper tiers, respectively. The corresponding figures for 1.4 explosives are 50 tons and 200 tons.

It will be noted that at present there are no regulations explicitly addressing the use of fireworks. Government argues that such activities already are adequately controlled by general health and safety legislation such as the Health and Safety at Work Act 1974 and the Management of Health and Safety at Work Regulations.

Importation Controls

Formal importation controls of fireworks into the United Kingdom were abandoned in the early 1990s, following the establishment of the single European market. Before this time, each consignment of fireworks in the United Kingdom had to be reported to the HSE and, if necessary, tested before sale. This level of control of fireworks has been superseded by reliance on classification of all fireworks in the United Kingdom, but sadly this change has been taken as a relaxation by many, with the consequence that many more poor-quality fireworks are now available in the United Kingdom than ever before. The proposed European Pyrotechnic Regulations (see later) may address many of these concerns but potentially also introduce new ones—particularly for the dwindling number of actual manufacturers that remain in Europe.

Consumer Legislation

In the United Kingdom, two major pieces of legislation control the sale of any goods—and thus fireworks—to the general public. These are the Consumer Protection Act (CPA) and the General Product Safety Regulations (GPSR). In essence, both pieces of legislation state that a product must be safe for the consumer to use without the need for any specialized training.

In 1996, following a season in which two deaths were directly related to firework misuse, the U.K. government introduced emergency regulations banning the sale of certain items to the general public. They were revised and reissued as permanent regulations—the Firework Safety Regulations 1997—in late 1997. The biggest aspect of these regulations was the banning from sale to the general public of any shell or maroon shell. Unfortunately, the definition of who was a "professional" was far from clear, thus allowing the purchase and use of shells to be very vague and open to abuse.

In 2003 an enabling bill—the Firework Act—was introduced into U.K. law by the unusual route of a successful "Private Member's Bill," a device that does not enjoy the direct support of the government but which attracts support from fellow Members of Parliament across the political parties. As it stands, the Firework Act changes nothing in the United Kingdom, but it does introduce mechanisms by which subsequent regulations can be developed. Likely candidates for early regulation include provisions to ban certain fireworks for consumer use, to limit the times and dates when fireworks may be sold or used, and to tighten the importation of fireworks into the United Kingdom.

It surprises many people that there is not at present any mandatory qualification for professional or amateur users of fireworks in the United Kingdom. The reason for this is mainly historical: Fireworks have been available to the public for hundreds of years, and a large industry has grown up around the supply of fireworks to amateur groups, particularly for the November 5th celebrations. It has been estimated that there are some 20,000 organized displays in the United Kingdom over the celebration period, ranging from small events, such as celebrations at sports clubs, to large nation-scale events. It would be impossible for all of these displays to be fired by "true" professionals, and as a consequence many local organizations have formed groups able and willing to fire these smaller displays on an annual basis. Future regulations made under the Fireworks Act may further restrict the sale of certain items to amateur firers.

The British Pyrotechnists Association (BPA) recently introduced a training and certification scheme for professional firework firers in the United Kingdom (see http://www.pyro.org.uk), and the Association of Stage Pyrotechnists (ASP) introduced a similar regime for users of theatrical items (see http://www.stage-pyro.org.uk).

Future Legislation

Much of U.K. legislation is now driven by directives originating from the European Union (EU). At the time of this writing, the European Commission is developing proposals for a "Pyrotechnics Regulation," which will have tremendous implications for the structure of U.K. domestic legislation and that in other European states. The proposed Regulation proposes mandatory adoption of the European Standard for Consumer Fireworks (EN 14035—see later) as well as introducing standards for professional fireworks and other pyrotechnics.

EUROPEAN STANDARD FOR FIREWORKS

Work on a pan-European standard for fireworks has been ongoing for many years, and it finally seems that the European Standard for Fireworks will be complete in 2005. The European Standard is very similar in concept to the British Standard (see previous editions of *Fireworks: Principles and Practice*) and again defines four categories. The major changes in the proposed standard are:

- Each firework type is the subject of a separate standard
- Additional fireworks (over the British Standard) are defined
- Changes in safety distances
- Changes in labeling requirements
- Type testing (in addition to batch testing)

There are, however, problems with implementing such a broad standard across Europe. Each EU country historically has permitted some types of fireworks for consumer use, while prohibiting other types. Consequently during the development of EN14035, many counties have successfully argued for their own derogations (called "A-deviations"), which further restrict the types of fireworks available in their states. For instance, handheld fountains are not permitted in the United Kingdom, but are permitted throughout most of the rest of Europe, and in general, the British Standard does not permit friction-ignition items, whereas these are common throughout the rest of Europe.

There also are some anomalies that have been highlighted in the drafting of the European Standard, one of the most amusing of which is the proposed requirement that all Roman candle batteries and mines should be buried to at least half their depth in the ground. The Scandinavian countries pointed out the impracticalities of this in frozen ground.

It will be difficult to argue that a firework type that is safe for use in, say, Germany, is unsafe for use in the United Kingdom. The proposed Pyrotechnics Regulation indeed makes this situation worse because it fundamentally states that no barriers to trade in approved goods should be erected by Member States of the European Union. In this way, there is a real possibility that a regulation designed to increase safety will actually have the opposite effect!

FIREWORK LEGISLATION ELSEWHERE IN THE EUROPEAN UNION

As time progresses, it is the intention of the European Commission to harmonize legislation throughout the EU. The introduction of the Seveso Directive (implemented in the United Kingdom as the COMAH regulations), the general adoption of ADR, the European Standard for Fireworks, and the proposed Pyrotechnics Regulation will address many areas of legislative control. However, there still are areas left uncovered (e.g., use) as well as areas where Member States have decided to implement European Directives differently. It is a commonly held belief (especially in the north of Europe) that where the northern states both implement and enforce European Directives fully and sometimes to extremes, the southern states implement in order to appear "good Europeans" but often fail to enforce the consequential domestic regulations fully or at all!

Gathering information also is difficult. During the early stages of the proposed European Pyrotechnics Regulations, it was stated by one National Competent Authority that there were only four incidents involving fireworks in that country in one year (the United Kingdom by comparison reports approximately 1,200 accidents to persons involving fireworks in any 12-month period). Only after protracted investigation was the figure revealed to be an average of four fatalities per annum (whereas the United Kingdom has not suffered a fatality since 1996). Without common reporting standards, it is difficult to judge if the perceived problems of firework use are common across Europe and therefore might conceivably be resolved by European-wide regulation, or whether they are regionally restricted, presumably as a result of national custom and practice. Of course, one accident is one accident too many, but we do need to maintain a sense of proportion: More people are seriously injured from Christmas trees and their associated decorations in the home than are affected by fireworks. It also is undoubtedly true that however "safe" a firework is inherently, the most common cause of accident is the user deliberately or negligently ignoring the instructions printed on the firework itself!

FIREWORK LEGISLATION IN THE UNITED STATES

Modern firework legislation in the United States is broadly similar to that in the United Kingdom, the United States having adopted the UN recommendations in the late 1990s. In addition, the DOT assigns an "EX number" to each firework device (or type), indicating that it is approved for transport. Fireworks must be examined by an authorized testing laboratory, or the manufacturer must certify that a new firework device complies with the provisions of the American Pyrotechnics Association standard APA 87-1 in order to be approved in this way.

The latter approach is similar to the U.K. default classification regime and simplifies the process of classification. In 1987 the APA developed a standard for the construction, labeling, and chemical content of fireworks that incorporated the requirements of the DOT and the CPSC. The standard also contained a list of "prohibited chemicals" that could not be used in fireworks for the U.S. market, as well as a list of "standard chemicals" that were recognized as having been used with an excellent transport and storage safety record in the United States for many years.

The APA Standard recognizes that many firework mixtures are similar in their chemical composition, and thus to repeatedly test blackpowder, red star composition, and so forth would be unnecessary and a waste of time and resources. Provided that the weights of compositions with specified fireworks do not exceed those specified in APA 87-1, a new firework can be placed into UN hazard division 1.4G without the need to have full UN tests performed. In addition, an article that exceeds the weight limits for assignment to a 1.4G classification may be assigned to hazard division 1.3G without laboratory testing, provided that it contains only standard chemicals.

In addition to submitting a formal application along these lines, giving a complete breakdown of the types and amounts of composition contained in the firework, as well as a detailed diagram of its construction, the manufacturer must complete a thermal stability test (48 hours at 75°C) performed on the article or its components.

The U.S. system is quite flexible; applications may be as broad or as narrow as the applicant wishes. This system certainly has been of enormous benefit to the U.S. fireworks industry.

From the previous discussion of the relationship between packaging and classification it will be seen that this is a potentially flawed method of assigning classifications, but nonetheless it is a system that, for the vast majority of cases, works well.

Both the DOT and CPSC assess penalties for failure to comply with their rules. DOT fines typically are $2,000 to $20,000 for infringements such as boxes not being properly marked or labeled or for shipping an unapproved device. The CPSC can assess fines up to $1,250,000 for failure to meet their rules, but typically the fines have been in the range of $20,000 to $75,000.

Operator Licenses

In the United States, operators usually are trained by a firework company, and a firework license is issued by that company for the operator. There is some mutual recognition of these licenses, but in other cases these licenses may tie an individual to working with the issuing company.

FIREWORK LEGISLATION IN CANADA

Canada has laws similar to those in the United States. Indeed, the United States has added the Canadian Explosives Research Laboratory to the list of authorized testing facilities. Classification is that of the UN scheme, and all fireworks must be authorized by the Explosives Branch of Natural Resources Canada. In addition, all aspects of storage, importation, manufacture, sale, and use of fireworks are controlled by the Explosives Branch. All authorized fireworks sold and imported into Canada are listed in the Canadian List of Authorized Explosives, and certain types of fireworks (e.g., flash crackers, trick fireworks) are banned completely.

Operator Licenses

Canada has one of the most highly developed training schemes for professional users of fireworks, consisting of two basic levels that determine which firework types and calibers may be used by the license holder. The training course specifies mortar types, dimensions, safety distances, firing procedures, and other requirements, as well as detailing accident and emergency procedures.

THE FUTURE OF FIREWORK CONTROL

There is a general need for greater harmony of legislation. The UN recommendations on the transport of dangerous goods now are widely adopted and have gone a long way toward facilitating the easy transport of fireworks throughout the world. Aspects of firework legislation not covered by the requirements of the UN recommendations, however, are much less harmonious. The most pressing need is for mutual recognition of the authorization of fireworks between countries and for the setting of internationally accepted standards for firework safety. The European Standard is progressing toward this goal. In this way, the ideal of a manufacturer producing to one worldwide standard may one day become a reality.

INTERNET INFORMATION AND LINKS

It is an unfortunate fact that many web site addresses change over time, particularly those administered by large organizations such as the United Nations. The author maintains a list of direct links to the web sites mentioned in this chapter, which can be found at http://www.davas.co.uk/fpp_links.html.

8

MIXING AND CHARGING

MIXING

Firework compositions usually are mixed in a special building. Indeed, it is common practice to employ several buildings: one for sulfur mixtures, one for chlorate compositions, one for ammonium perchlorate, and one for magnesium. Needless to say, even if the operator did not use chlorate and sulfur in the same composition, he or she would not use the same sieve for the two materials either. Sieves usually are made of copper or brass and sometimes are earthed against static electricity.

Sieving usually is a straightforward matter with gunpowder-type mixes, but special care must be taken with chlorates and perchlorates, particularly when fine metal powders also are in the mixture. Explosions or fires do not often occur, but the possibility of an electrostatic spark or friction caused by a fingernail or other tool scratching the sieve must not be overlooked.

Where hand mixing is employed, large circular sieves about 60 cm in diameter frequently are used, the mixture being sieved onto a large sheet of paper. Frequently use is also made of a nest of square sieves, one on top of the other (Fig. 8.1).

On top of the second and third sieve is a sheet of cloth to retain the material that has passed through the sieve above. The mixture can then be hand-mixed on the cloth before allowing it to pass through the sieve.

As a general rule, sieves of 14 and 25 mesh are used for mixing. Chlorate mixtures usually are passed three times through a 25-mesh sieve, while gunpowder/sulfur mixtures usually pass through a 14 and a 25 mesh. Rocket mixtures may be sieved once through a 14-mesh screen and three times through a 25-mesh screen. It is usual to sieve the potassium nitrate or chlorate (i.e., the oxidizer) first on its own before adding the other materials. This is very necessary because most of the

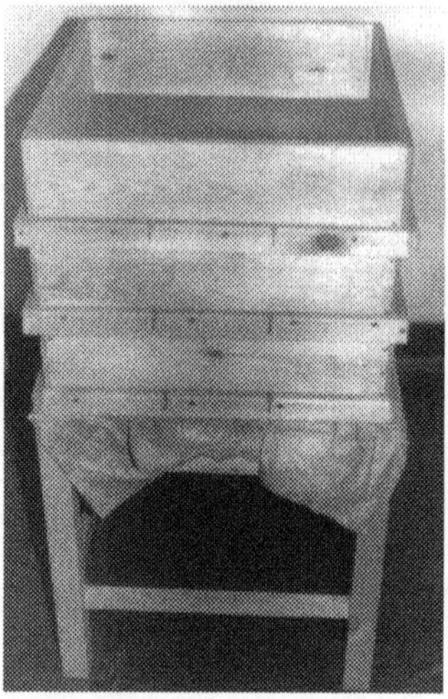

Figure 8.1 Nest of sieves

oxidizers tend to agglomerate on standing unless a flow agent has been incorporated. The word *saltpeter* (Greek: *petros*—a rock) perhaps is evidence of this. Since a good deal of pressure is needed to force lumps of oxidizer through a sieve, it is very unwise to have fuels mixed with it at the same time, particularly with chlorate mixtures.

There has been an increasing tendency over the past few years to use mechanical mixing methods that operate under remote control. The sieved materials are placed in a special drum made of metal or compressed fiber. After the drum has been securely closed, it is placed in a special apparatus that will cause the drum to turn and mix the contents. Earlier models simply turned end over end, but many people now use an apparatus that holds the drum at 45° to its axis and makes it rotate in a circular fashion (Fig. 8.2).

The inside of the drum frequently is fitted with metal baffles but, in some cases, balls of wood, porcelain, or rubber are placed in the drum along with the composition.

The whole mixing operation usually takes place in a concrete mixing bay made up of three solid concrete walls, no front, and a light roof extending over the top of the building to keep out rain. The mixing operation usually also is connected to a gate by electricity so that the mixing process will stop as soon as the gate is opened. Therefore, no one can be inside the mixing bay during processing.

Open mixers fitted with paddles also are used occasionally, but these are suitable only for mixtures that contain a fairly high proportion of water or other solvent or for those mixtures that are not particularly sen-

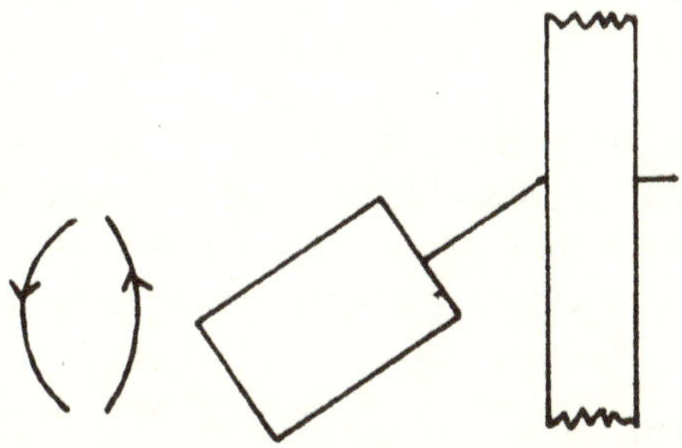

Figure 8.2 Modern mixing drum

sitive. In all these processes, it is important to earth the apparatus against static electricity. Firework makers seem to vary significantly in their preferences with regard to mixing methods. In some cases, dirty gunpowder and charcoal compositions are mixed mechanically to reduce the amount of dirt and dust and because they are less liable to explosion due to friction. For exactly the opposite reason, they also mix chlorate compositions by hand. However, some manufacturers would argue that chlorate compositions should be mixed by remote control because they are potentially hazardous. Mixtures of potassium chlorate and gallic acid are less popular now, but they certainly fall into this category and are best mixed in some out-of-the-way place. Mechanical mixing methods can be useful, but they also can be time-wasters in situations in which a few kilos could be mixed quickly on the spot by hand. With mechanical and vibratory mixing, there also sometimes is a tendency for coarser materials to settle out and rise to the surface or sink to the bottom of the mixture. In general, it is wise to avoid mixing more than seven kilos of ordinary composition at one time, and, after each batch of mixture is produced, it should be transferred to a suitable expense magazine situated a safe distance from the mixing shed. It is not wise to work with more than 500 g of flash powder. When mixing flash powder, use should be made of containers and bench mats made of carbon conducting material along with wrist cuffs connected to the earthing system.

The transportation of mixed powder has unique problems. Care must be taken to see that the rubber, wooden, or paperboard containers do not leak, and separate transporters should be used for sulfur mixtures and chlorate mixtures. One accidental factory ignition was caused by sliding a box of blue star composition along a transporter that had been carrying a sulfur mixture. The blue composition ignited but fortunately did not explode. Some companies divide their plants into a sulfur-containing section, often called the "bright" or "black" side, and a chlorate division, often referred to as the "color" side. There is much to recommend this practice.

CHARGING

Charging methods have slowly changed over the years, and fewer and fewer manufacturers use the old-fashioned technique of charging single items one by one, except for special exhibition work. More will be said later about charging methods, but the following techniques are available.

Hand Charging

It is fairly common practice to use a large block of wood as a base for ramming, about 40 cm square and about 75 cm high. In some instances, it also may be necessary to strengthen the floor of the building. Mallets usually vary in weight from 1⁄4 kilo to 4 kilos, depending on the nature of the work, and they usually are made of rawhide, box wood, or lignum vitae.

Drifts used for ramming normally are made of boxwood, or some of the new synthetic materials, and occasionally brass. Nipples normally are turned out of a piece of solid brass, aluminum, or stainless steel. Powder scoops should be made of copper and are easily made to suit the needs of the charger. The scoops often are semicylindrical and slightly smaller in diameter than the tube that is to be charged, but hemispherical or cone-shaped scoops fill and pour more easily (Fig. 8.3).

To make it easier to insert the composition, it often is convenient to place a small wooden or aluminum collar on top of the tube. Where only light charging is required, several tubes can be fixed in one frame (Fig. 8.4).

The nipples are used for tubes that require some constriction, such as fountains and drivers. In those instances in which the tube is "pulled in," the nipple merely serves as a base and fills up the end of the tube. Clay also can be used to choke the tube, and all that is necessary is to charge a scoopful of clay around the nipple before charging the composition. Mass filling methods are tending to revert to "pulled in" chokes because they are more convenient to handle. Nipples for fountains or drivers choke the tube down from a quarter to a half of the diameter.

Funnel and Wire

For charging narrow-bore fireworks, such as pinwheels, lances, and squibs, a funnel and wire often proves to be the quickest and most efficient method.

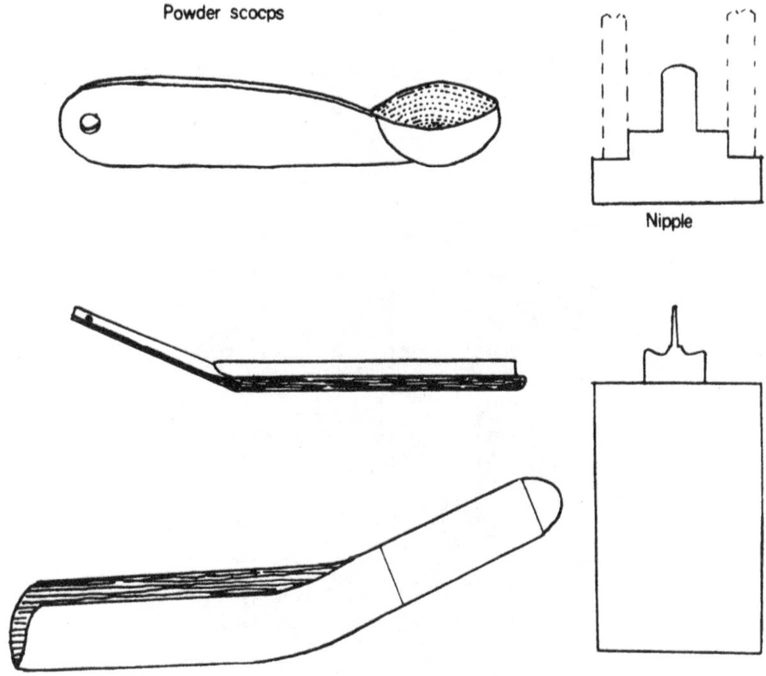

Figure 8.3 Powder scoops and charging nipples

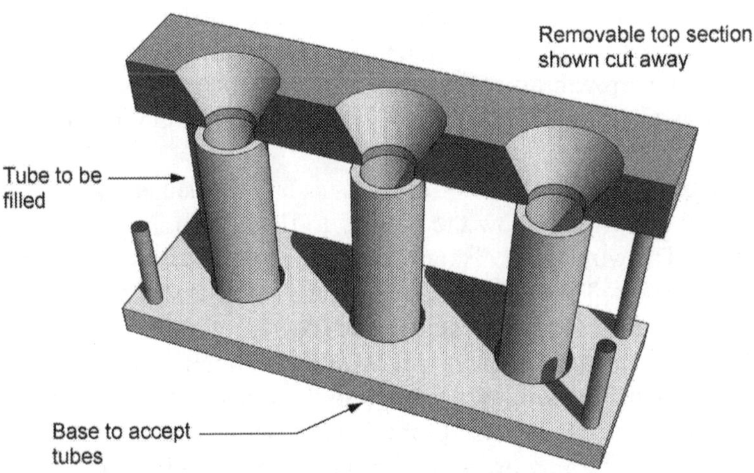

Figure 8.4 Single and multiple collars for charging operations

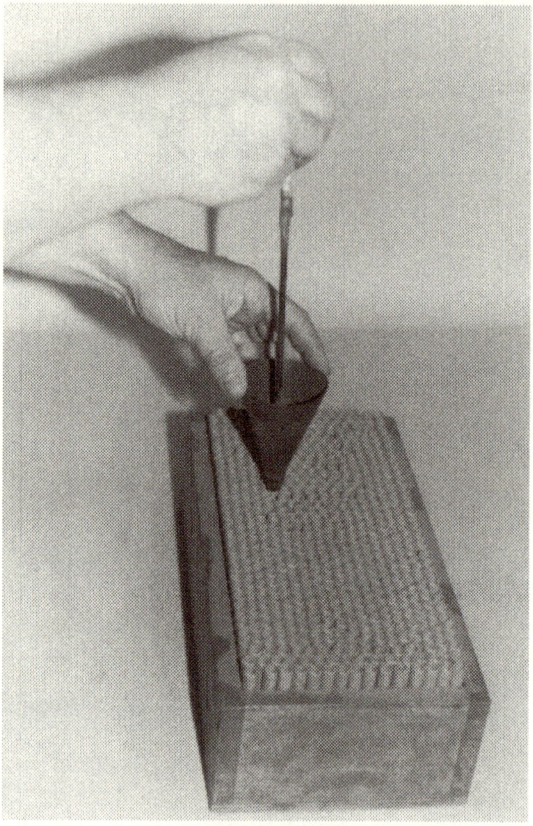

Figure 8.5 Charging by funnel and wire rod

The narrow end of the funnel normally is turned out of brass, with the remainder of the funnel made of copper that is soldered onto the brass. The narrow brass end usually just fits into the end of the tube to be charged and, after filling, leaves a convenient empty space for the prime (Fig. 8.5).

The brass or phosphor bronze rod is as large a diameter as possible, provided that it will allow the powder to flow down the side and into the tube. The wire usually is furnished at the top with a wooden or brass knob or handle. Tubes can be charged quite solidly by this method, and it does not take long to acquire some degree of skill. Compositions filled by this method frequently require the addition of a "flow agent," or the composition will tend to stick in the funnel. One percent tricalcium phosphate can be added for this purpose.

Many slow-burning fireworks, such as short, squat fountains, used to be filled upside-down. The tubes were covered at one end with touch paper and then placed with the touch paper downward in a large tray. The composition is merely pulled over the open tubes and shaken down by banging the container on the bench. The tubes are then closed up with a paper disc or cork. Needless to say, compositions that are filled by this method must be very slow-burning; otherwise a tube filled with a loose mass of fast-burning material may explode.

As nitrated touchpaper has a variable burning time, its use has been discontinued in favor of visco fuse, and the preceding process is consequently reversed.

Filling Machines

Firework manufacturers tend to develop their own machinery and, having spent a good deal of time and money on this exercise, they naturally are unwilling to share the information.

The most common English methods in the past have been those that used automatic ramming machines operated by a cam shaft. Naturally, the distance over which the ram can operate depends on the size of the cam, but large banks of these machines can be modified for filling fountains, pin wheels, small rockets, and other items. In some instances, quite heavy pressures can be exerted by counterbalancing the table on which the tube rests. The machines also are arranged so that they can stop when the tube is full.

Another type of filler uses a shaking table, which enables composition to be loosely charged by agitation. The method works well for slow-burning compositions, but there can be a tendency for the coarse and fine materials to separate during agitation (see page 314).

Screw feeding has been tried, but there appear to be many drawbacks and a tendency for the machines to become misaligned and churn up the inside of the tubes.

There can be no doubt that the use of compressed air mechanisms have been a great boon to the firework manufacturer, and they appear to be good deal safer than many other mechanisms.

Presses

The chief drawback to hand charging is the fact that the operator becomes tired as the day progresses, with the result that variations in quality are bound to occur. Presses, however, are more precise and controllable, giving good, consistent results. As a general rule, it is usual to press sufficient composition to produce a pressed increment equal in height to the diameter. Pressing loads vary from a few pounds up to 100 tons or more. Indeed, the burning characteristics depend, to some extent, on the pressing load. In the absence of any other readily available figures, interesting experiments quoted by Dr. David Hart (11) are shown in Table 8.1.

Table 8.1 Effects of Loading Pressure on Luminous Intensity

Loading Pressure lb/in^2	Luminous Intensity in Candles/in^2	Burning Rate in/mn
6,000	78,000	9.8
10,000	82,000	9.1
14,000	90,000	9.0
18,000	93,000	9.0

Hand presses are useful for work that requires pressing loads up to half a ton, and it is quite useful to fit the press with some kind of counterbalanced table to produce a consistent dead load (Fig. 8.6). Automatic presses are used for heavier loads, although it usually is necessary to protect the tube with some kind of mold (see Chapter 9). Pressing without a mold also is possible if the tubes are quite thick-walled, and in these circumstances as many as 120 tubes can be pressed at one time. The rams usually are fixed to the top of the press or are arranged to slide along rails (Fig. 8.7).

Granulated composition frequently is charged into the tubes by means of a measuring tray with a sliding base.

Loading pressures frequently can present problems, mainly because gauges can be inaccurate and vary from one press to another. Naturally, it also is necessary to take the diameter of the ram into consideration

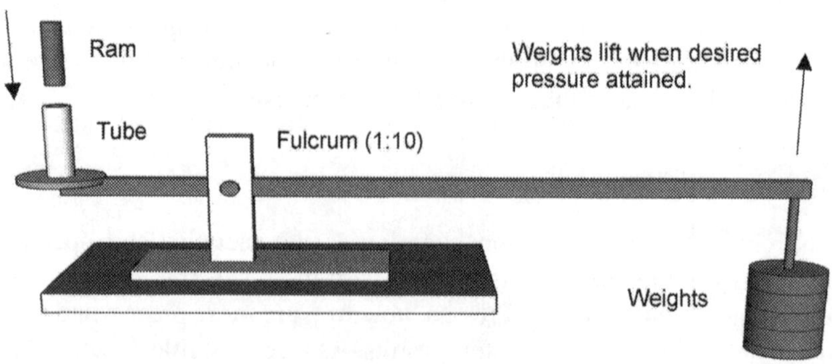

Figure 8.6 "Dead load" press mechanism

Figure 8.7 Press for 24 tubes (Courtesy Kimbolton Fireworks)

when calculating tonnage. Conversion from English to metric systems is as follows:

1 lb = 0.454 kg	1 kg = 2.205 lb
1 in = 2.54 cm	1 cm = 1/2.54 in
$1 \text{ lb/sq in} = \frac{0.454}{(2.54)^2}$	$1 \text{ kg/sq cm} = 2.205 \times (2.54)^2 \text{ lb/sq in}$
$= 0.0704$ *kg/sq cm*	$= 14.2$ *lb/sq in*
Thus, multiply lb/sq in by 0.0704 to convert to kg/sq cm	Thus, multiply Kg/sq cm by 14.2 to convert to lb/sq in
1 Tonne = 0.984 tons	1 Ton = 1.016 tonnes

A reasonably good fraction for covering the above pressures is 5/71, i.e.,

lb/sq in × 5/71 = kg/sq cm
kg/sq cm × 71/5 = lb/sq in

The drift of modern legislation has been to try at all times to separate the worker from the explosive. The ideals are laudable, but they also can be very impractical and expensive. Some forms of protective clothing are so hot and cumbersome that people hate to use them and do less work. Others seem to be as dangerous as they are protective. Face visors, which are favored in the United Kingdom, would actually direct flames and hot gases to the face if the fire were at bench level. Fireproof (Probanised) overalls normally are used when working with loose powders. Leather aprons are useful fire protection along with gloves, but the main need is to cover as much of the skin as possible.

In all these situations, the answer clearly is between those in industry who do not take enough care and the legislators whose mission seems to be to leave no stone unturned in justifying their existence. The fact that they contribute to the decline in Western competitiveness is clear to most people except themselves.

The fact that fireworks are inexpensive and must remain inexpensive means that compromises are vital. This is not the case with signals, which often employ sophisticated engineering to provide the essential reliability of the product. They also must perform well in adverse weather conditions.

The mixing of flash powder is discussed in a later chapter, but it may well be that a risk lies in the hydraulic pressing of some firework compositions in molds. An accident occurred from pressing almost a kilo of mealed gunpowder with titanium in a mold. The ratio was 70:30. Later trials determined that this was almost the worst sort of mixture with regard to sensitivity (48). Blackpowder was in its most energetic form, and the tube was inside a steel mold. The worker died because there was no effective barrier between the press and himself.

Although there is no guarantee of anything by way of safety, there are safer ways of doing things. In this case:

1. Do not use a mold. Free pressing with a thicker tube is always better, accepting that this is not always possible.
2. Use simpler mixtures (i.e., a mixture of nitrate, sulfur, and charcoal, which is less energetic and less expensive). Also add 0.5% paraffin oil to ease the pressing and lessen the dust.
3. Have a wall of concrete or steel between operator and press.

Many modern workshops are based on the principle that it is better to have the press in one room and the operator on the other side, with at least 0.3 m of highly reinforced concrete in-between. The hole in the wall is large enough to operate the tooling, and the door is closed by compressed air during the pressing. If the door is steel, the operation cannot be observed without a mirror system. However, 8-mm polycarbonate screens are very useful.

In an ideal world, presses could be loaded and set on fire to see what kind of effect is produced. It might well be every safety officer's dream to film dozens of such experiments at someone else's expense. However, in one experiment the author burned half-charged 18-mm gerbs outside the building.

The half-filled gerbs had the press rams firmly clamped on top of the composition, and an electric squib was inserted on top of the composition of one tube. The composition itself was blackpowder, potassium nitrate, sulfur, and sponge titanium. The tubes were free-standing and not in molds.

To our surprise, there was no explosion. The flames shot up the side of the ram, ignited other tubes, and eventually burned through the sides of the tubes. If the squib had been buried in the composition further down the tube, it might have caused an explosion, but in any event an ignition would not be likely to take place there during pressing.

Priming

Fireworks usually are primed, which consists of applying gunpowder paste to the mouth of the tube to ensure that ignition takes place. Mealed gunpowder containing about 5% dextrin can be used after it has been damped with water. Some firework makers pour a slurry of prime on the top of the firework and embed a piece of quickmatch into this. Only a small quantity of prime should be used; otherwise a large amount trapped inside a thick paper capping can be almost explosive. In fact, it often is better to use a prime consisting of potassium nitrate, sulfur, and charcoal in gunpowder proportions; it is much less fierce. Water-based prime can cause paper to swell and spoil clay chokes. This is avoided by using powder mixed with a solution of nitrocellulose in acetone.

Capping and Labeling

Fireworks for display purposes generally are capped. This means that about three turns of brown wrapping kraft paper or white sulfite paper are pasted around the end of the tube. The paper overlaps the end by about 35 mm and is used to tie in the quickmatch pipe where several fireworks are to be fired together. The paper is pasted only at the edge and along the side of the paper that is in contact with the tube. Fancy colored labels also are pasted on the outside edge only; otherwise they dry out on the tubes in a very unsightly manner.

Factory Sites

In some parts of the world, legislation has been very weak and little attempt has been made to regulate the danger in firework factories. In the United Kingdom, legislation in 1875 attempted to regulate manufacture and remove factories from domestic sites and other places where people congregate. It is inevitable that further legislation was taking place at the close of the twentieth century. However, the enactments of 1875 have served the United Kingdom well. There have been accidents, but they generally have been contained so that the effects have been limited. People in other parts of the world have been less fortunate. There has been loss of life and large-scale devastation of some plants. The 1875 U.K. Acts emphasized the obvious:

- Limitation on the amount of explosives in a workshop
- Limitation on the number of people in such a building
- Providing more than one escape route and ensuring that escape routes are not blocked with various objects
- Making sure all explosives are in covered containers
- Having the minimum quantity exposed so that there is a better chance of escape
- Wearing cotton or leather clothing and avoiding all synthetic materials, which could carry electrostatic sparks
- Covering as much of the skin as possible to minimize burns

The United Kingdom has been somewhat preoccupied with the wearing of rubber or leather overshoes in danger buildings in order to avoid grit being carried into buildings. There is some wisdom in this where sensitive chlorates are involved, but there is no necessity with quite a lot of mixtures, and some overshoes can be a positive hindrance to making a fast exit! Wet mats outside the danger areas are useful for cleaning shoes. Electrical switches are best on the outside of danger buildings. It is wise to have a copper earthing line in the danger buildings. Machines and carbon conductive matting can be connected to this. Lightning conductors are a matter of choice. In some cases, lightning rods have been thought to attract lightning. In the United Kingdom, it is more common to vacate danger buildings in these circumstances. In earlier times it was considered best to make workshops of lightweight wooden construction. Recent practice, however, has been to use concrete and save space. The wooden structure needed to be 23 m apart, but this could be reduced to 11 m if a screen or mound was placed in-

Figure 8.8 Modern concrete workshops (Courtesy Kimbolton Fireworks)

between. The modern system tends to use concrete walls, well reinforced with steel. This wall is built on three sides with a lightweight roof and front or other patterns with blowout panels. The walls project at the sides and at the top to contain the effects of fire or explosion (see Fig. 8.8).

In the United Kingdom, it would be normal to work with about 7 kilos of dry firework composition in a room or as little as 250 grams of perchlorate flashpowder at any one time. Rather, more wet star composition would be normal.

The distance of such a danger building or magazine from another building would depend on the quantities involved (e.g., the workroom block would be 50 m from a store containing 2,000 kilos of fireworks of UN category 1.3G or 600 kilos of 1.1G). Each magazine or store would be 20 m from the next building.

Magazines usually are built of concrete or brick, but it also is quite common to use freight containers lined with wood. This wooden lining helps absorb moisture and reduce the temperature just a little. In the United Kingdom, it is common practice to use specially made steel stores. These stores are easily movable, and in situ they are bolted down on the inside to a concrete plinth. They also are earthed with a copper strip that is connected to a copper plate some distance away. The assumption is that such an arrangement will produce a Faraday cage. The doors of these stores are close-fitting to prevent unlawful entry, and escutcheon plates of very hard metal around the key holes prevent the locks from being drilled out. The wooden matchboard lining is fixed with copper nails. There is an important drawback to such steel con-

Figure 8.9

Figure 8.10

Modern movable steel stores
(Courtesy Kimbolton Fireworks)

tainers in that they require painting regularly, particularly if you have an inspectorate that worries about rust. See Figures 8.9 and 8.10.

Stars are quite dangerous in large quantities, so it probably also is necessary to have strongly built stores and driers. The main danger with stars is the airspace around them, which contributes to their burning speed. It is more than likely, for example, that a pile of loose powder will simply "blow" when ignited. However, a large pile of stars is likely to detonate. Star driers usually are only just large enough to utilize the hot air for the best purpose. Wooden or aluminum shelves support wooden trays for the stars. The heat source can be an oil-filled electric heater or some other type of heated pipe. Some manufacturers use hot air.

Air circulation can be a problem. It must be sufficient to allow humidity to escape but not so much that it allows heat to escape unnecessarily. Hot stars also are more sensitive. A fire in the United Kingdom was caused by a worker scraping a tray where warm chlorate stars had stuck to the bottom.

A good idea for a star drier is to use a reinforced concrete cross with four small chambers around it (Fig. 8.11).

The floors of danger buildings can be covered with linoleum, modern industrial concrete coatings, or a coating of old-fashioned asphalt.

Good housekeeping is essential in a firework factory. They do say that cleanliness is next to godliness, and in the firework trade this is most important. Waste must be swept inward and placed in covered containers to be taken away for careful burning. Floors and working surfaces must be washed daily.

Figure 8.11 Concrete driers (Courtesy Kimbolton Fireworks)

Where it is possible, it is advisable to divide the factory into areas. "Clean" areas where powder is used should be marked and painted. It also is normal in the United Kingdom to separate chlorate/perchlorate buildings from the sulfur/blackpowder buildings. This is mainly because the admixture is prohibited by Orders in Council. It is a wise precaution. We look forward to the day when chlorate and sulfur admixtures will be outlawed everywhere.

The close of the twentieth century has seen a somewhat curious reversal of some of the conditions that led to the 1875 Explosives Acts. They do say that "what goes around comes around," but it was the laxity of the period that led to the strict rules for firework premises that produced those Acts. The Faversham Society quotes two delightful letters of the period concerning the report of a leaking cask of gunpowder to Major Majendie, a diligent man who had been appointed HM Inspector of Gunpowder Works in 1870, but a man who was not prepared to turn a blind eye to anything. Later he was to become Colonel Sir V. D.Majendie RA(KCB). The Company replied to his letter as follows (ref. 60):

> Sir, We have had forwarded to us from our London Office your letter of the 29th ultimo, giving us details of the enquiry you had made as to a broken powder cask which had been observed by the Police on the 5th ultimo.
>
> We had heard of the incident in question, and had made it the subject of grave consideration with all concerned, and we have now assembled all our coopers and headers-up of casks, and in the presence of our Manager and Foreman have read your letter to them and enforced all its points in the most emphatic manner.
>
> We pay the highest wages to coopers, we use only the best oak staves, and we endeavour to retain in our employ as coopers only men who have a

sense of responsibility and character, on purpose to ensure the best casks that can be made. We are taking additional measures to ensure still further inspection and supervision in this department, and we can assure you that no expense and no trouble shall stand in the way of our attaining as nearly perfection as possible in this important branch of our business.

We have the honour to be, Sir, Your obedient humble Servants

John Hall & Son

Today Sir Vivian must be turning in his grave because much of the work he established has been undone by default. As the established factories have gone out of business as a result of competition from imports, the trade has continued by people who started in a small way with material that they had purchased rather than manufactured. The authorities probably genuinely believed that the real danger was in manufacture rather than storage, so things have slipped. In the United Kingdom, 10 manufacturers post–World War II have now disappeared, but businesses legal and illegal have increased by the hundreds. Old military stores are stacked with imported material and it grows year by year

An accident took place on four storage containers on May 13, 2000, at Enschede in Holland. The resulting fire and explosions produced 22 fatalities and 900 wounded. The whole urban area was devastated, causing 500 million Euros worth of damage. The cells had been licensed for 159 tons of 1.4G or 147 tons of 1.4G with 2 tons of 1.3G.

What the cells actually contained remains unproved (80). It was not the only accident in recent times in which containers of brisant fireworks were cooked up by external fires.

There has been a learning curve for all concerned about the explosive nature of high-energy fireworks such as shells and salutes, but a study of the Dutch reports makes it clear that more than one regulatory agency was involved and that a degree of "buck passing" took place. The 1875 Act had a vital purpose. Modern attitudes that all will be well with an Approved Code of Practice and nonspecialized policing have diluted a serious business.

9

CONTAINERS

All fireworks require some kind of container. In the past, tubes have been made out of paper or even wood, but paper is still the most common type of container to be used and is still the least expensive. Nevertheless, the past few decades have seen many changes in technique, and in some situations in which high pressures are exerted, it has been necessary to use metals or resin-bonded tubes instead of paper. Signal flares frequently are pressed into steel tubes, but in those situations in which it is necessary for the tube to burn away with the composition, plastic materials or thin aluminum has tended to replace paper. Plastic and aluminum also have tremendous advantages over paper, which tends to swell or shrink, depending on its water content. Since most of these thinner tubes have to stand up to large pressures during filling, they must be protected with a mold to prevent the tube from splitting. Molds usually are of two types.

The first type of mold consists of a block of metal into which holes are bored of such a diameter that they will just admit the tubes to be pressed. The inside surface of the molds must be highly polished. A plate on the underside of the mold keeps the tubes in place during the pressing operation, and when this is completed, the plate can be moved to one side to allow ejection of the pressed tubes.

The alternative method is to use a split mold. This method employs three metal segments that completely surround the tube during pressing. The segments are held in position by a collar of some kind, and there is a taper on the outside of the segments to allow the collar to be removed easily (Fig. 9.1).

Earlier works have expanded at some length on the various ways of rolling paper tubes. No purpose would be served by repeating this information, so only the general principles are outlined here.

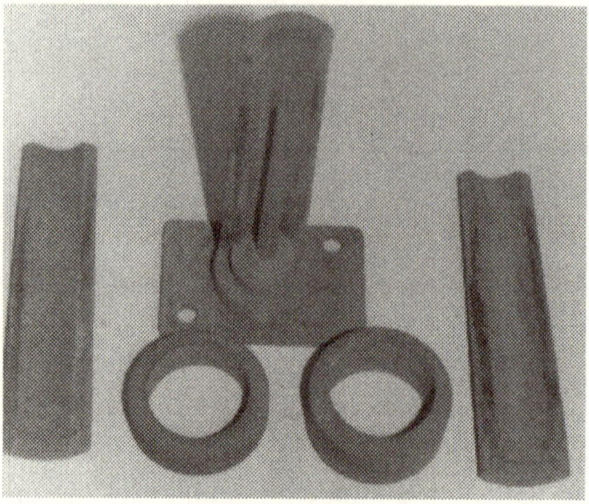

Figure 9.1 Mold

Commercial adhesives are many and various, and much depends on the type of paper to be employed. Casein- or dextrin-based adhesives frequently are employed for strong tubes made with kraft paper, but a starch or methyl cellulose type of adhesive can be used for thinner or cheaper papers. Unfortunately, tubes made with dextrin adhesives tend to absorb moisture in damp conditions.

A good inexpensive paste can be made by mixing one part of wheat flour into a smooth cream with sufficient cold water. Hot (not boiling) water is then added to make the total volume of water up to 10 parts. This mixture is then boiled until it thickens. A small percentage of preservative (boric acid) should then be added to prevent bacterial action; otherwise there will be mold formation and rotting in damp conditions. In the past manufacturers added alum as a coagulant, but this is neither necessary nor desirable.

It is not difficult to roll good tubes; it is just a question of the time needed to acquire the necessary skill.

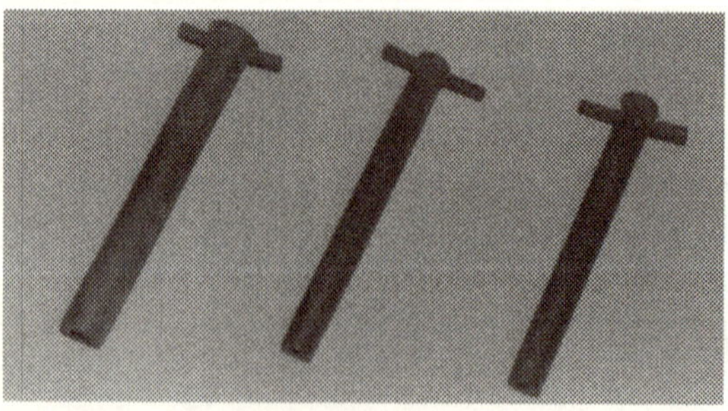

Figure 9.2 Drifts

Figure 9.3 Rolling narrow bore tubes

Formers in the smaller bores usually are made of brass or aluminum; larger sizes invariably are brass tubing (Fig. 9.3). Wood, of course, is quite useless for wet straight rolling, but it can be used for cones or other forms that easily slip off the former. The larger formers frequently are fitted with a wooden handle at one end (Fig. 9.4).

Long narrow bore tubes sometimes can be difficult to get off the former. The best way to deal with this is to fix a spike on the edge of the bench and drill a hole through the end of the former. After slotting the hole over the spike, the tube is lightly held as it is pulled off the former (Fig. 9.5).

The rolling bench should be made of heavy solid timber with a non-absorbent surface that can be easily cleaned. Marble or synthetic resin is very good for this purpose.

The choice of paper naturally is important, but it usually is not as crucial as might be expected. Strong tubes that must withstand pressure or hammering usually are made with unglazed kraft paper. The weaker tubes for Roman candles, which take only a light ramming, invariably

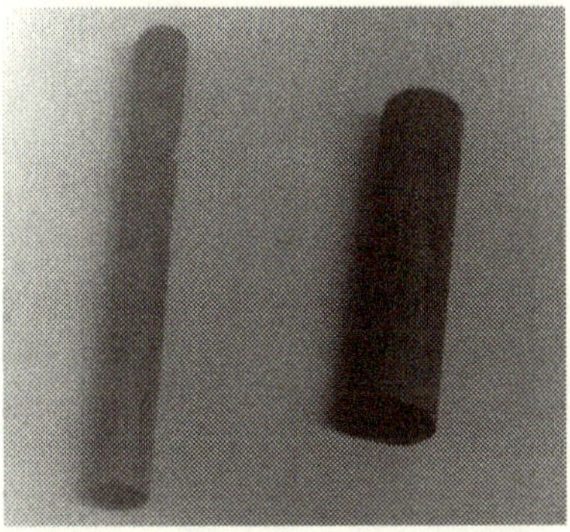

Figure 9.4 Formers for wide bore tubes

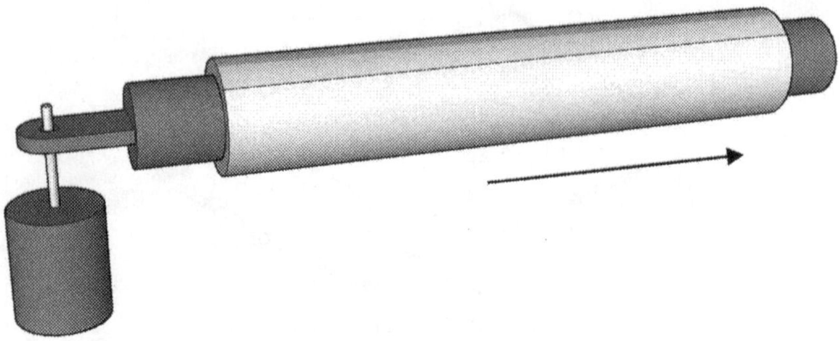

Figure 9.5 Using a spike to remove a long tube from the former

are made of inexpensive, unglazed, common brown paper (or "sugar" paper, as it is sometimes called).

Paper has a grain, so it is obvious that it must be cut in such a way that the "roll up" is with the grain and not against it. Gently bending the dry paper in the hands quickly shows that it folds more easily one way than the other and must be cut accordingly.

To obtain tubes with thick walls, it is far more convenient to roll several sheets of paper together at once rather than attempt to roll up a single long sheet of paper. It also often is more convenient to use a large sheet of thin paper on the outside and one or more liners on the inside of a somewhat thicker paper. By using this method, it will be seen that it is easier to stick down the outside paper, and by using a fairly short "roll up," there is less likelihood of making a bad tube.

The procedure may be best illustrated by an example. Assuming that a tube of 18 mm inside diameter, 45 cm long, and having a wall thickness of about 5 mm is required, two pieces of paper would be needed for each tubes.

For example,	1 piece	0.014 gauge	45 cm by 53 cm
	1 piece	0.025 gauge	45 cm by 29 cm

If, for example, 12 tubes are to be made, the 12 outside (014) papers are laid on top of each other, a quarter of an inch apart (Fig. 9.6). The quarter inch edges of the paper that are laid bare are then liberally covered with paste and scrubbed with a wire brush to roughen the edge of the paper to make sure that it will stick down easily on the tube. After this is completed, the batch of papers is turned over and the process is repeated on the reverse side without disturbing the positions of the

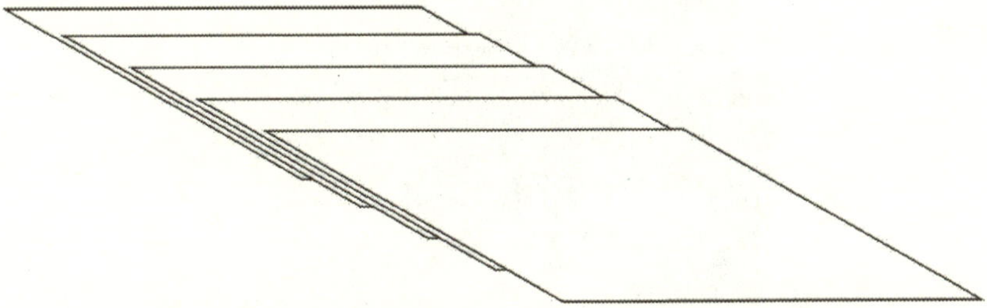

Figure 9.6 Paper ready for pasting

sheets of paper. As a result, each outside paper has its outer edge pasted and scrubbed up, but on opposite sides of each sheet one edge will be on the inside of the tube and the other on the outside of the completed tube.

To roll the tube, one of these outer sheets is placed on the bench and covered with paste. Next, an inner (025) sheet in placed about 12.5 cm up the sheet and also is pasted. The former is then placed on the paper; the paper is bent up over the former and checked to be square. After ensuring that the edge of the paper is biting closely under the former, the tube is rolled up, easing the liner a little during the process if necessary. The edge must be well stuck down, and the tube should be rolled on the former three or four times before it is removed. Sometimes a wooden board with a handle is used to roll the tube on the former (Fig. 9.7).

When thicker tubes are required, more than one liner can be inserted, although it will be necessary to space the liners about 12 mm from each other to give a well-balanced wall thickness. The paper should always be uniformly pasted or there may be the possibility of uneven drying. Drying must be a slow, gradual process for very wet tubes, a week at room temperature being better than a shorter period at a higher temperature. Wet tubes that have been rapidly dried with ordinary adhesives frequently end up banana-shaped. During the rolling process the tubes become so firmly stuck to the former that they cannot be

Figure 9.7 Rolling board

Figure 9.8 Choking machine

removed. This is because the rod has become dry, but this can be easily overcome by placing plenty of paste on the former and the edge of the paper where the former is to lie. Tubes that are used for making lances, pinwheels, or match-piping are not pasted all the way through, the paper being merely pasted on the edge.

Sometimes it is necessary to constrict tubes that are used for rockets, serpents, and fountains. The constriction usually is produced by a machine that contains six plates arranged similarly to those that control the aperture of a camera.

Alternatively, a piece of piano wire or stranded wire can be wrapped around the tube so that the tube is constricted when gentle pressure is applied. Choking machines also can be used to constrict dry tubes provided that the wall is not too thick, but if the hand method is used, only wet tubes can be choked. Choking machines also are used to close up the ends of small, thin-walled tubes. Choking machines are shown in Figures 9.8 and 9.10. As choking is a somewhat time-consuming process, it is common practice to constrict rockets and fountains with a clay washer. The tube is fitted onto a nipple made of brass, stainless

Figure 9.9 Tube rolling. Early twentieth century (Brock's Fireworks)

steel, or aluminum, and a scoopful of clay is hammered or pressed around the nipple before the composition is charged into the tube. At the end of the operation, the clay washer is left at the end of the tube after the spindle is withdrawn (Fig. 9.11). Cones and the swell heads of rockets are made by pasting a specially cut paper shape around the appropriate form (Fig. 9.13). Cones are made from a circular piece of paper with a cut across the radius (Fig. 9.12).

Wad cutters are used to cut the circular pieces of paper and for cutting the millboard discs that are used for making cylinder shells. Wad

Figure 9.10 Interior of a choking machine

Figure 9.11 Tube with clay washer

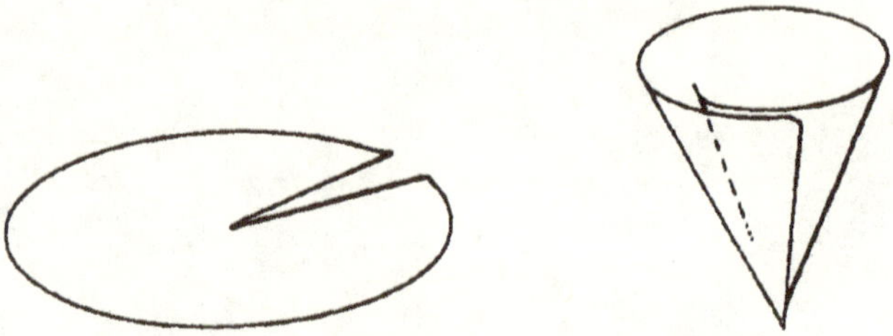

Figure 9.12 Paper circle for cone manufacture

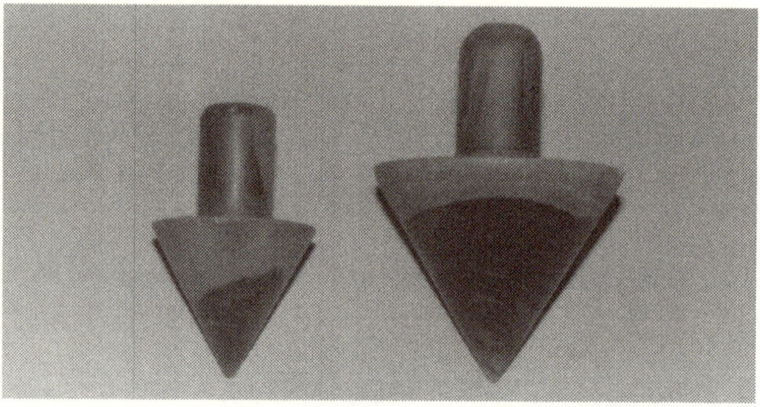

Figure 9.13 Formers for cones

Figure 9.14 Wad cutters

cutters are shown in Figure 9.14.

Spirally wound tubes are regularly finding their way into firework manufacture. In those fireworks that do not place a great strain on the tube, they are quite useful, but as a general rule they are not as strong or efficient as the older convolute-wound types. Unless the inside of a spirally wound Roman candle tube has been carefully executed, it is common to find that the pieces of paper also are ejected with the stars.

More accurate details of tube sizes will be given along with the description of the firework concerned.

Roman candle tubes that are spirally wound can be particularly liable to produce the "machine gun" effect because gunpowder can slip down the gaps in the spiral winding. This can be prevented by using an overlapping thin paper liner as the final inside layer. There is no doubt that spirally wound tubes have improved greatly, particularly those that use the more modern high-performance papers.

The kind of paper that is used for fireworks varies enormously and to some extent is subject to the preferences of individual producers. It is still a comparatively inexpensive item, but in modern times supply becomes an increasing problem. In the middle of the past century, there were dozens of manufacturers and merchants, but they have slowly disappeared, and the manufacturer cannot always get what he needs in a quantity that is economical. Ribbed kraft paper with a weight of 39 grams per meter is useful, but not if it must be bought by the ton. Kraft paper is widely used in the industry, perhaps at 50 g for capping or wrapping, and there are some uses for 70 g paper, and a paper of 110 g might be used for placing under sieves for mixing.

Colored poster papers of 50 to 70 g are used for rolling lances and portfires, but even these now seem to come from very limited sources. A more unusual paper is the pure kraft, which is used for papering shells the old-fashioned way. Paper of 0.25 mm at 225 g per square meter sometimes is called manila and also is used for making labels, but it has a great wet strength; it can be wet with paste and, after soaking for a short while, can be rubbed up until it is like chamois leather. This paper is wrapped around old-fashioned cylinder shells after stringing, and within a few days it sets very hard indeed. In experiments with Japanese-type shells, it was found that this paper works well and reduces the number of layers to be pasted.

In practice, manufacturers use what is available, and friends in some places use the high-quality kraft paper that is used for making cement bags and such.

10

STARS

In view of the fact that stars feature in many types of fireworks, their manufacture is described first. An endless variety of effects may be obtained with the chemicals now at our disposal, although only five types of stars are commonly used: (1) pumped (or cylindrical), (2) cut (or cube-shaped), (3) pill box, (4) pressed, and (5) spherical stars.

Before the composition can be made into a star, it must be moistened with some kind of adhesive so that it will set into a hard mass when it is dry. The amount of water that is added sometimes can be very crucial, and care must be taken not to add too much or too little. It is usual in Europe and the United States to mix a percentage of dextrin to the dry star composition and then add the required amount of water. From 3% to 5% dextrin normally will be adequate. It is unwise to use more than 5% unless it is absolutely necessary because dextrin is very hygroscopic, and stars containing large amounts of it tend to go soft in damp conditions.

Sometimes it is useful to use a mixture of alcohol and water for damping star mixtures; for one thing, it reduces surface tension, making it easier to dampen lampblack mixtures, for example. It also is easier to dry the stars. On the other hand, dextrin does not dissolve in alcohol, so no more than 1⁄3 alcohol to 2⁄3 water must be used or the adhesive power of the dextrin will be inhibited. Gum arabic also is used as an adhesive. A 7% solution in water will be adequate, but it can be time-consuming to make up the solution, which must be used the same day. When gum arabic solutions are allowed to stand, they ferment and become acidic and consequently would be very dangerous if used to make colored stars. Therefore, if it is essential to use gum arabic for some reason, it should be used only on the gunpowder side of the factory.

The other important method of gaining adhesion is to use alcohol only with shellac or accaroid resin. It is common practice to have some of the finely powered resin in the composition and merely damp the mixture with alcohol, damping a small quantity at one time because it dries out quite quickly. Ethanol and methanol are commonly used, along with isopropyl alcohol, which contains no water. It also will be discovered that shellac is the best resin to work with because it is much less sticky than accaroid resin.

Small pumped stars usually are dried in a gentle heat of about 35°C for two days. Larger stars and very wet stars will take much longer, of course. Drying cupboards usually are heated with hot water pipes, special oil-filled heaters, and the like, and some manufacturers use vacuum ovens. In the summer it often is convenient to leave trays of stars out in the open air during the day, although out of direct sunlight. Not every country has the luxury of a hot dry climate, but chlorate stars are thought to be affected by ultraviolet radiation.

PUMPED STARS

Pumped stars were used more than any other type of star in the West, mainly because they can be manufactured easily and quickly.

A single star pump is very useful for making small quantities of stars. It consists of a short cylinder of copper or brass, fitted with a brass plunger and a wooden knob. A stud in the side of the plunger and a slot cut into the outer sleeve complete the apparatus. When a star is to be made, the plunger is raised to the top of the slot, the pump is pressed into the loose composition, and the star is ejected. Some manufacturers use six of these pumps fixed in a row on a wooden handle. The principle is the same except that instead of having slots cut in the sleeve of the pumps, the plunger is spring-loaded. The plungers normally are in the raised position, but the springs allow the plungers to move down the sleeves to eject the stars, afterward returning to their raised position (Fig. 10.1).

The other common method of making cylindrical stars is to use a star plate. Large quantities can be made by this method, although the stars often tend to be rather soft unless they are made carefully. The star plate consists of a sheet of brass furnished with a number of holes of the same diameter as the star. Underneath this plate is another plate filled with spigots, each spigot fitting into the base of holes in the star plate. The two plates are fixed into the bench, and the bottom plate of spigots has some arrangement attached to it so that the spigots can be raised to eject the stars. To make the stars, the damp composition is pulled over the surface of the plate and tamped into the molds by another plate filled with spigots. When the molds are full, the excess composition is

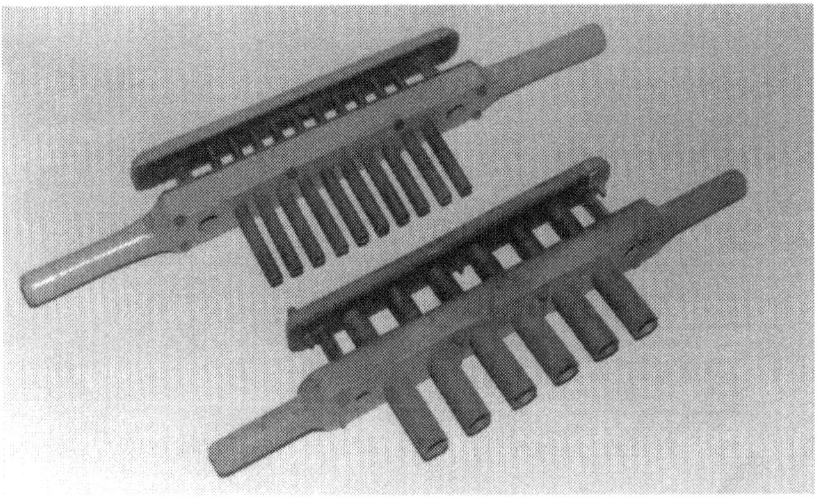

Figure 10.1 Star pumps

removed with a scraper; the base plate of spigots is raised to eject the stars, which are then transferred to a tray.

Roman candle stars must be very fast- and fierce-burning; otherwise they frequently blow out of the tube without igniting. For this reason, they often are made with potassium chlorate. Potassium perchlorate can be used, of course, but a considerable percentage of charcoal also must be added, and the stars always look smaller because there is not as much flame as with the chlorate. The following are good compositions for cylindrical Roman candle stars:

Red	Weight (%)
Potassium chlorate	70
Strontium carbonate	15
Accaroid resin	10
Dextrin	4
Charcoal 150 mesh	1

Green	Weight (%)
Barium chlorate	53
Potassium chlorate	28
Accaroid resin	10
Charcoal 150 mesh	5
Dextrin	4

Yellow	Weight (%)
Potassium chlorate	70
Cryolite	15
Accaroid resin	10
Dextrin	4
Charcoal 150 mesh	1

(continued on next page)

Blue	A Weight (%)	B Weight (%)	C Weight (%)
Potassium chlorate	68	65	67
Paris green	22	—	—
Basic copper chloride	—	11.5	—
Copper (II) oxide	—	—	13
Colophony resin + 60 mesh	6	—	—
Dextrin	4	5	5
Chlorinated PVC or Parlon	—	7	5
Lactose	—	11.5	—
Accaroid resin	—	—	10

White	Weight (%) A	Weight (%) B
Potassium nitrate	51	60
Sulfur	18	10
Mealed gunpowder	15	5
Antimony metal powder	10	—
Antimony sulfide	—	20
Charcoal +150 mesh	3	—
Aluminum dark pyro	—	1
Dextrin	3	4

Pumped stars usually have diameters of 8 mm or 9.5 mm, 14 mm or 16 mm, 25 mm, 35 mm, or 40 mm. Roman candle stars are slightly less in diameter than the bore of the Roman candle tube. Thus, for a 16-mm bore candle, the star would be 14 mm in diameter and approximately the same in length. Gold stars normally are made with charcoal or lampblack, or a mixture of the two. Dr. Shimizu gives details on page 388 of golden streamers made only with charcoal, and the following also will be found to be quite useful:

	Weight (%)
Meal gunpowder	54
Charcoal 150 mesh	20
Lampblack	13
Antimony sulfide	6
Dextrin	7

	Weight (%)
Meal gunpowder	66
Lampblack	23
Antimony sulfide	8
Accaroid resin	3

The star containing dextrin is damped with a mixture of two parts water to one part alcohol. The star containing accaroid resin is damped only with alcohol.

The following star is a modification of one quoted by Weingart and is said to be Japanese in origin.

	Weight (%)
Lampblack	49
Potassium chlorate	13.5
Potassium perchlorate	13.5
Potassium nitrate	15
Dextrin	6
Shellac 120 mesh	3

The composition should be made quite wet with alcohol and water like the aforementioned lampblack stars, but they require a long time to dry out.

GOLD STARS

It is not usual to employ lampblack in modern manufacture because of its extreme dirtiness and problems of obtaining suitable material. The reader is referred to Chapter 4 on charcoal, as there also are supply problems with this material. The following compositions are typical:

	A	B	C
Gold Stars	Weight (%)	Weight (%)	Weight (%)
Potassium nitrate	57.5	53.5	55
Sulfur	7.5	6.5	15
Charcoal + 150 mesh	35.0	—	—
Charcoal pine + 150 mesh	—	31.0	—
Charcoal mixed	—	—	25
Dextrin	—	—	5
Gum arabic powder (additional)	2.0	—	—
Glutinous rice starch	—	9.0	—

A is a good, all-purpose small gold star using the fine commercial blended charcoal.

B is a Japanese willow effect. Pine charcoal from Japan is better than much that is available from Europe.

C is a comet mix. The charcoal is a blend of that which passes 150 mesh and some softwood or pine charcoal of about 40 mesh, to the manufacturer's choice.

A good comet star is simply a mixture of two parts mealed gunpowder and one part charcoal with a suitable binder. To get a good sharply defined gold comet in a shell, possibly with a burst of flashpowder, a gunpowder-type mix will be required unless the manufacturer has suitable ball milling facilities.

GLITTER STARS

The well-known glitter, twinkling, or tremolante effect often presents problems, and some manufacturers have been puzzled to find that they are unable to obtain the right effect even though they may have the correct formulation. The problem once again is very closely tied up with the particle size of the ingredients and the method of production of the composition. If the material is wetted with water or some other mixture, this also can influence the final results. The old hands in the trade have always been aware of two things, even if they did not understand the chemistry:

1. That the glitter was produced from a bubbling molten mass; and
2. That adding charcoal spoiled the effect.

Dr. Shimizu has written about these matters (p. 390 and Refs. 27 and 78), and it is clear that the main reaction involves the formation of potassium sulfide around particles of aluminum and that, as this is spattered around in the air, the coated aluminum particle is able to ignite with a flash. Clearly, the addition of extra charcoal would cause reduction and spoil the melt (see Chapter 6).

The old manufacturers in Europe concentrated very much on mixtures that contained gunpowder, sulfides of antimony and arsenic, sulfur, and lampblack. The lampblack was thought to contribute to the fused mass, but storage and the influx of moisture caused chemical reactions that disturbed the glitter. It is more than likely that in the unheated magazines, the alkaline carbon black reacted with the aluminum. The situation is not a lot better with the use of sodium oxalate, but the addition of boric acid has changed this.

The author has reason to believe that the use of boric acid is distinctly English in origin and that the dissemination of this information took place in the first edition of this book. In mixtures cited by Dr. Shimizu (see Chapter 6), he is of the opinion that it is better to do without boric acid if possible; it all depends on the mixture, and one certainly would need boric acid in a barium nitrate/aluminum sparkler slurry.

An example for academic interest of the earlier U.K. mixes would be:

	Weight (%)
Mealed gunpowder	45
Antimony sulfide	16
Potassium nitrate	18
Aluminum + 120 bright	6
Dextrin	5
Barium nitrate	5
Lampblack	5

Who knows what kind of lampblack was used in these mixtures? In the course of time, it was eliminated, and the next two mixtures would be fairly typical:

	Weight (%)		Weight (%)
(A) Mealed gunpowder	50	(B) Mealed gunpowder	16
Potassium nitrate	11	Potassium nitrate	35
Sulfur	5	Sulfur	4
Charcoal +150 mesh	—	Charcoal + 150 mesh	5
Barium nitrate	—	Barium nitrate	15
Antimony sulfide	18	Antimony sulfide	14
Aluminum "bright"	8	Aluminum "bright"	6
Dextrin	5	Dextrin	5
Wheat flour	3	Wheat flour	—

Looking at the matter in a more scientific way, a number of people have quite reasonably assumed that all that is really needed is to take the basic materials and keep the matter as simple as possible in order to avoid the problems associated with the variations in too many ingredients.

In commercial terms, one major problem comes from the use of manufactured blackpowder. It now is very expensive, and most manufacturers need to use as little as possible and simply use the basic ingredients. Nevertheless, the author is convinced that some of the best glitter effects—particularly those that are to be viewed at a distance in the big firework displays—do need to use manufactured blackpowder, and some of that blackpowder preferably should be fine soft grain powder (U.K. grade NPXF). This raises the issue that glitter mixes should not be sieved in fine sieves. The best screens to use are about 18 mesh (840μ). Some amateurs have reported that ball milling of glitter mixes destroys the effect (it also risks destroying them).

The grade of aluminum used to be especially important. The author has no doubt that the best effects were made with old-fashioned flake aluminum ("bright") powders, which were made in stamp mills and about 120 to 200 mesh. Some ball-milled powders are satisfactory and some are not, and the same applies to atomized aluminum. In general, atomized aluminum works well, and some U.S. powders seem to be more reactive than some European grades. A formulation provided by the late Bill Withrow is a little unusual in its proportions and probably was formulated as a sparkler slurry:

	Weight (%)
Potassium nitrate	37
Barium nitrate	16
Antimony sulfide	10
Sulfur	10
Charcoal + 150 mesh	10
Dextrin	10
Aluminum atomized	7 (+53μ)

Some manufacturers would consider this a high proportion of dextrin and not suitable for damp climates. For most formulations, 3% to 7% would be the norm, and it is arguable whether dextrin helps or hinders these glitter formulations.

The reader should study some of the recent writings about the glitter phenomenon, particularly those written by Winokur (27) and Shimizu (29). Many different substances have been used in the trials, and it is clear that fine iron (III) oxide has become one of the favorites. However, it should be made clear that there can be a world of difference between observing a beautiful glitter fountain at 5 meters and a glitter star moving through the air at a few hundred meters. In the two cases, the formulations may be quite different. In practice, the author has found (and perhaps he has not looked widely enough) that barium carbonate or strontium oxalate serves both purposes very well with the materials at his disposal. Some quite splendid effects can be obtained, but like so much of firework manufacture, it is very much a mixture of chemistry and cooking, and once again we are brought face-to-face with the problem of the materials. It is essential to maintain the quality control of the ingredients. In practice, this often is a question of particle size, and most manufacturers at some stage realize that a commercial vibratory sieve is a sine qua non.

It is more than likely that the first glitter stars to be made were those made with sodium oxalate. In some regards they are the easiest, and they work quite well using atomized aluminum.

The following are typical:

Yellow Glitter	A Weight (%)	B Weight (%)	C Weight (%)
Mealed gunpowder	70	—	60
Soft grain gunpowder	—	27	—
Potassium nitrate	—	32	—
Charcoal + 150 mesh	—	6	—
Sodium oxalate	10	10	7
Antimony sulfide	8	8	21
Aluminum flake "bright"	7	—	6
Aluminum atomized + 350 mesh	—	7	—
Aluminum flitter 30–80 mesh	—	—	1
Sulfur	—	8	—
Dextrin	5	—	5
Gum arabic	—	2	—
Boric acid (about 0.5% additional)			

Composition C has a large (and expensive) quantity of antimony sulfide, but it does have an unusual glitter that has a slight strobelike quality. The stars should not be made too large and are for use in shells. A and B are suitable for all types of stars.

The best white glitters are made with blackpowder, although good ones also are made with the basic raw materials of gunpowder. Not everyone would agree, but if manufactured gunpowder is not used (i.e., certain burning characteristics are established for this main ingredient), then unexpected variations will appear from time to time in bulk manufacture. The author certainly has experienced problems when the nitrate has not been fine enough and when too much water has reacted with atomized aluminum.

White Glitter	A Weight (%)	B Weight (%)
Mealed gunpowder	26	24
Soft grain gunpowder	40	40

(continued on next page)

Antimony sulfide	14	12
Strontium oxalate	8	—
Barium carbonate	—	12
Aluminum flake "bright"	8	5
Aluminum flitter 30–80 mesh	—	2
Dextrin	4	5

If the glitter is quickly dried and the grease-containing flake aluminum powders are used, boric acid is not vital.

By using atomized aluminum powders in these white glitter mixes, some quite interesting variations can be obtained for viewing at a short range, but the upper and lower limits of the mesh sizes must be closely defined.

In fact, with the use of atomized aluminum, the type of "spritzel" obtainable does rather depend on the size of the particle, which in practice means the mesh size. However, as with every theory and practice surrounding this effect, so much depends on the material being used. Commonly used U.K. aluminum has been made by pouring molten aluminum into an airstream and the final particles are spindle- or sausage-shaped. A gerb made with fine dust that passes 350 mesh will give a very definite large glitter effect. However, if the fines are sieved out and the material used is between 120 and 200 mesh, then the gerb will produce a rather delicate flower, which we might call a "flitter." In practice, some flitter gerbs are made with a mixture of 120 to 200 atomized aluminum and flake aluminum flitter between 80 and 120 mesh.

When such formulations are used in stars, the distance at which the effects are observed frequently are too great; consequently, the formulations for flitter stars in shells or Roman candles must use an even coarser atomized aluminum.

As yet another example, a typical candle star of 30 mm called Silver Rain produced a silver tail based on gunpowder, barium nitrate, flake aluminum and flitter, and so on. However, at 200 m it was difficult to distinguish it from a titanium star made with the constituents of gunpowder and a titanium passing 40 mesh. Since the titanium star was cleaner to make, the Silver Rain star was all but abandoned.

A well-known formulation consists of approximately

Gunpowder	77%
Barium carbonate	4%
Antimony sulfide	2.5%
Aluminum atomized 120–200 mesh	12.5%
Dextrin	4%

But the finer the aluminum, the more like a silver streamer it becomes at a distance

A Roman candle flitter comet may contain less meal powder, more antimony sulfide, and a goodly percentage of potassium nitrate with atomized aluminum to obtain a goldish stream with a large excess of silver flicks.

It is worth mentioning that the use of calcium silicide was quite popular in former times. A silver streamer star that is appropriate here was made as follows

Potassium perchlorate	54%
Barium nitrate	11%
Accaroid resin	11%
Calcium silicide	16%
Aluminum bright flake	5%
Charcoal	3%

There always is a slight odor when such mixtures are damped with water, but usually there is no heating and the star can always be made with alcohol.

At the moment we have only theories of the chemistry of the glitter effect, but there is an excellent summary of the differing opinions written by Dr. Clive Jennings-White in the *Journal of Pyrotechnics* (88).

STROBE STARS

If there are problems with glitter stars, the situation is even worse with strobe stars. The author asked one well-known European manufacturer how he got along making strobe stars and was told that it all depended. Depended on what, the manufacturer was not sure. Some days the stars were good, some days they were less good!

Many people have looked into the manufacture of good strobing stars (29) and many have had a degree of success, but there are considerable problems for the consistent manufacture of large quantities.

The effect is essentially based on barium nitrate and magnesium with or without ammonium perchlorate. It seems clear that the Chinese were the pioneers of the firework strobe effects, particularly because postwar Chinese thinking has made great use of magnalium. The basic idea is the same as with glitters in that particles of metal are brought to ignition in a smoldering molten mass. The addition of small amounts of "catalysts" can produce flashes with a very clearly defined frequency, such as one or two flashes per second. Such flashes of one or two hertz are ideal for fireworks.

The first mixtures were based on barium nitrate, fine magnalium, and sulfur. The percentage of sulfur is quite high, and the stars are not easy to ignite. The stars are made quite small, say, 5-mm spheres, and must be very heavily primed with a hot igniter. A good green effect also can be made by adding a suitable chlorine donor. At first hexachlorobenzene was used, but now that this is becoming abandoned, Parlon or Alloprene can be used, albeit with difficulty.

The main question is how to make the stars into pellets, bearing in mind that the addition of dextrin can upset the chemical balance and prevent strobing, and the use of water-based solvents can cause heating with some forms of magnalium. The other problem is that organic solvents can be extremely inconvenient when working with commercial quantities. Red gum in an alcohol or nitrocellulose solution can be difficult to work with and to clean up. The delicate balance of the mix, the grades of chemicals, the chemical reaction, and the amount of solvent make the whole process quite idiosyncratic.

Mixes without sulfur are by far the best, and the reader is directed to the articles by Dr. Shimizu on Japanese practice (29,34). Very good strobes can be made with ammonium perchlorate, magnalium, and a sulfate, but again care must be taken to look after the chemistry because ammonium perchlorate and magnalium react quite dangerously under normal circumstances.

Many people are looking for good strobe effects that can utilize simple materials and be made in commercial quantities that have a good shelf life. It will be interesting to see what happens.

CRACKLING STARS

There can be no doubt that the Chinese discovery of crackling stars has been a source of great interest in the West over the past years. The effect is a source of delight in its use, both in small fireworks and at the large displays.

Most of the fireworks made with these microstars come from the Far East, not only because of the price advantage, but also because Western legislation makes it difficult to work commercially with lead compounds where a regular blood test is required.

Readers who wish to know more about the background to these mixtures should refer to *Pyrotechnica* (XIII, XIV, XV). Dr. Shimizu's work

(XIII) establishes that any of the three lead oxides can be used along with atomized aluminum of 60 to 100 mesh and magnalium of something between 60 and 200 mesh. An optimum magnalium alloy could be Mg/Al 31/69 if it were commercially available.

The following basic combinations have been explored:

Crackling Stars	A Weight (%)	B Weight (%)	C Weight (%)
Magnalium	11	17	15
Lead oxide, Pb_3O_4	89	44	—
Potassium nitrate	—	4	—
Sulfur	—	4	—
Copper (II) oxide	—	31	10
Bismuth (III) oxide	—	—	75

The next and most important discovery was by Dr. C. Jennings White, who found that bismuth oxide not only works well but also is virtually nontoxic (*Pyrotechnica* XIV). There always is a snag, of course, and in this case it is the high price of bismuth.

The stars are made into small grains of 0 mm to 4 mm and probably are best made with nitrocellulose dissolved in an organic solvent, such as amyl acetate or acetone. Dextrin also can be used, but this would be a snag if the stars were to be incorporated into the mixes of other stars that are bound with a water-based adhesive. A technique is described by Dr. Shimizu in Chapter 6.

It is important to remember that these stars are quite sensitive to impact and friction (*Pyrotechnica* XV). Tests conducted by Tony Cardell in the United Kingdom on the bismuth formulation bound with nitrocellulose established the following results:

Rotter Impact Test. The figure of insensitiveness is 35.

Comparisons	0–30	very sensitive
	30–90	sensitive
	90+	comparatively insensitive
Examples	Blackpowder about	100
	RDX	80
	Blackpowder/titanium	60
	Lead azide	50

The Rotary Friction Test gave a result of 0.9.

Comparisons	3 or less	very sensitive
	3–6	sensitive
	6 or more	comparatively insensitive
Examples	Lead azide	0.07
	Gunpowder/titanium	2.25 possibly
	RDX	3.0
	Gunpowder	+ 6.0

It is difficult to make silver pumped stars that are safe. Very good stars can be made with potassium chlorate and aluminum, such as the following:

	Weight (%)
Potassium chlorate	56
Aluminum "bright"	19
Aluminum dark pyro	19
Dextrin	6

but they are sensitive and dangerous. A much safer and more beautiful star can be made with potassium perchlorate and aluminum, but these often refuse to ignite and are best made as pill box stars. The following is an example of an inexpensive electric streamer without chlorate, but it lacks the brilliance of the chlorate/perchlorate stars.

	Weight (%)
Potassium nitrate	45
Aluminum dark pyro	30
Aluminum atomized 120 mesh	10
Sulfur	10
Meal gunpowder	5
Dextrin	about 5%

Silver stars for mines and shells quite frequently now are made with magnalium and atomized aluminum. They ignite well and ignite with normal priming methods.

Potassium perchlorate	42%
Accaroid resin	8%
Magnalium 120 mesh	20%
Aluminum atomized 120 mesh	15%
Aluminum atomized 300 mesh	15%

The reader is also referred to Dr. Shimizu's discussion of silver star effects on p. 390.

CUT STARS

Dr. Shimizu has described the manufacture of cut stars, and little more needs to be added since this corresponds more or less with Western practice. Considerably more water must be added to cut star composition so that it will beat down into the frame in a tacky, pliable mass like pastry. If it is too dry, the mass will not cut into cubes and just crumbles. After cutting, the stars usually are agitated in a bowl of meal powder, which coats the exterior. This has long been the practice, and it is wise to have some dextrin mixed in the blackpowder so that it adheres well to the wet star.

Reference to Chapter 6 will make it clear that it is not easy to coat stars that are cut or cylindrical after they are dry. As dextrin is invariably used, the water/slurry is very quickly absorbed and the stars are quickly deformed. For example, it might be a nice idea to place stars in a star tumbler, spray them with water, and then add the priming composition. Once that is tried, the operator soon learns that speed is essential.

Some people maintain that the sharp corners improve the chances of ignition of these stars, but this is doubtful. The gunpowder coating undoubtedly improves ignition, but it also increases the danger because of the sulfur and the chlorate being in close contact. Many pumped star compositions are suitable for cut stars, but the following are specifically used for the cut variety:

Red	Weight (%)
Potassium perchlorate	70
Strontium carbonate	15
Accaroid resin	9
Charcoal 150 mesh	2
Dextrin	4

Yellow	Weight (%)
Potassium perchlorate	70
Sodium oxalate	14
Accaroid resin	6
Shellac 80 mesh	6
Dextrin	4

Green	Weight (%)
Barium chlorate	72
Accaroid resin	12
Charcoal 150 mesh	8
Dextrin	4
Barium carbonate	4

MODERN FORMULATIONS

There is much to commend the original Chinese technique of using magnalium instead of magnesium for making high-temperature stars. These mixtures certainly are more stable.

There is a certain amount of room for maneuver so that variations can be obtained by using more perchlorate and less magnalium or the other way around using more magnalium and less perchlorate. The remainder of the components of the mixture can remain as normal (i.e., a percentage of coloring agent, halogen donor, accaroid resin, and binder).

The following would be typical:

	(%)
Potassium perchlorate	30–60
Magnalium	15–30
Strontium carbonate	10–25
Parlon	6–14
Accaroid resin	5–8
Dextrin/rice starch	4–5

Many people believe that there is wisdom in passifying the magnalium with potassium dichromate if water is used. It has been suggested that this is primarily a Western rather than an Eastern view, but also that it might give rise to the formation of small amounts of acid in certain mixtures. There is wisdom, therefore, in using carbonates as coloring agents. Barium nitrate still gives the best green, however, and copper carbonate is not very effective.

The various chlorine donors seem to work better in different mixtures, and the system has the advantage of allowing different shades and depths of color to be obtained. A mixture of copper and strontium produces a range of violet colors, but these combinations on the ground do not always give the same effect when viewed from a distance in the air.

Parlon remains a firm favorite as a color donor, but it frequently produces a large amount of ash.

PILL BOX STARS

Although pill box stars are rather more time-consuming to manufacture, some of the best effects can be obtained with them. Not only do

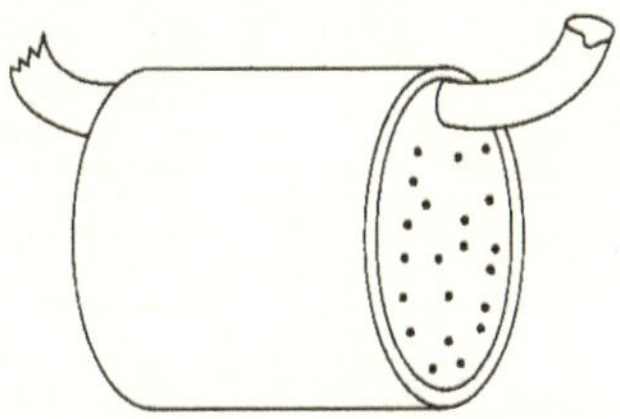

Figure 10.2 Pill box star

the stars have a longer burning time—usually about 5 seconds—but also they always ignite because of the piece of quickmatch that is embedded in the composition.

The stars are made in thin-walled paper spools of various sizes, a common size being 10 mm inside diameter, 14 mm long, and with a wall thickness of about 1 mm. The match is inserted into the tube, and the damp composition is firmly pressed into the tube with the thumb and first finger. The match extends over the outside edges of the tube by about 14 mm (Fig. 10.2).

The compositions are somewhat slower-burning than pumped stars, the following being fairly typical:

Blue	Weight (%)
Potassium chlorate	70
Paris green	20
Shellac 60 mesh	10
(add alcohol)	

Blue	Weight (%)
Potassium chlorate	39
Ammonium perchlorate	29
Copper carbonate (basic)	14
Accaroid resin	14
Dextrin (add water)	4

Red	Weight (%)
Potassium chlorate	64
Strontium carbonate	19
Accaroid resin	13
Dextrin (add water)	4

Green	Weight (%)
Barium chlorate	48
Potassium chlorate	17
Accaroid resin	17
Barium nitrate	11
Barium carbonate	4
Dextrin (add water)	3

Amber	Weight (%)
Potassium perchlorate	60
Sodium oxalate	26

Silver	Weight (%)
Barium nitrate	55
Potassium nitrate	13

(continued on next page)

Shellac 60 mesh	14	Aluminum dark pyro	21
(add alcohol)		Dextrin	6
		Sulfur	4
		Boric acid (add water)	1

Silver	Weight (%)	**Silver**	Weight (%)
Potassium perchlorate	64	Potassium perchlorate	64
Aluminum "bright"	14	Aluminum "bright"	24
Aluminum flitter 30/80	14	Aluminum dark pyro	4
Shellac 60 mesh	8	Shellac 60 mesh	8

(Damp these stars with a 10% solution of shellac in alcohol)

Green/Silver	Weight (%)	**Red/Silver**	Weight (%)
Barium chlorate	25	Potassium perchlorate	47
Barium nitrate	25	Strontium carbonate	20
Aluminum "bright"	19	Accaroid resin	10
Potassium chlorate	13	Aluminum "bright"	10
Accaroid resin	7	Aluminum flitter 30/80	10
Dextrin	5	Magnesium + 200 mesh	3
Barium carbonate	4	(add alcohol)	
Charcoal 150 mesh	2		
(add water)			

Modern thinking would consider the green/silver to be dangerous. Purple stars can be made by substituting about half the Paris green of the blue star composition with strontium carbonate. Needless to say, Paris green is no longer used in fireworks.

Yellow Illuminating	Weight (%)
Barium nitrate	68
Cryolite	10
Aluminum dark pyro	11
Sulfur	5
Accaroid resin	5
Boric acid (add alcohol)	1

Two very old-fashioned but beautiful pill box stars were:

	Crimson Weight (%)	Lilac Weight (%)
Potassium perchlorate	57	60
Shellac + 60 mesh	14	10
Strontium nitrate	29	—
Strontium carbonate	—	5
Copper oxychloride	—	5
Mercury chloride (calomel)	—	5
(damp with alcohol)		

The use of calomel today almost makes one gasp considering both the price and attitudes to toxicity; in the case of mercury, calomel used to be taken internally in medicine. It is now clear that mercury vapor is dangerous, and methyl mercury in the food chain from the sea was a disaster. The question is, where are the uninformed to draw their lines on toxicity?

The use of ammonium perchlorate in pill box stars used to be quite common in the United Kingdom and was exploited elegantly by the Wells family. Using the material commercially has its problems in factories where chlorates are regularly handled, so good housekeeping is vital. The stars are quite slow-burning, but they do tend to go damp when they come into contact with potassium nitrate (but not barium nitrate), so it is wise to avoid using normal quickmatch for making pill box stars in this case. The layering of stars to prevent reactions of this type is much too specialized for commercial work.

In a tragic flooding of the River Thames, the Wells factory was flooded. When the water receded and the magazines dried out, they exploded—was this a reaction between goodness knows what—seawater, ammonium perchlorate, aluminum, barium chlorate, potassium chlorate?

Beta naphthol is now abandoned and considered to be carcinogenic, but based on early military mixtures, it was used a fuel to make a beautiful green star with barium chlorate and a blue star made with ammonium perchlorate, potassium perchlorate, and copper carbonate. The two colors in pill box stars together in a shell made the blue look more blue that it was. In the past the author has used it in competitions.

The green composition was

Barium chlorate	85%
Beta naphthol	8%
Dextrin	7%

PRESSED STARS

Some stars are pressed dry in a hydraulic press or they can be pressed on a modified tablet press. Roman candle stars can be made by the latter method, but the scope is limited to certain gunpowder-type mixtures. Chlorate stars are very sensitive for pressing in this way and usually become so hard when they are pressed that they will not ignite. Gunpowder, potassium nitrate, and sulfur mixtures press quite well, but charcoal is not easy to press, and stars containing a large amount of charcoal tend to break up as they are ejected from the molds.

Most of the pressed stars are magnesium stars used principally for signaling in Very cartridges. The magnesium composition is pressed dry in a sleeve of paper or aluminum that must be supported in a mold. Nitrates mixed with magnesium and PVC can be pressed dry in this way, although there is a limited shelf life for these articles because the magnesium is uncoated. In some compositions, the magnesium is coated with linseed oil or drying oil and allowed to stand for some hours before mixing in the other ingredients. Polyester resin also is sometimes incorporated into the composition before pressing.

All magnesium stars require an igniter composition when the stars are pressed, so it is common practice to press an increment of igniter composition on top of the illuminating composition and finally an increment of some type of blackpowder. Igniters usually are mixtures of silicon powder, potassium nitrate, and some type of gunpowder, but mixtures of silicon, lead (IV) oxide, and copper (I) oxide also are used. According to the BIOS reports (12), the following were used during World War II:

	Weight (%)		Weight (%)
Red		**Green**	
Strontium nitrate	55	Barium nitrate	55
Magnesium	28	Chlorinated PVC	29
PVC	17	Magnesium	16

	Weight (%)		Weight (%)
Illuminating		**Yellow**	
Barium nitrate	55	Sodium nitrate	55.5
Potassium nitrate	10	Magnesium	17
Aluminum dark pyro	21	PVC	27.5
Sulfur	8		
Barium fluoride	6		

	Weight (%)
Igniter	
Meal gunpowder	50
Potassium nitrate	16
Barium nitrate	16
Aluminum dark pyro	10
Sulfur	8

The stars were pressed at about 10 tons on 24 mm.

Magnesium grades 3 to 5 are used in these stars, but it is necessary to experiment with the grades of magnesium to obtain the correct burning time. The red and green stars also make good pill box stars if 5% accaroid resin is added to the mixture, which is then damped with alcohol. English and American formulations normally incorporate a percentage of potassium perchlorate into the composition to obtain a higher candle power [e.g., Formula 62 in Ellern (5)].

	Weight (%)
Magnesium	30
Strontium nitrate	42
Potassium perchlorate	9
PVC	12
Laminac (polyester)	7

Dr. Becher (10) quotes the following star composition, which is of interest:

	Weight (%)
White	
Barium nitrate	50
Strontium nitrate	10
Aluminum dark pyro	25
Sulfur	8
Meal gunpowder	7

SPHERICAL STARS

Spherical stars originally were very much a specialty of Japan, but now are an almost universal. There are more ways than one to use the spherical star technique, and the degree of care that is taken in making this kind of star depends on its ultimate use.

Stars can be made very precisely in quite simple equipment, and the reader is referred to Dr. Shimizu's description on page 419. If the starting point is some kind of seed, there is no fast way that stars can be made. The tiniest sprays of water must be introduced to a handful of seed and then powder is added. The seeds must be constantly agitated with the hand to prevent their sticking to each other, and they are dried with only a light coating. This process is carried out over and over again until 100 grams of seed become very many kilos of stars.

The rolling and grading of stars in the Japanese style and the consequent production of stars of the same size requires much precision and dedication. The drums tend to be large and slightly apple-shaped, and in order to keep the stars rolling, the bulk of the large drums are empty. Where cut stars are made into round stars, the drum tends to be rounded but it also has a flat base.

Dr. Shimizu points out in Chapter 6 that there are distinct advantages in the use of glutinous rice starch for making spherical stars, not only as a binder but also because the final product is more waterproof than when dextrin is used. This makes it easier to add layer after layer, which is so necessary for color-changing stars.

In the West it has been more common to use single-colored stars, so it is not quite as necessary to take the same degree of care. Spherical drums are preferred by some, but others prefer the circular disc-shaped drums, which have a flat base and 90° side. Stars of variable diameter are quickly made in these drums using only powder and water. It is again important to coat, dry, sieve, and recoat rather than to try to make big stars too quickly.

Needless to say, dense powder rolls better than light mixtures, and at all times the drum will be at least one-third full to create the weight needed in rolling. Star pans usually are made of copper, aluminum, or stainless steel. If possible, it is better to have the turning machinery outside the building, but flameproof motors also are used.

Modern stars tend to be made with perchlorate and so are less sensitive than they used to be. However, care must be taken in the sieving process, and dust masks should be used.

The drying of stars must be carefully controlled. There is much to be said for drying in the open and out of strong sunlight, certainly when organic solvents are used. But it is important to dry additions of about 2 mm rather than try to make big stars with moisture left in the center. Indoor driers require a warm air current and, in some areas, a dehumidifier.

Provided that some simple basic rules are followed, the process can be quite straightforward, but beginners usually are in a hurry. Starting

Figure 10.3 Star making (Courtesy Kimbolton Fireworks)

Figure 10.4 A 100 tonne double-acting press for comet stars (Courtesy Kimbolton Fireworks)

with dry powder, it is better to start with a pan that has sides at 90° to the base. Small granules are made to begin with, and some larger lumps will be formed and need to be broken down. With patience these can be built up to about 3 mm with some drying taking place in the process.

Once the stars are about 3 mm, they can be built up more rapidly in the crucible-shaped pan, which also is good for making a start with cut stars. The water is sprayed judiciously, and the powder is gently added at the base of the mass of rolling stars. The Japanese system of adding toro to the dry stars requires much more experience; some of the powder is mixed with water to give a very pourable slurry that quickly covers the dry stars, but the amount of water varies with the mixtures.

A good star maker will have a drum full of stars of the same size, and when these reach the required size, a check can be made with a short length of angular aluminum. A handle is fixed on one end and a row of 20 stars is placed along it. Grooves at the side indicate the length of, say, 20 mm × 8 mm stars telling the operator whether to add more composition.

If the technique of rolling stars can be a problem, the priming of stars can be yet another difficulty. Priming dried stars is not always the best way, and it is common to see stars with chipped priming. There is an advantage in adding the first layer of prime straight onto the last layer of color before it dried. Mixtures of potassium nitrate, sulfur, and charcoal serve the purpose well, with the first layer having a lower percentage of nitrate, say, 55%, while the second layer is more akin to blackpowder proportions.

It is not a good idea to prime stars with manufactured blackpowder, and even mixtures of blackpowder and charcoal do not always work well. Inevitably, however, much depends on the kind of charcoal.

Even then, some stars do not ignite when the shell is violently burst with flashpowder or perchlorate mixtures and a much hotter prime is required. Silicon has been the additive of choice because it provides a hot slag. A well-known mixture is:

Potassium nitrate	40%
Silicon powder	40%
Gunpowder	20%

Many people favor a mixture of gunpowder and its constituents with about 10% fine ferrotitanium powder.

Very useful information on stars can be found in Barry Bush's contribution to the Hardt book on Pyrotechnics (78).

11

COLORED FIRES, BENGALS, LANCES, PORTFIRES, AND TORCHES

Colored flames are made with almost the same compositions as those used for stars except that fuels that are used to make stars burn rapidly (e.g., charcoal) usually are omitted. On the contrary, it is usual in these items to add materials that will retard the burning rate. Colored fires, torches, and bengal illuminations that are sold to the public usually are required to burn quite slowly and normally have the burning rate reduced by adjusting the fuel. The use of coarsely ground materials, or the cutting down or increasing of the amount of fuel, can have this effect, but it also is usual to add retardants such as sawdust, woodflour, starch, or flour. The addition of these materials also makes the composition less expensive, of course, but care must be taken with materials such as woodmeal because they may contain a good deal of water.

According to older textbooks, extensive use seems to have been made of tableau fires in the past. This is not very common now, mainly perhaps because piles of loose composition burn very rapidly and the same effect can be obtained from a smaller quantity of material in the appropriate container. Nevertheless, it has been a feature of English firework displays to burn heaps of red or green fire behind bushes. Smoke enhances colored flames because it tends to reflect light, and unless the colored fire is to be used indoors, no attempt should be made to eliminate all the smoke. Some who have tried to do this have discovered that part of the effect also was sacrificed at the same time.

Weingart and others have set out compositions containing picric acid in color production, but this is highly dangerous with chlorates, quite pointless, and expensive.

Containers for colored lights normally are made with a thin wall so that the tube will burn away with the composition in order to gain the best color effects. With these slow-burning compositions, the tube need only be dry-rolled and pasted at the edge; lances, for example, are made this way, but if the filling operation exerts any pressure, it may be necessary to use a wet-rolled tube; wet-rolled tubes with thin walls can be quite strong. When thick-walled tubes are used for colored fires, they must be quite short (e.g., 2 to 3 inches long,) or it will be found that the color deteriorates as the fire moves down the tube. The color also is destroyed by the yellow flame color of the burning tube. The best color effects are thus obtained with wide-bore tubes, 18 mm in diameter and upward, and with thin walls.

It is well known that the human perception of color can be erratic, and care must be taken when colors are produced in variety at close range. The human eye is more perceptive to reds than to greens, for example, a fact that is well illustrated when color-changing lances change from green to red, for then the changeover is easily noticed. Yet if the color order is reversed with a change from red to green, the green cannot be seen until it has been burning for some time. In addition, a red flame appears to be orange or yellow if it is observed for a period of time at close range. For these reasons, it is best to arrange colored lights so that the observers see the reflected light but not the source itself. Aerial fireworks utilizing greater distances do not present the same problem, although care must be taken with magnesium colors, which do not always mix well in close formation. Red and white, green and white, and possibly red and green magnesium colors will mix, but yellow must be used on its own, and to mix all four magnesium colors together would destroy the colors. Blue magnesium colors are not reliable because the combustion temperature and the magnesium destroy the color, quite apart from the fact that the usual blue chemical combinations with magnesium are unstable.

It must be remembered that the copper chloride molecule can exist only in low-temperature flames, so aluminum and magnesium will destroy the molecule. On the other hand, very beautiful high-temperature flames can be produced in red and green. Chemically, the displacement reactions of the electrochemical series also decree that copper will react with aluminum and magnesium, which have higher positions. Having said this, copper oxide is quite insoluble, so some

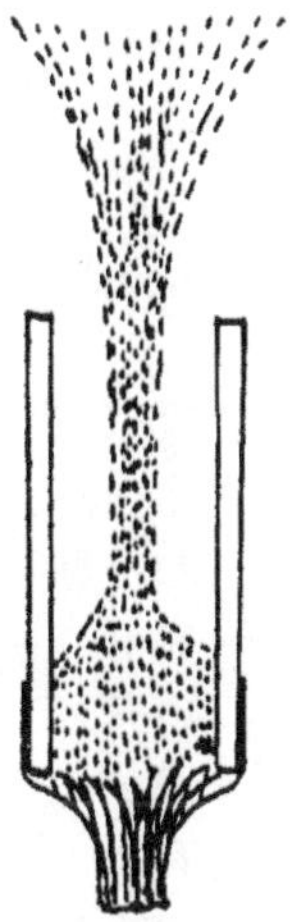

Figure 11.1 Composition being poured into a tube

interesting effects can be obtained without having to worry about chemical reaction; the fact is, however, that a good, deep, high-temperature blue may be beyond our reach.

COLORED FIRES

The following compositions may be burned as heaps of loose composition or gently tamped into short, squat tubes. As already stated, any length of tube can be used provided that the tube will burn away with the composition; in this case, the wall thickness of the tube would be about 1.5 mm. A short, squat tube should be 18 mm inside diameter, 75 mm long, with a 2-mm wall. The composition should be charged into the tube with a drift and hand pressure only, or a funnel and wire could be used. Many of these mixtures burn more easily and reliably when they are gently compacted or even just shaken down (Fig. 11.1).

	Blue			**Green**		
	A Wt (%)	B Wt (%)	C Wt (%)	A Wt (%)	B Wt (%)	C Wt (%)
Ammonium perchlorate	46	30	—	—	50	—
Potassium perchlorate	26	40	60	—	—	49
Basic copper carbonate	10	15	—	—	—	—

(continued on next page)

Red gum	15	15	—	12	15	15
Stearine	3	—	—	—	—	—
Copper (II) oxide	—	—	10	—	—	—
Barium chlorate	—	—	—	55	—	—
Potassium chlorate	—	—	—	33	—	—
Barium nitrate	—	—	—	—	35	31
Lactose	—	—	20	—	—	—
Chlorinated hydrocarbon	—	—	10	—	—	5

	Red			Yellow	
	A Wt (%)	B Wt (%)	C Wt (%)	A Wt (%)	B Wt (%)
Potassium perchlorate	—	66	68	75	75
Strontium nitrate	66	—	—	—	—
Strontium carbonate	—	20	11	—	—
Cryolite	—	—	—	10	—
Accaroid resin	—	14	11	15	—
Shellac 60 mesh	17	—	—	—	15
Woodmeal	—	—	10	—	—
Sodium oxalate	—	—	—	—	10
Potassium chlorate	13	—	—	—	—
Charcoal 150 mesh	4	—	—	—	—

	White		Silver
	A Wt (%)	B Wt (%)	A Wt (%)
Potassium nitrate	74	13	15
Sulfur	8	—	—
Orpiment	18	—	—
Barium nitrate	—	—	45
Potassium perchlorate	—	64	—
Antimony powder	—	13	—
Aluminum dark pyro	—	—	15
Aluminum 30/80 filter	—	—	20
Accaroid resin	—	—	5
Copal gum	—	10	—

BENGAL ILLUMINATIONS

This special type of flare candle is used for illuminating public buildings and is extensively used in Europe for castles, which particularly lend themselves to this type of illumination. In order to fulfill the necessary conditions, the composition should produce maximum color, burn efficiently but as slowly as possible (i.e., 40–60 seconds per 25 mm), and not emit too much smoke.

The candles are made in various sizes, but the large ones are about 50 mm in diameter, 30 cm long, and with a wall thickness of about 1.5 mm. The end of the candle is fitted with a wooden plug and a screw eye for attachment to the holder. The candle is fixed so it burns in a horizontal position so that the dross will not run down the side of the thin wall and accelerate the burning time. Finally, the candle is placed behind some kind of shield so that onlookers do *not* see it burning (Fig. 11.2).

The following compositions are typical. Paraffin oil sometimes is added as a retardant and to give some protection against moisture, which may be a problem in strontium nitrate mixtures that are not heavily compressed.

	Red		**Green**	
	A Wt (%)	B Wt (%)	A Wt (%)	B Wt (%)
Strontium nitrate	65	63.5	—	—
Barium nitrate	—	—	68.5	70

(continued on next page)

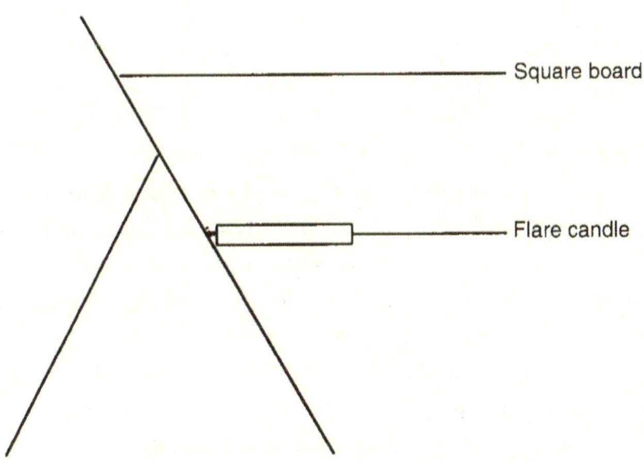

Figure 11.2 Bengal illuminations

Potassium chlorate	20	—	—	16
Potassium perchlorate	—	16	15	—
Shellac 30/200 mesh	15	—	—	—
Accaroid resin	—	9.5	15	13
Sawdust	—	9.5	—	—
Lampblack	—	1.5	—	—
Antimony metal powder	—	—	1.5	—
Paraffin oil	—	—	—	1
Approximate burning time in sec/25 mm	45	30	20	40

It is always the case with firework manufacture that there are many snags that practical workers discover only by experience. Strontium nitrate mixtures sometimes shrink if they contain water that subsequently dries out, leaving a gap between the composition and the tube. This naturally decreases the burning time.

Strontium nitrate mixtures are very unreliable if they are charged in paper tubes. The issue is even more severe in damp climates. The above Red mixture A is particularly liable to shrink if it has been kept for a short time. Dr. Shimizu pointed out in Chapter 6 that this is likely to be caused by the use of shellac, so red gum would be a better option.

LANCES

Large quantities of these lances are manufactured every year for use in making set pieces, as fire pictures, mottos, motifs, all made with these little tubes of colored fire. Lances usually are 8 mm or 10 mm in diameter and about 10 cm long. The paper usually is an 0.004 bond or poster paper cut into pieces 95 mm × 125 mm, and they are dry-rolled on a former that is countersunk at one end. The papers are placed on top of each other with just the outside edge and side uncovered. A pasted sheet of paper is placed on the bench, and the former is laid along the paper about 6 mm from the end. The paper is brought up and over the former and then rolled up to make the tube. The pasted edge, which is left overhanging the end of the former, is then pressed into the countersunk base of the former. This completely closes the base of the lance in a neat fashion, thus preventing the powder from trickling out. The diagrams in Figure 11.3 will help to make this clear. The following compositions are typical of those used for lances:

A funnel and wire are used to charge the composition, but the base of the funnel has a nozzle that goes about 10 mm into the lance tube. In

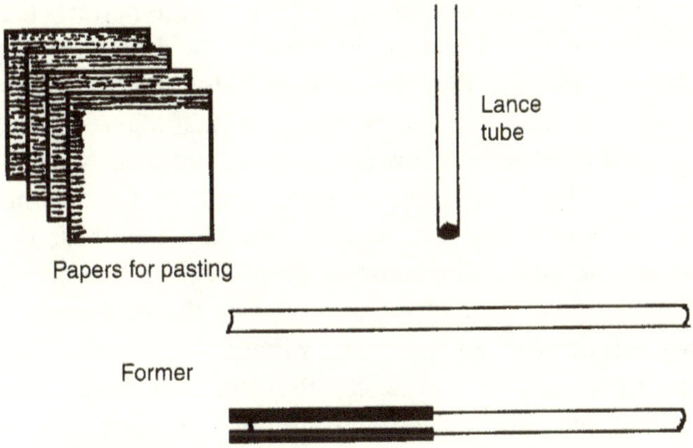

Figure 11.3 Apparatus for manufacturing lance tubes

this way, the lances can be filled to within 10 mm of the top. The next operation is to add the topping composition in a similar way, to a depth of 8 mm. The old manufacturers used to use the white lance composition as a topping, but it is not wise to funnel and wire a sulfur mixture onto a chlorate or perchlorate, and it would be very hazardous with barium chlorate. A simple mixture of potassium nitrate and accaroid resin serves the purpose quite well. The lance is finally primed with a paste of gunpowder, dextrin, and water.

Lance Compositions	**Red**	**Green**	**Blue**		**Amber**	**White**
			A	B		
	Wt (%)	Wt (%)	Wt (%)	Wt (%)	Wt (%)	Wt (%)
Potassium perchlorate	75	25	64	27	73	—
Barium chlorate	—	36	—	—	—	—
Barium nitrate	—	25	—	—	—	—
Strontium carbonate	12	—	—	—	—	—
Cryolite	—	—	—	—	14	—
Copper carbonate	—	—	—	11	—	—
Ammonium perchlorate	—	—	—	47	—	—
Paris green	—	—	32	—	—	—
Red gum	13	14	—	15	10	—
Lactose	—	—	—	—	3	—
Copal gum	—	—	4	—	—	—
Potassium nitrate	—	—	—	—	—	65
Sulfur	—	—	—	—	—	20
Antimony metal powder	—	—	—	—	—	10
Mealed gunpowder	—	—	—	—	—	5

As Dr. Shimizu points out on page 379, lances sometimes are a problem mainly because the flame is disturbed by the dross or they produce a long chimney-like ash that obscures the color and the brightness. In view of this, a good lance can be tested only at about 50 m distance when it should be observed burning in a horizontal position. A lance 9 cm long should burn for about 50 to 60 seconds. Lances made with potassium perchlorate are less likely to chimney than those made with the chlorate, and some care must be taken with the choice of paper. Spirally wound lance tubes have appeared on the scene now, but they are greatly inferior to the hand-rolled variety.

The use of spirally wound lance tubes makes it much easier to fill lances in large quantities. The tubes are more consistent in shape and can have a precise outside diameter, thus making them easier to charge in quantity in a mold. Bush in Hardt (78, p. 119) describes such a device and charging technique. However, it is still difficult to obtain precise filling, the snag being that without due care, increment by increment, more powder seems to get into the central area of the mass of tubes rather than on the outsides.

When we consider the economic benefit the West has in buying from China, it should be possible to buy very good quality lances, but they do not seem to exist. They look unattractive; are poorly filled and primed; and compositions, which should not be made with magnalium, do not have a good shelf life.

PORTFIRES

Portfires, which are used for lighting fireworks, are made in exactly the same way as lances except that the tube usually is longer and has a slightly thicker tube. Portfires normally are 9 mm inside diameter and about 38 cm long. The paper for the 38 cm size would be 18 cm × 38 cm. Portfires usually burn for three or four minutes, but they must not burn too slowly or they are liable to go out. The following compositions are satisfactory:

	A Weight (%)	B Weight (%)	C Weight (%)
Potassium nitrate	60	63	—
Sulfur	20	16	—
Meal gunpowder	20	11	—

Antimony sulfide	—	10	—
Potassium perchlorate	—	—	50
Barium nitrate	—	—	35
Red gum	—	—	4
Copal gum	—	—	11

Unfortunately, portfires drop hot dross during burning, and this frequently lands on the ankles: Many people use a color mix (e.g., composition C) to avoid this.

This composition very typically was made with shellac in former times for Miner's portfires. After careful hand-charging, the portfires were placed in an oven until the shellac reached its low softening point. After cooling, they were quite hard and solid.

To make them waterproof, the tubes then were painted with a black shellac paint. Until recent times these shellac lacquers frequently were used in the firework industry, a white one still being used to identify fuse tips.

The black paint was made by adding 3 parts of lampblack to 40 parts of methylated alcohol and then adding 32 parts of shellac. After standing and stirring for a day or two, the paint was ready. A similar "off white" can be made with 16 parts of titanium dioxide and 48 parts of alcohol and then adding 25 parts of shellac.

The burning speed of portfires can be a frequent discussion point, and there are those who boast that they can make the slowest-burning portfires. So can we all. Such a mix might be:

Potassium nitrate	75%
Sulfur	23%
Charcoal fine	2%

However, it is not much use if it is extinguished when you throw it on the ground. After all, a portfire is an aid at a display, and you throw it on the ground when you need both hands free.

The normal test for a portfire is to hold it vertically a meter above the ground and then drop it on the burning end. If it stays alight, it passes!

Gunpowder prime on lances or portfires can burn off without lighting the lance, particularly if it has shrunk during drying; it also is prone to happen with perchlorate mixes. This is overcome by using a mixture of the three components of gunpowder with some dextrin and silicon. It also is a possible reason why lances made with magnalium can suffer from this problem as a result of excess moisture at the junction of the prime and main composition.

BLUE LIGHTS AND STAR LIGHTS

These little items seem to have almost disappeared from the market. In effect, they were small lances 5 mm inside diameter, 15 cm long, and with a wall thickness of 1.5 mm. The compositions were quite slow-burning and produced either golden spur fire or molten sparkling drops of the type mentioned by Dr. Shimizu on pages 159 and 392. The following compositions were used:

	Blue Light		Star Light
	A	B	
	Weight (%)	Weight (%)	Weight (%)
Meal gunpowder	51	13	6
Potassium nitrate	35	62	62
Sulfur	14	19	19
Orpiment	—	6	9.5
Lampblack	—	—	3.5

The potassium nitrate sometimes was a mixture of fine powder and the more coarse crystalline powder. These fine tubes were not always charged with a funnel and wire; the composition was merely poured over the open ends of the tubes and shaken down in a similar way to the method used for filling English crackers (see page 315). Again, not surprisingly, many compositions filled in this loose manner function better than those that are carefully charged.

TORCHES

Weingart wrote a great deal about torches, but they seldom are used now in Europe. The reason for this could be that they are comparatively expensive, or that they are liable to deposit dross, make a lot of smoke, and are apt to cause accidents in crowds. The most commonly used torches today are made of a piece of Hessian or sackcloth that is rolled around a former and then dipped in molten paraffin wax. It is quickly removed from the former and subsequently fitted with a suitable handle. These torches are non-pyrotechnic, but they are safer and less expensive. Pyrotechnic parade torches and railway fusees are 18 mm to 24 mm in diameter and 20 cm to 50 cm long. The tubes normally are wet-rolled with a piece of thin kraft paper that makes about four to five turns on the former. Lance compositions are quite suitable for colored torches and, if necessary, they can be made less expensively by the addition of 5% to 10% woodmeal or 5% starch or wheat flour. The following compositions also are quite useful:

	A Weight (%)	B Weight (%)	C Weight (%)
Potassium perchlorate	52	—	25
Strontium nitrate	—	—	45
Aluminum 150 mesh "bright"	24	—	—
Aluminum flitter 30/80 mesh	20	—	—
Aluminum dark pyro	—	18	—
Dextrin	4	—	—
Barium nitrate	—	76	—
Sulfur	—	4	—
Petroleum jelly	—	2	—
Shellac 30/200 mesh	—	—	14
PVC	—	—	6
Linseed oil	—	—	1
Magnesium grade O	—	—	9

Parade torches usually are furnished with a wooden handle that is glued into one end, while the other end is primed with a modified gunpowder and touch-papered (Fig. 11.4). Signal torches frequently are primed with a mixture of potassium chlorate and fine charcoal that has been mixed into a paste with dextrin and water. A small wood striker on which is pasted a slurry of red phosphorus also is provided for ignition. More sophisticated signals have striker ignitions embedded in polythene units, with ignition being achieved by pulling the wire to release the striker.

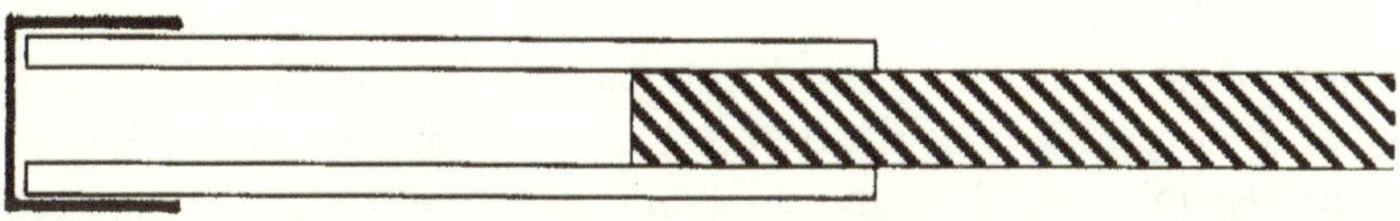

Figure 11.4 Torch or hand flare

FLARES

In some circumstances, illumination can be achieved almost as well with aluminum as with magnesium. The following compositions can be charged into wide-bore tubes with thin walls, but care must be taken when charging mixtures of barium nitrate and aluminum because these mixtures are more sensitive than one might think. It is, in fact, much wiser to press them into tubes on an arbor press.

	A Wt (%)	B Wt (%)	C Wt (%)	D Wt (%)	E Wt (%)
Barium nitrate	56	—	—	22.5	50
Strontium nitrate	—	60	42	—	—
PVC	21	15	12	15	—
Magnesium grade O	—	—	30	35	—
Potassium perchlorate	—	—	9	22.5	—
Polyester	—	5	7	5	—
Magnesium grade 4 or 5	16	20	—	—	—
Aluminum dark pyro	—	—	—	—	35
Potassium nitrate	—	—	—	—	15
Montan wax	7	—	—	—	—

Formulas A, C, and D are from BIOS (12) sources and Ellern (5), the two latter being from the American patent literature.

Compositions containing barium nitrate and aluminum also are somewhat difficult to ignite, and it is always necessary to add a suitable ignition mixture, such as those mentioned on page 228, or a hot-burning antimony white star mixture would work equally well.

Magnesium flares are rather more complex to manufacture, mainly because they depend so much on the correct preparation of the materials and they require pressing under heavy pressing loads. Ellern and others have published many typical formulations.

WATERFALLS

In view of the fact that waterfall compositions are closely related to the rest of the fireworks described in this chapter, they are described here.

The general principle of formulation is to ensure that there is an excess of aluminum present to guarantee that the burning material will fall to the ground with the appropriate sparks. The tube usually is a wet-rolled tube consisting of four or five turns of an 0.010 kraft paper on a 18 mm to 24 mm former. The tube is 15 cm to 30 cm and naturally must burn away with the composition or the essential dross will not fall to the ground. Old formulations were made with potassium chlorate, for example:

	Weight (%)
Potassium chlorate	72
Aluminum "bright"	28

This was damped with a 10% solution of shellac in alcohol, mainly to control the aluminum dust, but the small amount of well-distributed fuel also improves the burning and holds the composition together. Naturally, it is much safer to use potassium perchlorate, and in this case a slightly larger proportion of aluminum is required.

	Weight (%)
Potassium perchlorate	50
Aluminum "bright"	25
Aluminum flitter 30/80 mesh	12.5
Aluminum flitter 5/30 mesh	12.5

The usual method of charging is to cover one end of the tube with a thin paper "drum head" or a piece of muslin. A small quantity of igniter composition is placed in the tube first, followed by successive increments of composition that are lightly tapped down with a mallet and drift. Bundles of tubes with a maximum of 7 kilos of composition are damped at one time. Finally, the tube is closed with a disc.

Many people prefer to use barium nitrate as an oxidizer, and although this is safer, it also is inferior because there is a tendency to get a large glare at the mouth of the tube and a weaker drop effect. Dr. Becher (10) gives the following composition, which is fairly typical.

	A Weight (%)	B Weight (%)
Barium nitrate	52	52
Potassium nitrate	8	8
Aluminum dark pyro	16	22
Aluminum flitter 5/30 mesh	21	14
Meal gunpowder	3	—
Aluminum flitter 30/80 mesh	—	5

In this particular instance, the increments of composition should be pressed dry with a hand press and not charged with a mallet. The igniter composition also is quite essential with this particular mixture. It also is necessary to adjust the aluminum to fit the particular grades used; a bright powder can be used equally well instead of pyro, but this will necessitate the adjustment of the flitter.

All the shower sticks are finally primed on the outside of the drum head with meal powder, capped, and prepared for use as waterfalls. Some people insert a piece of wooden dowel into the end of the shower stick and then nail this onto a length of board that is fixed so that the shower sticks burn in a horizontal position. In England it is common to fasten a piece of thick string on the end of the shower and use this to fasten the showers onto a board. In this way the showers actually move during burning (Fig. 11.5).

Good waterfall effects also can be obtained by using a bank of iron or titanium gerbs of 24-mm bore, fixed to burn in a horizontal position. It has been mentioned elsewhere that some mixtures are very difficult to ignite and may require a special igniter composition. Mixtures of barium nitrate and aluminum in stars and waterfalls are a case in point. It is usual to use an igniter that produces a hot slag such as:

	Weight (%)
Potassium nitrate	40
Silicon 300 mesh	40
Gunpowder	20

Sulfurless gunpowder can be used with mixtures that contain a chlorate. With a binder, the mixture also can be used for stars.

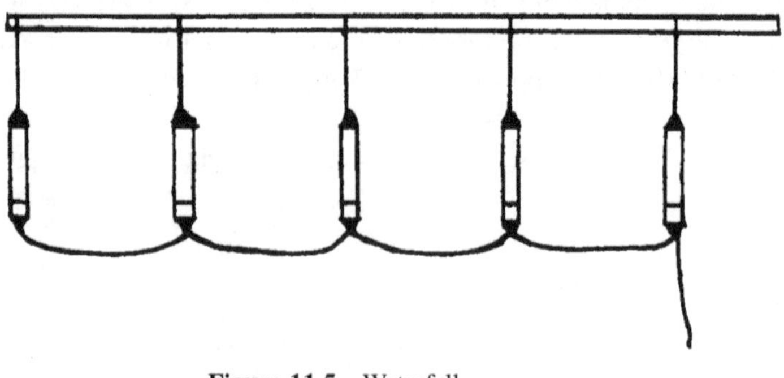

Figure 11.5 Waterfall

12

ROMAN CANDLES, COMETS, AND MINES

ROMAN CANDLES

There can be no doubt that it requires much practice to make really good Roman candles, and the best ones are still made by hand. So many factors play a part in the performance of the candle with the inevitable result that there is always something to go wrong. Each of the factors is therefore examined in turn and the possible snags are pointed out.

First, the tube is all-important, and for the purposes of this chapter, only one size of Roman candle will be referred to, the tube being 16 mm inside diameter, at least 30 cm long, and having a wall thickness of about 6 mm. The wall must be of this thickness or the tube will catch fire during the performance of the candle. The best tubes are hand-rolled, and it is quite essential that the inside lap of paper be completely stuck down or the powder will creep down the loose paper, cause a "blow-through," and eject the stars like machine gun bullets. The stars are made in a 14-mm former, which means that they just comfortably slide down the inside of the tube. In addition, as would be expected, stars made in a 14-mm former shrink very slightly during the drying process, just as a wet-rolled tube on a 16-mm former also shrinks during the drying process, with the result that both factors might play a part. Stars that are too tight are ejected so fast that they frequently are extinguished because they are projected too rapidly. Similarly, stars that are too slack often land on the floor still burning because they were projected only a few inches into the air.

The quality of the paper of the tube also is quite important. A good-quality kraft paper, for example, rolled with a good casein glue, provides a tough tube that hardly burns away during the firing of the candle. If a good-quality grain powder were used for such a tube, the stars would be ejected with increasing force as the fire proceeded down the tube and probably would be extinguished. The reason for this is because the tube is not burning away at the top and a high pressure is thus maintained. However, the use of an inexpensive paper could mean that the tube burns away at the top, allowing the gas to escape, with the result that the stars are not projected quickly enough. Clearly, the manufacturer uses a good-quality grain for poorer quality papers and vice versa. In practice, it is usual to use the less expensive paper. The 14-mm stars usually are about the same length, but the burning rate of the composition depends on the way the stars are placed in the tube; there are two methods.

The better method uses very fast-burning stars, such as those on page 211, and surrounds each star with a small amount of gunpowder to make sure that it will be ejected immediately. Since the star is ejected so rapidly, it is essential that it be fast-burning or it will be extinguished.

The second technique places no powder around the star but uses a much slower-burning star that actually burns for a second or two in the tube before it is ejected. Gas is lost up the side of the star, which is not ejected as high as those made by the first method. The latter method also produces candles that are more erratic in their performance because a good Roman candle should eject all its stars regularly to the same height, the stars burning out just as they turn over to come down again.

The Roman candle delay also plays a part because its burning characteristics are directly related to the grain powder charges under each star. Manufacturers naturally tend to have their personal preferences, some maintaining that it is essential to granulate the composition by damping it with water and subsequently drying it. If this granulation takes place for a definite purpose, such as making the composition dust-free and therefore cleaner to handle or easier to manipulate in the measuring boards for mass filling, this is fine, but otherwise it is unnecessary. Some argue that the fuse should contain a large proportion of gunpowder or sulfur or both, although it is not always clear why this view is held. The most significant factor would appear to be the need for a reasonably dense fuse that consolidates well, a condition mainly satisfied by the presence of sulfur, potassium nitrate, gunpowder, and not too much charcoal. The fuse ideally should produce a good show of golden sparks, but this is not necessary, of course, and there always is a tendency to produce too much smoke, as would be expected.

The following composition is quite satisfactory because it combines most of the characteristics mentioned:

	A Weight (%)	B Weight (%)
Potassium nitrate	50	55
Meal gunpowder	22	15
Charcoal 40/100 mesh	11	8
Charcoal 30/60 mesh	11	8
Sulfur	6	14

The manufacture of a 30-cm Roman candle, for example, would take place as follows. The tube (see the dimensions at the beginning of the chapter) is placed on a small nipple, and a charge of clay is placed inside. Using a drift that slides easily down the tube, the clay is firmly consolidated. A charge of English 4FA or FFF grain powder is next placed in the tube, followed by a star. A small charge of gunpowder or fine-grained powder is then placed on top of the star, and the tube is tilted backward and forward so that the powder will trickle down the side of the star. A scoopful of delay fuse is now placed in the tube. This is consolidated lightly with a mallet, and it should be noted that this is where the skill really lies. If the delay is rammed too lightly or too firmly, the candle will not function properly, with the result that the stars are ejected too fiercely or in quick succession. Only experience will teach the operator just how hard to strike the drift.

The fuse and clay scoops usually are semicylindrical, being beaten from a sheet of copper. Grain scoops are smaller, but are made in a similar way. A 16 cm × 30 cm Roman candle takes about six stars. The approximate quantities are:

Fuse scoop holds approximately 4 g
Grain scoops hold 0.45, 0.55, 0.60, 0.90, 1.20, 1.50 g
About 0.50 g of gunpowder is placed on each star

The smallest grain charge goes in the bottom of the tube on top of the clay, and then the charge increases as it approaches the top. Figure 12.1 shows a Roman candle in cross section.

In practice, these candles are not filled singly, of course. Display candles usually are filled in small bundles, whereas commercial items may be charged 100 or more at a time. American practice commonly appears to employ gang rammers, but this is unusual in Europe, where

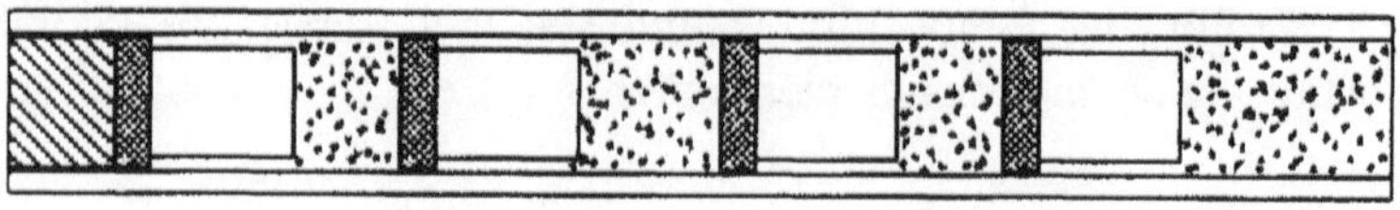

Figure 12.1 Roman candle

the candles are rammed singly or six at a time. The European method undoubtedly springs from a fear of the mixture of chlorate in the stars and the sulfur in the fuse plus friction from the gang rammers.

There is a considerable history of accidents charging Roman candles and widespread agreement that most of them were caused by chlorate-sulfur contacts. Traditional fuse is made with sulfur—as much as 26% in one U.K. mix—and one can imagine what would happen if a chlorate star became stuck halfway down the tube and was then forced into position. One accident in the United Kingdom was caused by an operator who added the chlorate stars and then forgot to add the delay fuse before he started the ramming operation.

It is much safer to use a delay fuse without free sulfur such as the following:

	Weight (%)
Potassium nitrate	52
Soft grain gunpowder	22
Charcoal 40 mesh	21
Red gum	5

Good-quality grain should always be used for making Roman candles. Homemade gunpowders are fine for many things, but consistency is required in good candle manufacture.

In mass-filling, the grain and the fuse are added by means of the measuring boards described by Weingart. Thick boards made of wood or synthetic resin are drilled with suitable holes just large enough to admit the required quantity of fuse or grain. Another similar board is then placed underneath, with corresponding holes to allow the powder to trickle through into the tubes. If the two boards are arranged so that they will slide over each other, the powder can be contained in the holes of the top board or allowed to run through into the tube by merely sliding one board over the other. The principle is not unlike that used by organ builders for making wind available to ranks of pipes. The diagrams explain the method more easily than a detailed description, and greater detail is pointless because every manufacturer will modify the method to suit his own purposes (Fig. 12.2).

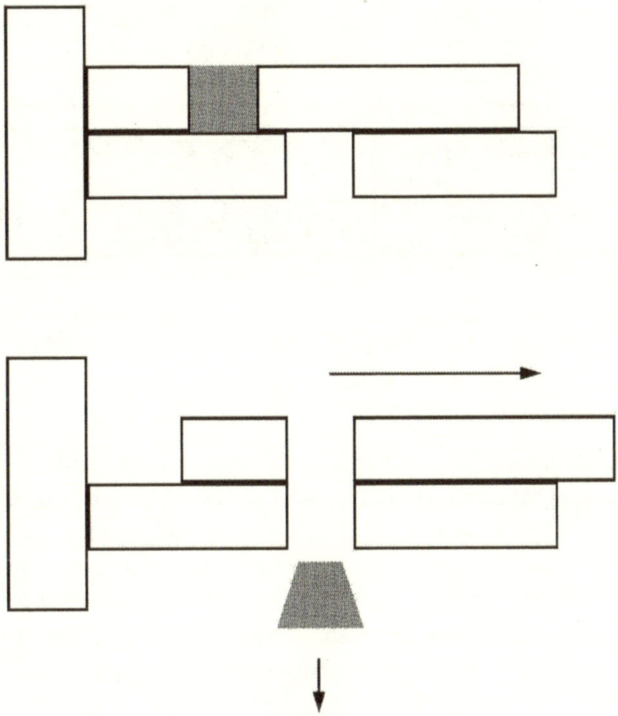

Figure 12.2 Sliding boards for charging powder

A good Roman candle is easily recognized. The fuse will burn rapidly and steadily, and each of the stars will come out at regular intervals with a sharp little crack, gain a good height, and just turn over before it is extinguished.

The reader is referred to an article by the author in *Pyrotechnica* XIV (29).

A wide variety of Roman candles are made, with inside diameters as small as 10 mm and as large as 60 mm. Most of the candles eject stars such as colors, glitters, aluminum, and lampblack stars, but more sophisticated versions eject pressed magnesium stars, whistles, humming stars, flash and sound, serpents, and large comets. The special larger-bore candles usually must be filled singly and require great care in their manufacture because they are such expensive items.

In recent years more sophisticated Roman candles have been made with felt wads as delays instead of the old candle fuse. The wads are very effective and clean to handle, but they are considerably more expensive. Economically, this method should be used only for tubes of approximately 25 mm in diameter and upward because candles of this

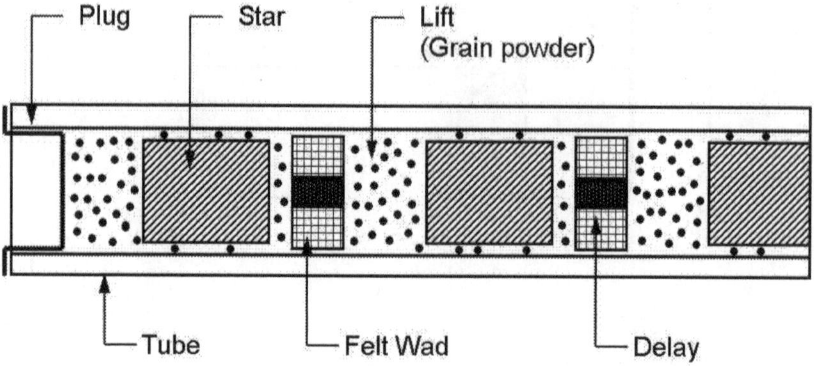

Figure 12.3 Section of a 30-mm candle

bore would require large quantities of candle composition, which in turn generates volumes of smoke and sparks. Densely pressed virgin wool felt makes the best wads, but much less expensive reconstituted scrap wool felts also are used. A 30-mm candle would require a felt washer about 20 mm high. The diameter of the wad must be about 3 mm larger than the diameter of the tube so that it is a tight fit when it is forced down the tube, thus producing excellent compression. In the center of the wad is a small hole containing a delay mechanism that fits tightly and comfortably.

The easiest form of delay is a self-consuming pellet such as a 14-mm nitrate-based white star. The star works very well, is easily and inexpensively placed in the wad, and should be a tight fit. The slight drawback is that this pellet requires a hole of about 12 mm, so there is some loss of gas and stars may be ejected a bit erratically. In all honesty, it must be said that the overall performance of this method is much more consistent than when loose candle composition is used.

The best delay to use is a short plastic tube charged with powder, such as a shell fuse. It need be only 4 or 5 mm in diameter, so there is much less loss of gas. Narrow-gauge Bickford can be used, but it is not satisfactory because it requires cross-matching for 100% performance; this is difficult to achieve on both sides of the wad, and the ramming of the wad can upset this cross-matching and cause a misfire. Most Europeans today use the excellent Portuguese 10-mm Bickford with a generous powder core. Thicker pieces of felt are required, but the results are excellent and consistent. In Spain, plastic cups with a molded fuse hole for 10-mm Bickford are being made for candles, but they are not 100% efficient.

These large candles are most useful for comets or bombettes and are made as large as 60 mm and used for multiple firing of bundles of

Figure 12.4 Roman candles for Thames Day in London

effects, such as whistles, serpents, and so forth. Such powerful candles require good tubes, and while the temptation is to use spirally wound tubes for economy, they frequently disintegrate and must be of exceptional quality; convolute tubes are much better.

The Chinese produce small inexpensive candles that are quite efficient and use granulated clay between the stars. A length of standard Chinese fuse goes down the inside of the candle. A charge of grain is placed in the bottom, then a star, and then a charge of granulated clay to make the delay time. The tube is continually charged in this way until it contains eight or more stars. The method works well and the price is good.

COMETS

Comets are single Roman candle stars in their simplest form, but they usually are gunpowder-type compositions because they are ejected at great speeds. The compositions are damped with dextrin and water or accaroid resin and spirit and then consolidated in a single star pump. Large star pumps are made of brass or a mixture of brass and copper and are arranged so that they can be charged with a mallet or placed in a press. The smaller comets of about 24 mm in diameter can be adequately charged with a mallet, but larger comets should be pressed to obtain greater consistency. When comets are being manufactured, it is usual to press more composition than is necessary, eject the excess, and then cut off this excess with a knife. Large 100-ton presses also are used (Fig. 12.5).

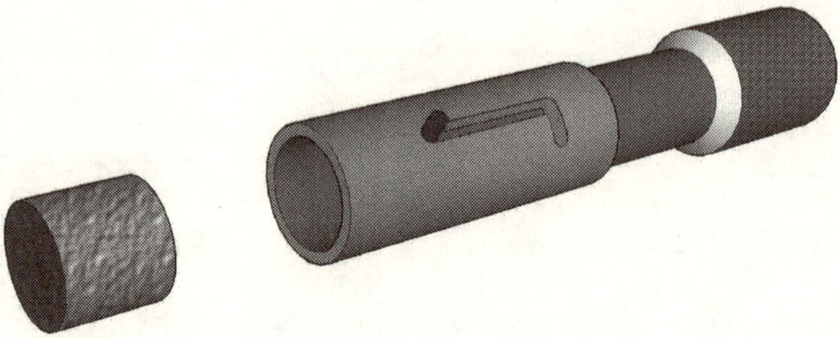

Figure 12.5 Pump for comet stars

The majority of the smaller comets are placed in cylinder bombs, but they also can be effectively used when fired from the ground in large numbers in quick succession. Comet batteries consist of steel plates on which are welded a dozen or more steel tubes about 20 cm long. Holes are drilled in the base of each tube to allow a piece of slow fuse to be threaded through each one. A charge of grain powder, possibly about 10 g, is loaded into each of the tubes, and then a comet star is placed on top of the charge. Finally, a tight-fitting wad is rammed on top of the star. The "wad" need only be a sheet of paper that can be consolidated with a wooden drift. Naturally, care must be taken not to ram the tube too tightly or there is a danger of breaking the insert. Large-bore tubes should never be rammed in this way.

Huge numbers of comet batteries are now reaching the West from China. Strawboard tubes suitably closed are threaded with Chinese fuse and then glued and wired into differently shaped batteries (i.e., a square of 16 or 25 tubes, or maybe a circle of 90 tubes, etc.). Some of the "cakes," as they usually are called, may extend to hundreds of shots and vary in diameter from 12 mm up to 75 mm or more.

Appearing with unusual names that in many cases seem to be literal translations of Chinese names, they may have a greater significance than we realize in the West.

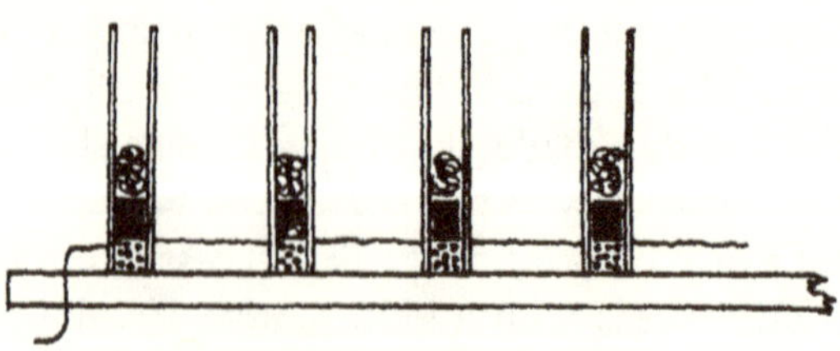

Figure 12.6 Comet battery

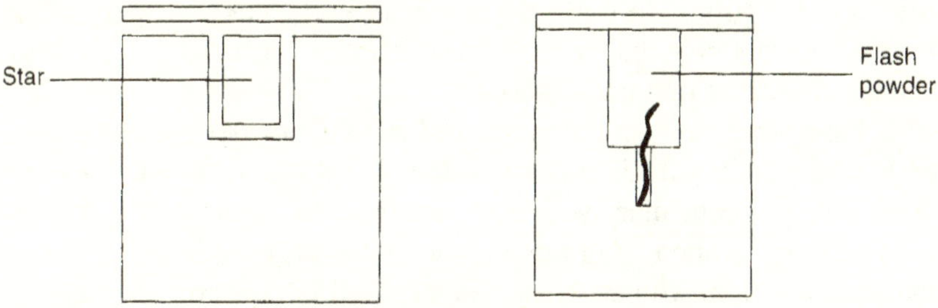

Figure 12.7 Comet stars

Larger comets of 4 cm or 5 cm in diameter are either solid or filled with colored stars. The star is formed with a chamber in one end of the star so that a colored star or stars can be placed into the cavity that is formed (Fig. 12.7).

The stars are finally covered with a pasted paper coat. The paper covers one end and three quarters of the length of the star. Solid stars normally are not covered. Crossette stars or splitting comets usually have a small charge of flashpowder instead of colored stars. These stars are very effective because the streaming comet suddenly bursts into many fragments.

Effective comet bombs can be made by pressing a star in the bottom of a paper tube and then filling the remainder of the tube with stars or other effects. The result is a spiral of fire terminating in a star burst. These items must be carefully manufactured; otherwise they are apt to "blow through" in the mortar (Fig. 12.8).

There can be no doubt that this technique has been dramatically exploited by the Chinese. There was no tradition of manufacturing multiple comet boards for commercial sale in the United Kingdom—that is, until they started to arrive from China. At first most of them seemed to be circular, with the inevitable consequence that they were called "cakes"; indeed, they are still called cakes, but this has given rise to

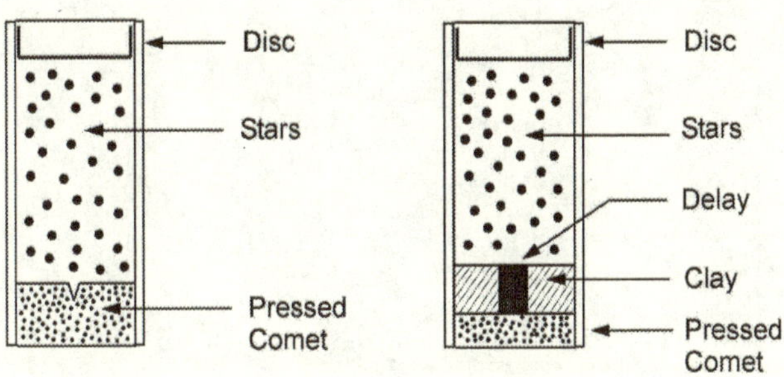

Figure 12.8 Small comet bomb

official descriptions. In these days, when pyrotechnics is more about law and bureaucracy than about fireworks, they basically are one-shot candles converted into multishots.

At the retail level, with attempts to limit the powder content of single Roman candles, it still will be possible for the general public to purchase some spectacular items. At the time of writing, it still was possible in the United Kingdom to buy multishots of maximum 30 mm diameter with a maximum gross weight of 10 kilos, but it also has led to some abuse and many complaints about noise. Many of the individual comet bombs have contained as much as 6 g of flashpowder; this loophole in the regulations has now been changed, but the damage to the reputation of the industry is incalculable. Once again, it has been caused by EEC rules that allowed people with no experience of the trade to bring totally unsuitable material into the United Kingdom and sell it to the public, but this is partly poor regulation. Chinese manufacturers must become more sensitive to what has been happening and only create reports in a smaller scale by the use of blackpowder.

Figure 12.9A Roman candle combination

Figure 12.9B Roman candle combination—fan shape

Looking at it more positively, the new multishots also have many benefits. They are easy to handle and produce a wealth of assorted effects even though they are based on a few simple types of units, small comet bombs (Fig. 12.8) where the fuse is embedded in a clay plug to provide a delay, comets, tourbillions, whistles, and mines.

Modern techniques in firing displays with electric ignition but also with computer-designed systems has meant that there has been an exponential rise in the demand for one-shot mines and candles. Minimum debris also is a modern requirement because many of these items are used in football grounds and stadiums. The sequential electrical firing of comets and mines has become a particularly impressive spectacle in firework demonstrations, giving the designer an additional dimension. Chinese ingenuity also has extended this technique to their multishot batteries in order to produce fan shapes, zig-zags, and backward and forward firings. What next?

FIREWORK MINES

The ordinary mines are simply a charge of gunpowder and stars at the base of a strong tube that acts as a gun. The "bags" of commerce, or *pot-au-feu*, as they are sometimes called, consist of about three turns of a medium kraft paper dry-rolled on a former that is about 12 m less than the diameter of the mortar. A 11.5-cm bag, for example, is 100 mm in diameter and about 15 cm high. The paper for this bag would be about 1 m long for the roll-up and about 30 cm wide. A gunpowder charge of about 50 g of grain powder is placed in a small bag, which is then tied to the end of a length of piped match. This bag is then placed into the bottom of the main bag, and stars, tourbillions, flash charges, comets, and so forth are then loaded on top of the powder charge to a depth of about 15 cm and the bag is tied off above the stars. Figure 12.10 illustrates the method more clearly.

Commercial mines usually are made in wide-bore tubes about twice the height of the diameter. The tubes are fitted with a strong base of paper or wood, which is securely glued into place, and a weakly fitting lid that will blow off easily. Sizes vary from quite small tubes of about 24 mm in diameter up to quite large tubes of 12.5 cm to 15 cm in diameter. Mines of this type normally are fitted with a central delay fuse consisting of a long narrow tube that contains no clay at the base, but is merely primed so that the fire can transfer to the powder charge. Roman candles also are used as delays. Special mines called "Devil

Among the Tailors" consist of a mine with a central Roman candle and three other candles fixed to the outside of the tube. The device thus produces a fine display of colored stars that terminate in a mine burst of crackers, stars, and serpents (Figs. 12.11 and 12.12).

Use of the word *mine* has been normal among the firework fraternity. With hindsight a better description could have been found since mines of the military kind have caused untold human misery.

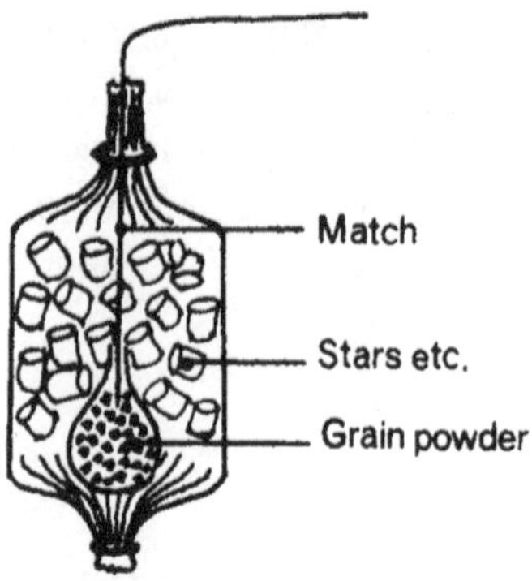

Figure 12.10 A "bag" or mortar mine

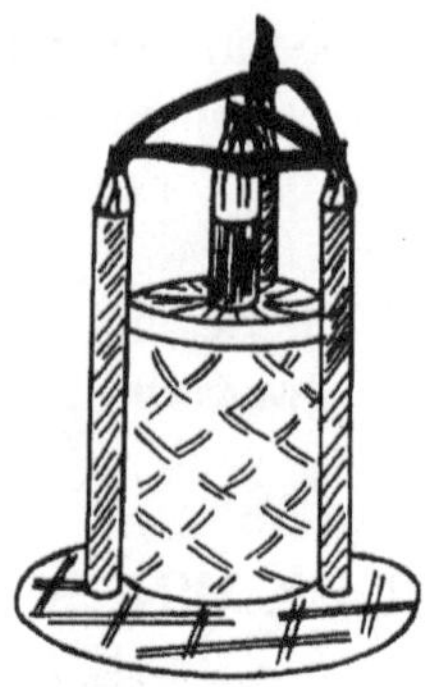

Figure 12.11 Devil Among the Tailors

Figure 12.12 Devil Among the Tailors. Early twentieth century (Brock's Fireworks)

13

NOISEMAKERS

FLASH AND SOUND

Explosive fireworks more than anything else have been the cause of serious accidents, and they probably have done more damage to the firework industry than anything else. Flash crackers in their various forms were on sale to the public for far too long and, happily, have been forbidden in many European countries. Gunpowder "bangers" are still sold to the public, but they are mostly of interest only to schoolboys who frequently make a nuisance of themselves by throwing them at people.

It also is a tragic fact that many young people cannot resist the urge to create explosions, and in their enthusiasm they choose unstable and highly sensitive mixtures that they charge into unsuitable containers in highly dangerous ways. Fatalities caused by sodium chlorate mixtures being charged into containers made of metal (of all things!) must be legion.

There can be no doubt that even among firework manufacturers, compositions vary greatly in their sensitivity, but so much depends on what is required of the mixture.

Flash compositions of some types will cause explosions in the lightest of containers—sometimes even in just a few turns of paper—but it also happens that these compositions are extremely brisant and sensitive. Horrifying mixtures of potassium chlorate, pyro aluminum, sulfur, and barium nitrate have been employed and should be avoided at all costs. Mixtures of the perchlorate, sulfur, and bright aluminum are safer and appear to be used extensively in the United States and Japan, but even these would be considered dangerous by many of us in Europe. In fact, the more common European technique is to use a strong paper tube with a composition consisting simply of potassium perchlorate and dark pyro aluminum. For example:

	A Weight (%)	B Weight (%)
Potassium perchlorate	66	70
Pyro aluminum	34	30

Sometimes barium nitrate is used as the oxidizer. For example:

	Weight (%)
Barium nitrate	68
Pyro aluminum	23
Sulfur	9

	A Weight (%)	B Weight (%)	C Weight (%
U.S. mixes			
Potassium perchlorate	67	63	60
Aluminum	17	27	25
Sulfur	16	10	—
Antimony sulfide	—	—	15

Magnesium flash usually is:

	Weight (%)
Potassium perchlorate	38
Graphite powder	5
Fine magnesium	57

These mixtures are not quite as easy to ignite, but they normally do so with a Bickford fuse and easily ignite with quickmatch and gunpowder primes.

It also must be pointed out that dark pyro aluminum is a dark gray impalpable powder with no visible silver particles. In fact, it does not even look like aluminum powder. Many so-called pyro powders are much coarser than the European dark pyro powders.

The original flash compositions were made with potassium chlorate and aluminum, and barium nitrate occasionally was added. These mixtures in themselves were very sensitive to impact and friction, and a quantity of less than 1 g could be quite devastating.

The so-called fine "bright" flake powders were used before World War II, but they were not always as ignitable as one might expect. Flash crackers made for general sale usually had a piece of black match in the tube to prevent the surprisingly large number of failures.

In Europe it has been usual to make the flash composition with the dark pyro aluminum powders. German pyroschliff has always been of good quality, although it is more than likely that the powder also is being made increasingly outside Europe for both economic reasons and stricter controls on manufacture. In at least one case, European manufacture was stopped by bureaucrats who considered it to be too great a risk, with the result that jobs were lost and the material was imported.

In the United States, atomized aluminum is used in some flash mixtures, but it also seems that the United States has better grades of these powders. It certainly would not be possible to use easily available atomized powders made in the United Kingdom.

Many amateur firework organizations in the United States have demostrated the extreme brisance of flashpowder by detonating something like 500 g in a small wooden shed. It is a sobering experience to see it detonate, unconfined. Even if a human were to survive inside a workshop, the effect of the pressure on the ears and lungs, for example, is devastating.

The smallest amount of flashpowder should be mixed at one time, and 250 g should be regarded as the maximum in a covered container with the smallest amount exposed.

Mixing is a serious issue and probably is best done by using free-flow potassium perchlorate newly sieved, adding the metal powder onto the perchlorate on a sheet of paper and rolling the two together by lifting opposite corners of the paper. The mixture can then, if necessary, be placed on an earthed sieve and passed through by tapping the side of the sieve. It must never be pushed through the sieve.

The bench should be covered with an earthed sheet of carbon conductivity matting and, if desired, the wrists of the operator can be earthed. Some companies who make large quantities of flashpowder mix the powders in small mixing drums made of this carbon conductive material, the machines being earthed.

The whole issue of sensitivity to static electricity is a complex and highly specialized field, but there should be an earthing system in all danger buildings. So much depends on the climate—the climate of Arizona is not the same as that of the United Kingdom.

Nevertheless, it is a sobering fact that while accidents do not often happen under careful control, familiarity does breed contempt.

Additionally, it usually is not one factor that leads to an accident; it is fate that sometimes brings as many as three or four factors together at one moment in time.

It is clear that if such small quantities of flashpowder are to be handled, there must be a very high cost factor involved in bulk manufacture and there must be some way around this. In practice, containers are filled with two different sets of charging boards. A measure of sieved free-flow oxidizer is placed in the tube first, followed by a measure of metal powder, and the tubes are sealed. The filled tubes are then taken to a gentle tumbling machine, which turns them end over end at the other side of a concrete barrier. The mixing is not as efficient, but it works.

The explosions normally are made out of a paper tube that is 1 mm thick and upward and which is closed at each end with clay or corks. A piece of Bickford or other fuse penetrates the tube in some manner and is firmly glued into position. The containers are filled about one-half or two-thirds full of flashpowder, which is charged loosely and not compressed. Tubes made of metals or brittle materials should never be used because of the danger of flying debris. The barium nitrate mixture is less fierce than the perchlorate composition, and in this case the tubes are filled quite full. Magnesium compositions also are sometimes used, but naturally they are less stable and rather more sensitive.

Maroons sometimes are filled with FFF grain powder, but they require strong tubes. In former times, and occasionally still, the cores were wound with strong twine and then glued on the outside.

Photoflash mixtures made with potassium perchlorate, barium nitrate, and atomized aluminum have much slower burning speeds and require a very strong tube to produce a loud noise.

WHISTLES

The oldest whistles were made with potassium picrate. Picrate whistles are very shrill and can be very entertaining in Roman candles with their black tails.

Potassium picrate normally is not manufactured by picric acid manufacturers because there is little use for this material, although the principal English manufacturer did, in fact, produce small quantities for the firework trade. Dr. Shimizu describes the manufacture on page 401.

Picric whistles are not popular with firework makers mainly because nobody cares to work with them. The salt stains the fingers and clothing a bright yellow, and the taste is bitter and unpleasant.

It is well known that picric acid and its salts are sensitive to impact, and great care must be taken when charging potassium picrate. In fact, it is much wiser to consolidate the material with a small hand press. Lead picrate is particularly sensitive, so lead-covered surfaces must be avoided.

Strong narrow bore tubes (e.g., 6 mm inside diameter, 32 mm long, and with a 6-mm wall) can be charged quite full, although the material must be well consolidated or it will explode. If neat potassium picrate is used without any additives, the tube must be filled to the top and not primed. Priming also causes explosions. The material purchased in the United Kingdom could be charged into whistles without any additional material. On the other hand, material made by the author has been known to explode. Since the material is not used anymore, the reason is academic, but it may well be the way it is neutralized. The short, full tubes should therefore have about 3% stearine or asphalt added to the potassium picrate, the former being more pleasant to handle.

Other types of whistles operate in such a manner that it is essential to have an empty space over the composition because the empty tube acts as a kind of resonator. Potassium picrate, as has already been stated, will work in short tubes without any additive, but it must be diluted in long tubes. Up to 20% of potassium nitrate or up to 15% of stearine or powdered asphalt can be added for this purpose.

Whistles should be only lightly capped, with no more than two turns of thin kraft paper or they are liable to explode.

Picric whistles are suitable for rockets, Roman candles, and ground pieces, but it might be unwise to place them in shells lest the lifting charge detonate the picrate.

Many people regard potassium picrate as an exceptionally hazardous material, but there can be no doubt that many accidents have been caused by the principal alternative—a mixture of potassium chlorate and gallic acid.

A mixture of three parts of potassium chlorate to one part of gallic acid normally is used for these whistles, although it appears that American formulations include a small percentage of accaroid resin. Potassium perchlorate cannot be used with gallic acid. The mixture is extremely sensitive, and the greatest care should be exercised during mixing and charging; explosions have taken place during mechanical mixing, and fires have occurred during sieving and charging. The mixture is charged into tubes from 6 mm to 10 mm in diameter, to a depth of 24 mm in about four increments. The top 35 mm of the tube should be left empty. Two pieces of quickmatch are inserted into the empty space before the tube is capped. An additional disadvantage of this mixture is the high price of gallic acid.

In recent years there has been a tendency to make whistles with a mixture consisting of 70% potassium perchlorate and 30% sodium salicylate. Again, in the United States, they add a small percentage of accaroid resin, and this is not necessary. In fact, this mixture is hygroscopic, and whistles should not be kept for long periods, particularly in unheated magazines over the winter. Oddly enough, the addition of accaroid resin to this mixture seems to increase its tendency to absorb water, but the reason for this is not clear. Polyester can be added to the salicylate composition, but it does not give complete protection from moisture. Potassium benzoate also can be used in place of sodium salicylate. This type of whistle can be pressed or hand-charged, but the potassium perchlorate and most of the sodium salicylate must pass a 120 mesh sieve. The exact particle size is something of a problem, and batches of sodium salicylate vary somewhat in their performance.

Work on whistles originally was done by Maxwell (49), but from a firework point of view great credit goes to a Turkish pharmacist, Selcuk Öztap. Two excellent articles in *Pyrotechnica* (XI, XIII) give a wealth of information on the kind of articles that can be produced with this mechanism, which is related to the strobe effect (see Chapter 6).

Öztap originally worked with salicylates and was original in that he also added vaseline or oil and a small amount of iron (III) oxide to the mixture. The following mix is representative:

	Weight (%)
Potassium perchlorate (170–200 mesh)	76
Sodium salicylate (170–200 mesh)	20
Vaseline or paraffin oil	3
Red iron oxide fine powder	1

The ingredients are mixed by warming to 40°C to 45°C.

Sodium benzoate has been employed in whistles, but potassium benzoate generally is favored in commerce. The Chinese seem to prefer potassium hydrogen terephthalate. Variations in tone can be produced with these whistles. Benzoate whistles, with a central spindle-shaped hole (but no choke) like a rocket, make an extraordinary rattling noise and have been commercially available in Europe for some years. As with rockets, the mixture must be carefully charged and the spindle must not be too large or the items are likely to explode.

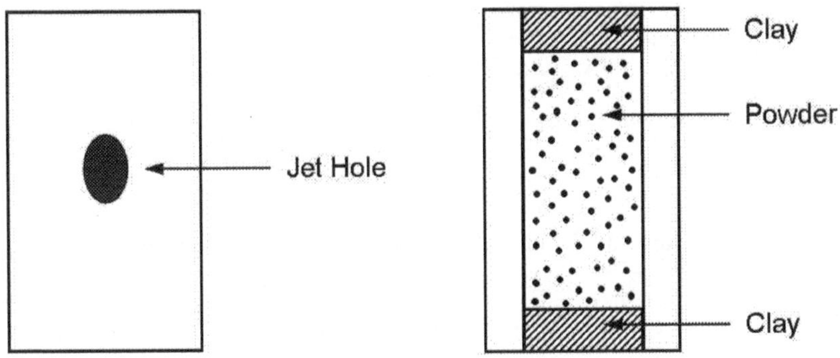

Figure 13.1 Firework hummer

This strange rattling noise, generally described as a "croaker," "screecher," or "knatter," is being used for general entertainment at firework displays or for scaring birds. Titanium also can be added to the mixture or to the ordinary whistles for a silver tail. It must be borne in mind, however, that adding titanium increases the sensitivity. All these mixes are fiendishly explosive, particularly when pressed into molds, and should be pressed behind a barrier.

The Chinese are using these mixtures pressed into tubes in similar ways to the hummer described later to obtain either a screecher or a humming effect. The sound can be positively flatulent in tubes with clay chokes.

Potassium 3,5 dinitrosalicylate mixed with potassium nitrate has been used to produce a very effective whistle in small-diameter tubes (3 mm in diameter).

HUMMING FIREWORKS

Humming fireworks are constructed in a manner that will allow a jet of gas to issue out of a tube at such a speed and at such an angle that the tube will be caused to rotate on its own axis. This causes a humming noise in the air.

The smallest hummers are made from narrow-bore tubes about 8 mm internal diameter and with walls 6 mm thick. The tubes are about 32 mm long and must be blocked with clay at each end. Inside the tube, between the clay plugs, a very fast-burning mixture of gunpowder and aluminum is carefully charged (Fig. 13.2).

The purpose of the aluminum is merely to slow down the meal powder to prevent an explosion and to produce some silver sparks.

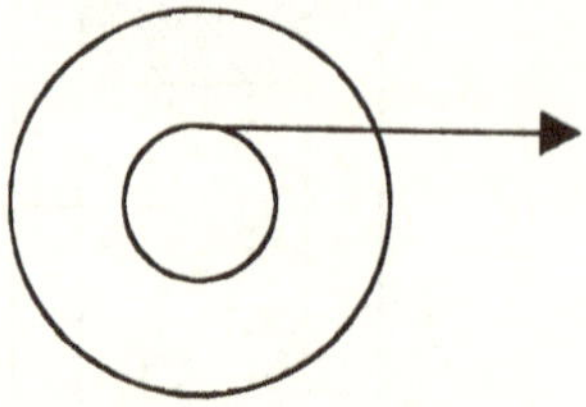

Figure 13.2 Angle of the jet hole of a firework hummer

Seven percent to 10% of fine titanium or silicon can be used for the same purpose.

Larger hummers with an inside diameter of 16 mm often employ a little charcoal in addition to the gunpowder and aluminum and frequently have two holes. The gas jet issues out of the tube from a small hole pierced through the wall in the center. The size of the hole is, of course, related to the burning speed of the powder.

It can be observed from Figure 13.2 that the angle of the hole is such that the tube will be made to spin in flight. Needless to say, it is better to pierce the hummer tube because drilling is rather hazardous. For ignition, a piece of quickmatch is placed in the pierced hole and then secured with a piece of tapc. It is possible to drill the empty tubes first and then charge a piece of match into the composition during the filling operation.

The best hummers are made with a long steady stream of gas. Ideally, this is done with a polymerized Bakelite resin containing potassium perchlorate and some aluminum. This was the subject of a patent by the late R. G. Hall of Brock's Fireworks and a further development of trials in Germany during World War II. The Bakelite resin in the form of a syrup is mixed with the manufacturer's recommended quantity of accelerater. Potassium perchlorate is added to produce a putty-like extrudable mass that can be extruded into the tubes. The mass dries into a hard solid that burns with a steady production of gas.

There was a time when it looked as if Brock's patent would indeed revolutionize the manufacture of fireworks. The main idea was that these wet extrusions would be safer and would reduce the danger of handling dusty powders. In the intervening years, a number of people, including the author, have done extensive trials, but it is not ideal.

Looking at it commercially, the resin is not inexpensive and in recent times the supply of resorcinol is not as plentiful as might be desired. An equally important factor is that formaldehyde is not pleasant to work with and invariably attacks the skin, which inevitably suffers, particularly in the cleaning process after extrusion.

Some resins are not miscible with cold water, so it can be a problem cleaning tools and dealing with effluent. The author learned the hard way by unthinkingly trying to wash a hand pump extruder in hot water, which immediately started to polymerize the resin!

Once the tubes have been charged, the setting time depends on the temperature and the amount of alcohol in the syrup, but clearly there is an advantage of curing at a constant higher temperature; this also can give a degree of porosity, which could be both an advantage and a disadvantage.

Tubes with an internal diameter of less than 25 mm (and preferably smaller) give the best results, but shrinkage is common and a pressure buildup can cause fire to strike down the inside of the tube. Air locks in the rather stiff putty-like material also can cause burning time problems, particularly with rockets.

There is no easy way to make stars with these mixtures. In the early period good colors were not very easy to obtain apart from yellow, which also comes from the nature of the material.

The conclusions can be that traditional mixtures are the easiest to handle, but small rockets, small tourbillions, hummers, portfires, and fountains can be quite successful. One great help is that uncoated iron is actually protected by the resin and will keep for long periods.

14

ROCKETS

Rockets are one of the oldest pyrotechnic devices, and their history can be traced from the Middle East, where they were used as weapons. It is believed that they were first used for display purposes in India and China. The manufacture of rockets was well known in Europe during the early Middle Ages, and the origin of the word *rocket* is thought to be based on the similarity in appearance of a rocket on its stick to the round piece of wood used to cover the point of a lance in mock combat. This was known as a "rockette," an Italian diminutive of "rocca," a staff.

All rockets, from the smallest firework to the giant satellite carrying Saturn, have four basic components in common:

- A case or rocket motor
- A "choke" or venturi
- A propellant charge
- A flight-stabilizing device

Rockets are reaction motors. On ignition, the propellant charge produces gas at a high temperature, so the internal pressure of the gases in the rocket body is raised. By allowing the gases to escape through a narrow opening, the pressure at this opening, or nozzle, falls and the velocity of the gases is increased. At any time, the momentum of the rocket must equal the momentum of the escaping gases so that, very simply, we have the following equation:

$$\text{Mass of rocket and unburned propellant} \times \text{Velocity of rocket} = \text{Mass of escaping gas} \times \text{Velocity of gas.}$$

Rocket bodies or cases can be made from a variety of materials. Simple firework rocket bodies are almost exclusively rolled from a good-quality kraft paper. After cutting to size, the paper is pasted and tightly rolled on a cylindrical former and allowed to dry. It makes little difference whether this operation is done by hand or by machine except that when the cases are machine-rolled, a slightly inferior quality of paper may be used because the tighter machine rolling gives greater mechanical strength to the finished case. Care should be taken to ensure that the inside surface of the finished case is smooth and regular and that the inside edge of the paper is firmly pasted down. This is important because it is possible to dislodge this edge during the charging operation and to turn it over into the composition. This would seriously affect the consolidation of the propellant charge and most likely would lead to the bursting of the rocket on ignition. The cases should be thoroughly dried; that is, they should be allowed to reach an even moisture content in equilibrium with the atmosphere in which they are to be charged. Failure to observe this simple precaution can cause swelling or shrinking of the case after charging and subsequent malfunctioning of the finished rocket. For signal rockets, which have to stand more vigorous handling and storage conditions, the cases may be made from phenolic resin–impregnated papers. After rolling, the resin is cured or polymerized in a heated oven, and this process gives a rigid case of great mechanical strength. Metal cases of steel, or aluminum alloy, also are used, and these often are varnished on the inside surfaces to give greater adhesion between the propellant and the case wall and sometimes to prevent undue heat transfer along the case. Good-quality, thin paper linings, which may or may not be varnished, also are used for the same purposes.

The "choke," or venturi, is an important part of the rocket. The pressure energy of the propellant gases is converted to kinetic energy by this device. In firework rockets, the "choke" is quite crude and often is formed by constricting the rolled case while it is still damp. The constriction, which may be from 25 mm to 35 mm from the end of the case, is then tied with string to prevent its opening out on drying. A slightly more refined "choke" is formed by placing the dried case on a metal step that has a nipple of the required "choke" dimensions.

A scoopful of dry clay is poured into the case and pressed or rammed into the case until it is solid. The case and its clay "choke" are then removed from the nipple. Nipple sizes and dimensions will be given later. In signal- and line-carrying rockets, the "choke" may be no more than a cylindrical hole bored in a metal plate, which in turn is roll-crimped or riveted to the rocket body. This is a common form of "choke" in which blackpowder-type propellants are used. When more sophisticated propellants are employed, the nozzle and venturi should be designed so that the maximum thrust is given by the burning propellant charge. Economic factors and the desirability of mass production of nozzle units invariably force the designer to compromise, and the result, while satisfactory, is not always the best that is technically possible. A full mathematical treatment of nozzle and venturi design can be found in the many technical books on rocketry (22).

The traditional propellant charge for the firework rocket is either:

- Gunpowder
- A mixture of mealed gunpowder and charcoal
- A simple mixture of potassium nitrate, sulfur, and charcoal
- A mixture of mealed gunpowder and calcium carbonate

To obtain consistent results with any of these compositions, a few simple but very necessary precautions should be observed. When grain gunpowder is used alone, the grains should be small, regular, and dust-free. They should be quite dry and show no signs of white crystals on the surface when examined under a powerful lens. The potassium nitrate and sulfur should be of a good commercial quality, and the particle size should be closely controlled. Samples passing 120 British Standard Sieves (B.S.S.) and retained on 160 B.S.S. are ideally suited for the mixed type of composition. The moisture content of the ingredients should be as low as possible, and this level should be consistent. Any drying out of the composition after charging invariably would lead to cracking of the charge and to increased pressure buildup and bursting of the rocket on firing. Charcoal is the most difficult ingredient to control. It is imperative that the charcoal be kept in a dry, warm store if the regularity of the burning rate of the finished composition is to be maintained. The particle size should be sufficiently large to give a good glowing "tail" when the rocket is fired. A charcoal passing 40 B.S.S. and retained on 80 B.S.S. is useful for small rockets, and a charcoal containing 10% passing a 15 to 20 B.S.S. and retained on 40 B.S.S. gives a good display in larger rockets. The ash content of the charcoal should be consistent from batch to batch, and the type of wood from which the charcoal is made also should be specified. Generally, hardwood charcoals give compositions that burn more slowly than similar compositions made from softwood varieties.

When a particular formula has been found to give a satisfactory rocket, subsequent batches of composition can be checked for consistency by column test. A weighed amount of composition is pressed in equal increments into a gun metal tube about 10 mm in diameter until the column of composition is about 50 mm in length. This length is controlled by a mark on the ram.

The measurements given are purely arbitrary, and any reasonable measurements can be used, but once the weight of the composition, the number of increments, the diameter of the tube, and the length of the composition have been determined by the manufacturer, they should be strictly adhered to. The column of composition is fired by a piece of quickmatch, and the burning time is checked by a stopwatch. A variation in the burning time of 1.5% is allowable. The gun metal tube is washed in warm water, then dried, and is ready for the next test. This test is quick, reliable, and easily carried out. Any discrepancy in mixing or in the purity of the ingredients shows up as an increase or a decrease in burning time.

Modern propellants for signal rockets may be of the double- or triple-base nitrocellulose powders. These are mixtures of nitrocellulose with nitroglycerine, or of nitrocellulose with nitroglycerine and nitroguanidine. More frequently, plastic/perchlorate propellants are employed because these allow the use of simply designed motors. These propellants are capable of being molded to any desired shape in order to give the correct burning characteristics. Typical formulations would contain polyisobutene, ammonium perchlorate, and ammonium picrate (13).

There are three basic types of solid propellant rockets. In the first type, the composition is pressed into the case in a solid mass and is ignited at only one end. The propellant is said to burn in a "cigarette-" or "end"-burning manner.

It will be seen that only a limited burning surface is available for gas production, and therefore a fairly quick-burning composition is needed to sustain an adequate internal gas pressure. In firework rockets, these compositions usually are compounded of mealed gunpowder and fine charcoal. The charcoal usually is present in proportions varying from 20% in rockets of 18-mm bore to 10% in the smaller 12-mm diameter rockets. Fine-grained gunpowder also is used, and such rockets are quite powerful for their size, but they lack a spectacular "tail" and, in the author's opinion, are not as interesting as rockets that contain charcoal. In these rockets, the "choke" is of clay, as already described, and the diameter usually varies between 2 mm and 4 mm, depending on the diameter of the rocket and the type of composition used. This end-burning type of rocket usually is pressed with the case supported in a metal mold to prevent collapse of the paper case. The composition is put into the case in four to six equal increments, and each increment is pressed at 300 to 500 lb on the ram.

The weight of composition may vary between 5 g and 25 g per rocket, depending on the diameter. It is not possible to be more specific because the ingredients vary in moisture, particle size, and so on, from place to place and from different suppliers. Indeed, most manufacturers devise their own formula to suit the particular type of case and charging pressures and stick to this as closely as possible. In the end, firework rocketry is largely a matter of trial and error. Most signal rockets are of the end-burning type, but the burning surface frequently is modified in shape to give better burning characteristics. As the charge burns away, the chamber in effect will become larger, causing a fall in gas pressure. To overcome this disadvantage, the end of the charge is shaped in such a manner that the burning surface will increase and thus tend to keep the chamber pressure constant. Cruciform or concentric channels are the two simplest and most favored designs for this purpose.

The second type of rocket has a charge that is perforated along its long axis. The perforation takes the form of a long, thin cone, usually formed during the charging process. The choked rocket case is placed over a tapering hard steel spindle, which normally extends into the case for three-fourths of its length excluding the choke (Fig. 14.1).

The diameter of the base of the spindle is about 1/12 of its length, and this should be the dimension of the rocket choke. The base of the spindle

Figure 14.1 Rocket spindles and drifts

is supported by a solid concrete or steel post. For rockets of 24 mm diameter and above, the composition is divided into 14 equal increments. Each increment is poured separately into the case and consolidated by means of a hollow drift or ram that is subjected to a pressure of 2 tons. The case should be supported in a mold or tube to prevent collapse.

When the column of composition has reached the top of the spindle, the remaining increments are pressed with a stepped drift to form a plug with an axial hole. The finished rocket will have the appearance of Figure 14.2 in cross section.

Smaller rockets of 18 mm diameter and under should have the composition charged in 8 to 10 increments. Very small rockets may be filled lightly with powder and then bored by machine. This operation also serves to consolidate the composition.

Compositions for this type of rocket are of the sulfur, charcoal, and potassium nitrate mixture type. The proportions vary widely: 60% to 70% potassium nitrate, 15% to 20% charcoal, and 15% to 20% sulfur. It is not possible to be more specific in view of the problems mentioned previously, but as a guide, rockets of 24 mm diameter and over usually require composition containing 63% to 67% potassium nitrate, and this percentage increases as the rocket diameter decreases.

Rockets should be kept in a dry atmosphere; moisture absorption causes the composition to crack, thus giving rise to an increased burning surface and explosion on ignition. Both types of rocket are ignited by means of a length of match inserted through the choke or by means of fine-grained gunpowder that adheres to the choke by means of a small quantity of thin shellac varnish. With both types of rocket, the flash of ignition should be sufficient to cause instant ignition of the burning surface; otherwise slow "takeoff" and erratic flight will result.

Stabilization of firework rockets is achieved by means of a stick. Many rules for stick dimensions have been given over the years, but the author has found that it is only necessary to ensure that the center of gravity lies a little to the rear of the head. The stick holds the rocket in a stable position during takeoff and tends to counteract deviation dur-

Figure 14.2 Section of charged rocket

Figure 14.3 Multiple firing of rockets in Portugal
(Courtesty Antonio Rodrigues)

ing flight. Rockets with sticks will always tend to turn into the wind as a result of unequal pressure on the stick.

Finned rockets usually are used for signal purposes. Here, the rocket usually is given sufficient forward velocity to maintain stability during takeoff by means of a small charge contained in the launcher. Fin design problems are covered fully in the technical literature. Line-carrying rockets are stabilized by the harness and line and also are launched by a powder charge.

Firework rockets may carry a payload of stars, hummers, or other pyrotechnic articles. The load usually is carried in a lightly rolled paper case attached to the head of the rocket. A flash from the propellant charge ignites a few grains of powder, which eject and light the effects. The payload weight should be in the ratio –5 load to 3 propellant for a well-constructed rocket.

In the first edition of this book, the late R. G. Hall, formerly Technical Director of Brock's Fireworks, kindly wrote this section on rockets. It clearly is the work of a man of experience, and anyone in the trade accepts that there is little sense in committing anything more precise to paper. Once a manufacturer has decided on the physical dimensions of his rocket and has chosen a formulation with the material on hand, he or she must obtain the best result by changing either the powder or the dimensions. At this point, he or she must carefully check the burning speed of the powder in future batches. Ron Hall has made this crystal clear. However, since the publication of the first edition we have had so many complaints about the fact that precise details were omitted that it has been decided to give some additional general details notwithstanding the fact that burning speeds of powder still must be carefully controlled.

Rockets with a Perforation Down the Center (Fig. 14.2)

Tube ID: 19 mm OD: 30 mm Length: 145 mm
Spindle diameter at base: 10 mm Spindle diameter at top: 3 mm
Length: 120 mm
Thickness of clay: about 14 mm

		Weight (%)
Composition	Potassium nitrate	61
	Sulfur	5
	Charcoal 150	20
	Charcoal 40	14

The composition must be sieved twice through 30 mesh and twice through 20 mesh for uniformity and then pressed in about eight increments using drifts with increasingly smaller central holes until the top of the spindle is reached. One diameter is pressed on top of the spindle. Let us assume that there is not enough gas pressure. The rule, therefore, is to increase the percentage of potassium nitrate and possibly reduce the amount of charcoal. If the rocket is too fast, then the reverse would apply. The percentage of sulfur is less critical, but it should not be less than 5%. Rockets that are too slow can, of course, have the spindle lengthened by reaming out the composition at the top of the spindle. Needless to say, there must be enough pressed composition there. One way is to press too much on top and then bore a hole into this from either end or both ends. This frequently is done with rockets that are "long headed."

Some manufacturers add a small percentage of gum arabic to the composition and charge the powder slightly dampened. The powder consolidates and sets into a solid mass that does not easily "blow through." This often is done in warm climates, but in more temperate zones the drying out period creates problems for testing. In one remote part of Spain, the author saw rockets being made in this way and charged solidly in the first instance. The filled tube then was placed in a modified lathe and a hole was drilled through the clay into the powder.

The method is quite neat for small rockets, and a few tests with each batch quickly determine how long the hole should be. The cutting edge must be on a narrower shank to allow the loose powder to escape. The author shuddered at the sight of loose composition everywhere and hoped there would be no foreign matter about to cause an ignition. Loose increments are quite a problem because they are liable to make the rocket explode. Great care must be taken when pressing large numbers of rockets at the same time. The 18-mm rockets mentioned in this article were pressed six or eight at a time. It was the custom in the United Kingdom to charge numbers of tubes up to about 16 mm. with "pulled in" chokes by simply constricting the tube with a choking machine (page 204). The rocket then was spiked with an arbor press or a machine. In general, this is fine, but if loose composition breaks away in the venturi, the rockets frequently explode. Bad handling of such rockets is a problem, and in poorly made specimens it was sufficient just to bang them against a table to make them explode. Chinese rockets were made this way with a red fire composition that contained magnalium. They should have ascended in the air with a red tail. By the time they reached the United Kingdom, they had turned into flash crackers of great ferocity that almost smashed the bottles in which they were fired.

The variation in the burning speed of commercial blackpowder and the fierceness of the powder means that rockets made with this material and spindle-type perforations are even more delicately balanced. The blackpowder is slowed down with either charcoal or precipitated calcium carbonate, but careful manufacture is essential.

If there is a final comment to be made on traditional rocket manufacture, it must be that the relationship between the speed of the powder and the area of the burning surface is paramount. The amateur simply cannot make small batches of composition and expect the rockets to perform consistently, but it is amazing how many people are willing to try it! One well-known U.K. manufacturer solved the problem by simply getting the gunpowder manufacturer to make its rocket powder to a given specification and burning speed. One such powder was made with 68% potassium nitrate, 23% sulfur, and 9% charcoal.

There can be no doubt that in Europe Portugal is the country with the most consummate skill in rocket manufacture. It has been a kind of open secret that the Portuguese had the best rockets, but somehow it was not realized because they were not exported very often. Yet the growth of the European competitions revealed their quality and elegance.

To witness a display of Portuguese maroon rockets is quite amazing. It is not only the exact timing and the rhythm, but also the precise height at which they burst and then the pattern of bangs each"foguete: produces—anything from 3 to 50.

We had heard that the Portuguese rockets were large, but we had not really comprehended what was meant by this. A payload of 100 g with a stick of 3 m would be considered to be large, particularly when some countries do not allow rockets with more than 20 g total powder and stars.

Antonio Rodrigues, a longtime acquaintance in the Portuguese firework industry, made us realize that we had no idea of these sizes and that in the north of the country payloads of 200 to 500 g were common and that even larger ones certainly had been made.

The incredible photographs bear witness to these huge rockets where a large display might contain anything between 600 and 1,000. The rockets themselves contain the usual items that are found in shell breaks, but the breaks themselves are not as fierce as they might be in modern shells.

Typical sizes are:
Motor: 80 mm × 13 mm increasing to 180 mm × 26 mm
Stick : 1 m up to 5 m

Special sizes are:
Rockets with 2, 3, or 4 motors
Daylight rockets with from 3 bangs to 50 bangs.
Night rockets with bangs and stars, stars only, or other effects and other effects of either light or sound

We hope that "progress" will not remove some of the special nature of these rockets.

END-BURNING ROCKETS

The traditional rockets with the central hole may look beautiful and gain great heights, but they do take a long time to charge and there are unfortunate composition variations to cater to. At the end of World War II, Germany introduced the commercial blackpowder rockets with great success. These rockets do not perform as attractively as traditional rockets because the gold tail is reduced, and neither do they go as high, but they are easier and less expensive to make. This type of rocket also can carry a good payload and, provided that the head is well constructed, they can look better than some small shells. Small rockets often are made with paper tubes and clay chokes, but the majority are pressed in an aluminum tube. The use of precision aluminum tubes makes it easier to use molds and press large numbers at a time. One end of the tube is sealed with an aluminum plug containing a central hole. A typical example would have an inner diameter of 24 mm and a venturi of 7 mm.

While there is no central spindle, in practice there is a slight indentation in the powder of about 18 mm in order to give the initial thrust. The blackpowder used may contain a little graphite. Some signal rocket motors also have used blackpowder and calcium carbonate.

The aluminum tube conducts heat, of course, but the motor normally is forced into a paper tube after pressing, and this makes it easier to add a star pot and fix a stick. This type of rocket also has the advantage that a different type of composition (e.g., silver) can be pressed on top of the blackpowder. It is, however, better to start with blackpowder.

Rockets imported into the United Kingdom from China have caused considerable problems where there was no regulation controlling maximum powder weight. Salute rockets containing as much as 65 g of flash powder with a TNT equivalence of 20 g have been used for nefarious purposes, giving fireworks a bad name.

15

DRIVERS, SAXONS, AND TOURBILLIONS

DRIVERS

Wheels and moving fireworks are operated by rockets or, more commonly, by drivers. Drivers are stout tubes from 12 mm to 40 mm in internal diameter, choked down to about one-third of a diameter and charged with a fierce composition containing a high percentage of gunpowder. The composition is charged solidly in small increments and does not contain the central spindle that is characteristic of rockets. The tubes are choked with a clay washer or by "pulling in" the tube in a choking machine.

Small wheels are turned quite adequately with tubes, which are 12 mm or 16 mm in diameter and about 100 mm long. A 16-mm tube, for example, would be placed on a nipple with a central spiggot about 5 mm in diameter. A small amount of clay is first charged around the spigot to produce a washer, and then increments of composition are added and charged with several light blows. With fierce compositions, many light blows are better than a few heavy blows, and increments of composition should be sufficient to occupy one diameter in height. A charge of clay finally is rammed into the tube to close it, although not all tubes are clayed; some are simply fitted with a paper cap to enable fire to be transferred to a new driver when relays are required.

The following compositions are suitable for small drivers:

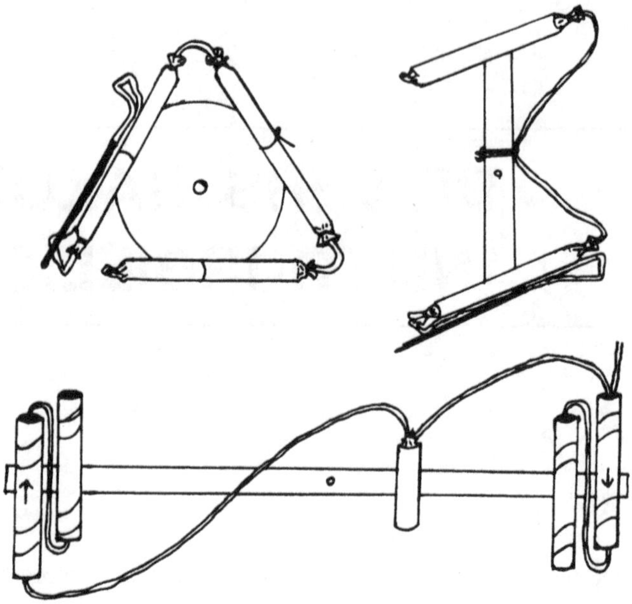

Figure 15.1 Small wheels

	A Weight (%)	B Weight (%)	C Weight (%)	D Weight (%)
Meal gunpowder	60	55	88	70
Charcoal 150 mesh	—	—	12	—
Charcoal 40/100 mesh	7	7	—	—
Potassium nitrate	25	23	—	—
Iron turnings 60 mesh	—	—	—	30
Sulfur	8	6	—	—
Titanium 20–40 mesh	—	9	—	—

The smaller drivers are used for a variety of different types of wheels, as Figure 15.1 shows. The older wheels were made with a central boss, three or four spokes, and a beech hoop, but these have been replaced in more recent times with plywood and hardboard constructions. Originally designed by Standard Fireworks in the United Kingdom, they now are widely made in China.

Large drivers are used for the bigger display wheels from 2 m to 5 m in diameter. The tubes are 24 mm or 40 mm in diameter and about 23 cm long. Compositions are a shade slower and often contain more charcoal than the smaller sizes. For example:

	A Weight (%)	B Weight (%)	C Weight (%)
Meal gunpowder	74	65	60
Potassium nitrate	10	5	15
Charcoal 40/100 mesh	6	—	5
Charcoal + 150 mesh	6	—	4
Sulfur	4	5	7
Iron turnings 20 mesh	—	25	—
Titanium turnings 40 mesh	—	—	9

SAXONS

These fireworks are used in combination with a small color case to produce small colored wheels that usually are part of a larger design. Thick-walled tubes are charged with composition, but the tube is closed with clay at both ends, and the fire issues from a hole that is cut into the side of the tube at right angles to the axis of the tube. The hole is near the clay plug that closes one end, but the size and position of this hole depends on the strength of the composition. The hole often is bored 12 mm or so from the clay plug so that the composition burns all around the hole when the tube is ignited, thus giving an extra large thrust at the beginning to get the wheel started. Unfortunately, it is a common sight to see sluggish saxons that do not seem to have enough power to get started, or they become stuck with a piece of match piping wrapped around them or have a nail that is too tight. Another snag with saxons is the fact that the hole in the tube becomes larger during the burning process, with inevitable slowing of the piece. Thick-walled tubes about 16 mm in diameter with a 4-mm wall or 12 mm in diameter with a 4-mm wall are used, and they are about 25 cm to 30 cm long.

Compositions should not be too hot (i.e., they should not contain too much potassium nitrate and sulfur) or they are liable to burn the jet hole away too much, with the resulting loss of force. Old mixes were rather hot-burning. For example:

	Weight (%)
Meal gunpowder	50
Potassium nitrate	25
Sulfur	25

The following are more typical and useful:

	Weight (%)
Meal gunpowder	50
Potassium nitrate	30
Charcoal 40/100 mesh	10
Sulfur	10

A little titanium can be added to this mixture, of course. Saxons are made in various forms, but are mostly double. A tube about 30 cm long has 12 mm of clay solidly rammed into one end and then is charged with 125 mm of composition; 24 mm of clay is then charged, followed by 125 mm of composition, and finally 12 mm of clay. The tube is now full. A hole to receive the nail is bored through the middle of the tube, through the clay, and then the jet holes are bored near the ends, but on opposite sides of the tube, as shown in Figure 15.2.

Saxons can be fired simultaneously from both ends, or they can be arranged to fire successively by inserting a piece of quickmatch in the base of the composition of one end, thus transferring the fire to the opposite end, as shown in Figure 15.2. Some manufacturers make the saxon in two separate units, joining the two together in the center with a paper tube or a piece of wooden dowel. Single saxons also are made. A small spacer tube should be fixed between a saxon and the timber work to prevent the saxon from striking the woodwork during the performance. Color cases usually are fixed on the side of the saxon to produce the fire rosette. They are timed to burn the same length of time as

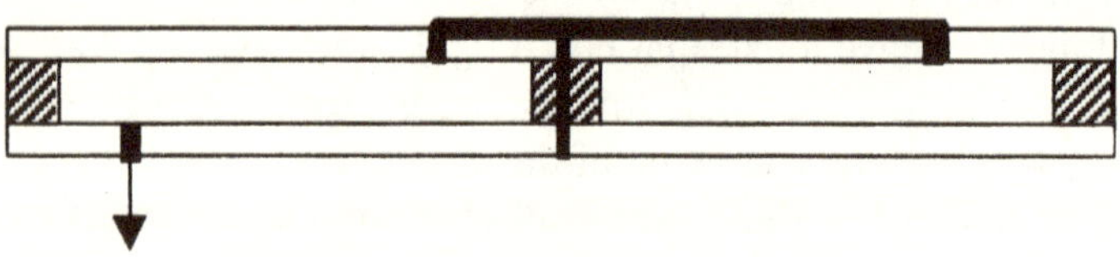

Figure 15.2 Double-ended saxon

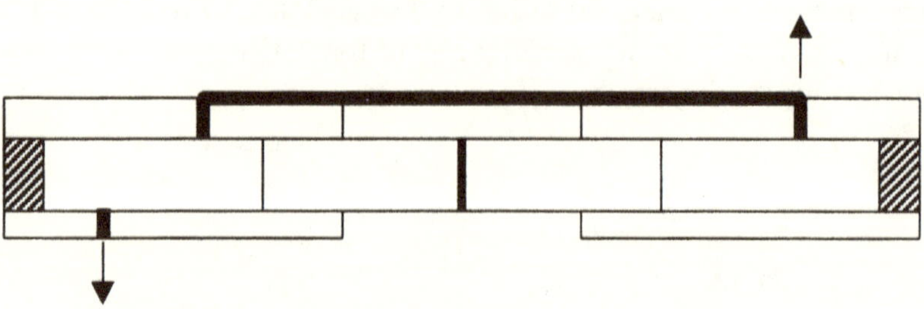

Figure 15.3 Two single saxons attached to a piece of wooden dowel

the saxon and require careful fixing with glue and paper or they are liable to fly off during the performance (Fig. 15.3).

TOURBILLIONS

The French word *tourbillion*, or "whirlwind," is still used to describe these fireworks, which have appeared in many forms and sizes. The older and larger forms were, in effect, a fierce type of saxon that first spun around on the ground and then lifted up into the air by means of two or four holes bored in the underside (Fig. 15.4).

A curved stick, approximately the same length as the tourbillion itself, was nailed to the center of the tube, but at right angles to the tube. The process of boring the six holes in a large tourbillion is quite tricky and must be carefully done. The holes in the side and base also must be exactly at 90° to each other or the performance will be erratic. A useful way to make the tube is to slip it inside a hollow metal or paper tube that already has the holes drilled in the correct positions, thus using this external tube to mark the position of the holes before drilling. It also is unwise to drill tubes that already contain composition; a much better method is to drill or punch the holes in the empty tube and cover them with a piece of paper or adhesive tape that can be removed after filling. This also applies to saxons, of course.

The larger sizes are made with a tube that is approximately 24 mm in diameter, 20 cm long, and with a wall thickness of 6 mm. The holes are about 4 mm, depending on the strength of the composition, but they

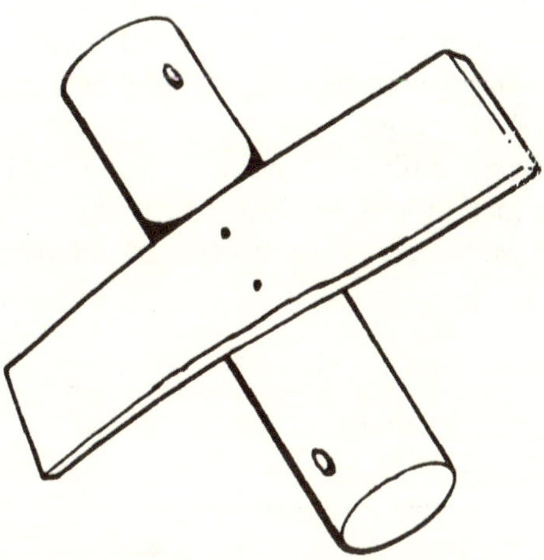

Figure 15.4 Tourbillion (underside view)

also must be carefully spaced so that there is approximately the same quantity of composition on either side of each hole. The following composition is a good one, although some adjustment of the diameter of the jet hole may be necessary.

	Weight (%)
Potassium nitrate	45
Meal gunpowder	35
Sulfur	5
Charcoal 40/100 mesh	15

Sticks usually are made from a piece of beech hoop of the type used for sieves and large wheels and are fixed to the tube with a copper nail. The tourbillion is matched in such a way that the match is ignited on the top in the center, transferring the fire first to the two side jets for spinning and finally to the lifting fire jets at the base. The match is fixed with tape and then covered with paper. Small tourbillions are made with a tube about 12 mm in diameter and about half full of composition. They are made to ascend with a single jet punched at about 30° or 45°, which both lifts the firework and spins it at the same time (Fig. 15.5). These small articles use much fiercer compositions with a larger percentage of mealpowder and, as before, the size of the jet hole is related to the burning speed of the composition. Wings usually are made of paperboard or plastic.

Closely related to this type of firework is the famous flying saucer (see Fig. 15.6). This piece has various names such as geyser, girandola, and flying crown, and consists of a circular horizontal wheel that spins at great speed on a spike at the top of a pole and then takes off into the air to climb several hundred feet. Some varieties ascend, descend, and finally reascend during flight.

This piece is very difficult to produce, but it basically consists of very fierce horizontal drivers that cause the unit to spin on the post and a few fierce vertical drivers or rockets that lift the piece in the air.

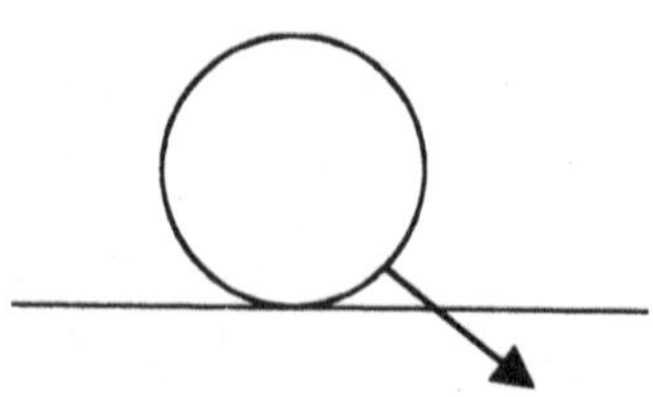

Figure 15.5 Position of the jet hole

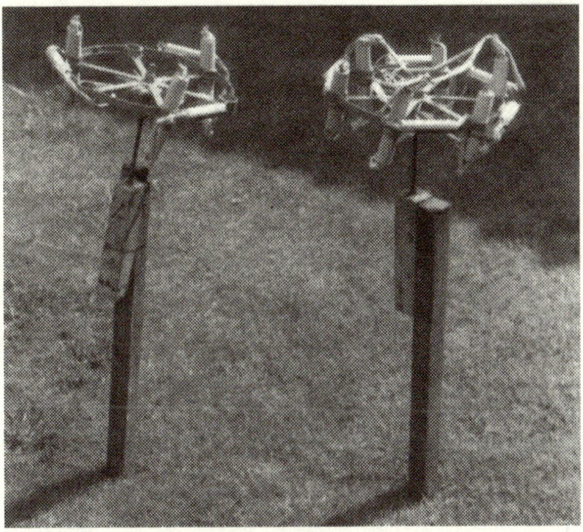

Figure 15.6 Flying saucer

No attempt will be made to describe the manufacture of these items because there are many pitfalls that only considerable experience can overcome. Those who attempt to make the pieces will find either that the drivers are not fierce enough to lift the piece or that they explode. In addition, poorly made drivers will burn incorrectly, sending the saucer into the air at the wrong angle and possibly falling into a crowd. The saucer usually carries a small shell that fires at the highest point of the flight. It also can be fitted with fountains or small pot au feu.

FLYING PIGEON

The line rocket or flying pigeon is a very popular item at firework displays and frequently consists of a wooden block with a hole through the center so that it can slide along a tightly stretched rope or wire. Four 16-mm rockets are fixed to the block, and four fierce drivers are attached to a small hoop fixed on the block (Fig. 15.7).

The first driver rapidly spins the unit on the rope, and as soon as it burns out, it ignites the first rocket, which transfers the piece to the other end of the rope. The burned-out rocket then ignites the next driver, and the process continues, backward and forward along the rope. Where ropes are used, the friction is quite high and large rockets of at least 18 mm bore must be used. Whistles and colors often are added to

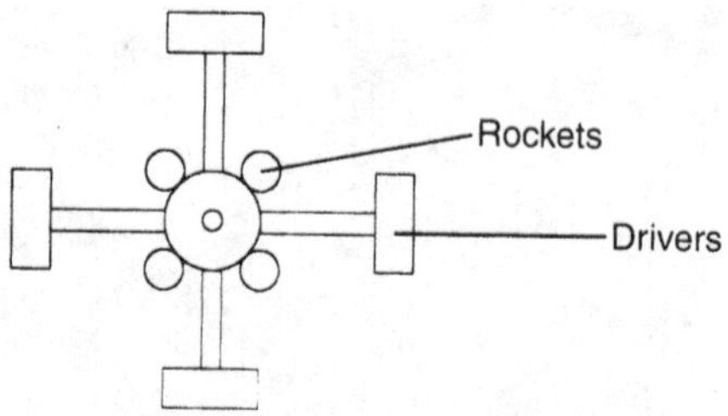

Figure 15.7 Flying pigeon

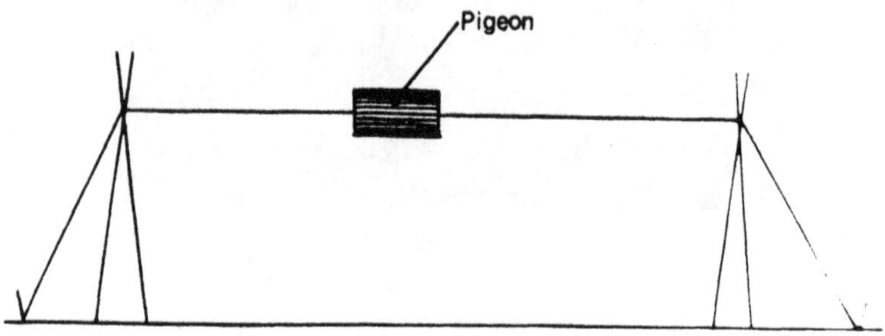

Figure 15.8 Flying pigeon

the pigeon. A tube must be placed on each end of the rope to prevent the pigeon from striking the post and thus being damaged. A good way to assemble the pigeon is to attach the rope to two firm stakes and then stretch it with two stretchers according to the diagram. The stretcher consists of two wooden poles loosely joined at the top with a nut and bolt (Fig. 15.8).

Smaller pigeon-type fireworks called rats consist of a relay of very fierce drivers that follow each other down the wire. They are charged with a mixture of gunpowder and iron, and the tubes are about 12 mm in diameter. See also Figure 1.7 (page 31).

16

SHELLS

The greatest skill of the firework maker can be exhibited in the manufacture of star shells, or "bombs," as they frequently are called in Europe. Shells are produced in huge quantities every year, but the quality varies from, at one extreme, the most ordinary types, which literally drop the stars into the sky in a disorderly heap, to the magnificent pattern shells at the other extreme. The older method of shell construction involved the manufacture of hollow paper spheres about 6 mm thick, but this was a costly and time-consuming business. Strips of paper were soaked in paste and then laid in layers in a hemispherical mold. A hemispherical former is used to press the paper into the mold in order to press the strips of paper down. When the shell halves are dry, the rough edges are trimmed with some type of lathe (Fig. 16.1). When the two perfect hemispheres are ready, a hole is punched in one of them to receive the fuse, and the two halves are glued together. The final process consists of pasting strips of paper, which are well soaked in paste, all over the outside. Long strips are used, usually stretching from the fuse hole to the base. Sometimes the paper sphere is made up from four hemispheres—two inner ones and two outer ones. The hemispheres are pressed out of sheets of pasted strawboard cut out in the following shape (Fig. 16.2).

Ideally, the thickness of the two inside shells should be equal to those on the outside so that when the halves are in different planes, the shell should break into at least four quarters to produce an even burst.

In practice, this even burst is not easy to obtain, and instead of all the stars being thrown out evenly from the center, they tend to fall downward in a kind of bouquet. The old method was even worse because, in effect, the shell broke into two halves and dropped the stars out in a heap.

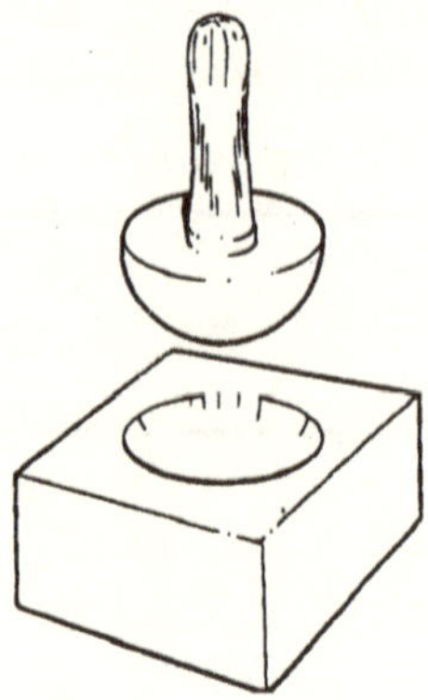

Figure 16.1 Shell mold and former

Figure 16.2 Paper cut out for shells

The formation of the shell wall is quite vital, and the pattern of the burst is entirely dependent on this in relation to the nature of the bursting charge.

The easiest type of shell to make is similar to the Japanese "poka." The sphere is loaded with stars, and 10 to 25 g of meal gunpowder is poured loosely among them before the fuse is glued in. In some cases, a mixture of gunpowder (85%) and fine charcoal (15%) is used instead, employing about twice the quantity one would use in the place of neat gunpowder.

Two methods are used to burst round shells. One method, which has been outlined already, consists of bursting a shell with a comparatively weak outside wall. The other method is to use the Japanese "warimono" technique, which requires the formation of an extremely hard exterior wall consisting of many layers of pasted kraft paper. This is adequately covered by Dr. Shimizu.

The shell maker thus has the choice of three methods of bursting round shells:

1. Employ a small gunpowder charge and a weak exterior wall to obtain the bouquet effect.
2. Employ a much fiercer bursting charge and a comparatively weak shell wall to obtain an even burst. The use of a fierce bursting charge involves the risk of destroying the stars or causing ignition failures. Flash powder sometimes is used for this purpose, but there are many drawbacks and dangers to the technique.
3. Employ a very strong outside shell wall and use a comparatively slow expanding bursting charge. Various grades of grain powder are used for this purpose in the West, but in the East they have their own distinctive techniques, as outlined in Chapter 23.

Round shells are used in various sizes, the smallest being made to fit a 50-mm diameter tube and the largest to fit a 1, 000-mm diameter tube. The most common sizes are 75 mm, 100 mm, 125 mm, 150 mm, and 200 mm, these sizes being fairly universal. Sometimes 300-mm and 400-mm shells are manufactured, but it is a byword among firework makers that their effect frequently is no better than a 150-mm or 200-mm shell. Dr. Shimizu reports that two 1,000-mm mortars are known to exist in Japan. It appears that this unusually large size is not regularly manufactured but has been a feature of an annual summer display in the city of Nagaoka. It is of interest that it has been said of these shells that they are not as attractive as one might expect.

Plastic ball shells have become very common in recent years, the molded plastic being made in such a way that the delay and lifting charge are all built into the same unit. They are less expensive and easier to fill, although in some respects paper still seems to be superior and plastic is rapidly going out of fashion.

Some manufacturers describe their shells in terms of circumference rather than diameter. Fortunately, the practice is not common and possibly is a device to deceive the uninitiated—300 mm sounds much better than 100 mm, but what a disappointment when you receive the shell.

After the shell has been filled with stars and blowing, the fuse is carefully glued into place, but care must be taken to see that the fuse is secure or it may be forced through into the shell by the lifting charge. Lifting charges vary, depending on the size of the shell, the quality of the powder, the length of the mortar, and the tightness of the fit of the shell. A 100-mm round shell would take about 40 g 4FA British pow-

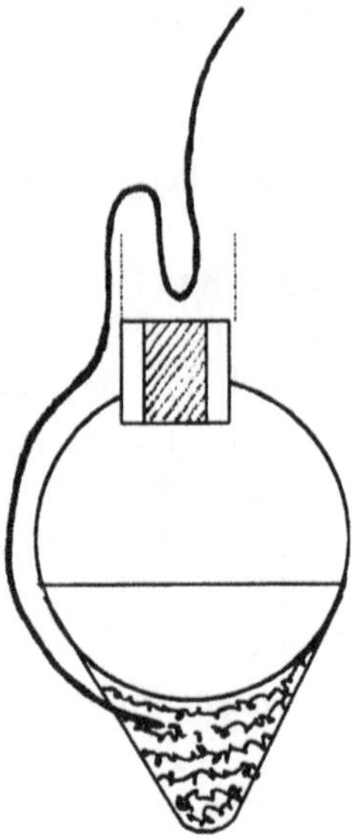

Figure 16.3 Cross section of a round shell

der, while a 150-mm shell would require about 100 g. The lifting charge usually is placed in a cone, although any substantial paper container would be adequate. The cone is particularly useful in situations in which mortars are to be reloaded because the cones fit into each other and do not take up too much room in the tube (Fig. 16.3). Finally, a length of piped match is tied into the fuse and continued down to the lifting charge so that the two will be ignited almost instantaneously. It is advisable to have two pieces of match between the time fuse and the lifting charge because a failure in this position could cause the shell to explode in the mortar. The shell fuse itself usually is wrapped with about three turns of cotton cambric to hold the match pipe because shells often are carried by their leaders and cambric will not tear as paper does. The black match at the end of the match pipe usually is covered with an empty lance tube for safety.

Three types of shell fuse are made. The older type consisted of a very hard tube about 18 mm in diameter, about 30 mm outside diameter, and about 40 mm long. A white fire composition such as the following

	Weight (%)
Mealed gunpowder	65
Potassium nitrate	25
Sulfur	10

Figure 16.4 Shell fuse

was charged into the tube in about eight increments of 3 mm when charged, each increment being rammed with many light blows until it is rock hard; 24 mm of composition burns for about 4 seconds. Solid filling is essential or there is a good chance of getting a "blow through." This fuse burns with a white light and is quite effective.

The second method uses tubes about 8 or 10 mm in diameter and about 65 mm long. Meal gunpowder, or better still a soft unpolished grain powder, is charged into the tube in the manner described previously until there is about 24 to 30 mm of solid powder. The remainder of the tube is filled with lengths of quickmatch that project about 12 mm outside the tube. The match is held in place through wedging it by breaking down the inside wall of the tube with a piercer or tying it into a piece of cambric that is glued on to the end of the tube (Fig. 16.4).

The shell moves in the air at great speed and turns end over end. These narrow bore fuses have been known (surprisingly) to burn out without igniting the shell. This happens particularly in completely airtight plastic shells. To avoid this, an indentation is made in the end of the fuse so that it will "spit" backward into the shell.

A third way to make fuses is simply to employ a piece of Bickford or other safety fuse. These fuses burn at the rate of about 3 seconds per 24 mm and vary in thickness from 6 mm to 10 mm. In England it was usual to glue two lengths of Bickford into a small wooden plug. Americans tend to use one piece of 9-mm Bickford fastened to the center of a mill board disc. The ends of the Bickford must be primed with gunpowder past or better still pierced and threaded with a short length of quickmatch, to ensure ignition (Fig. 16.5).

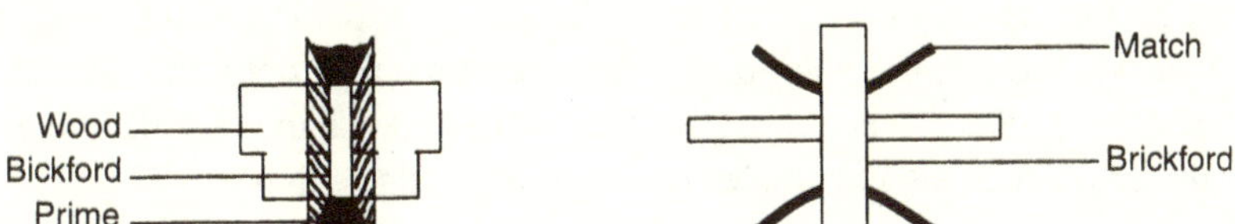

Figure 16.5 *Left:* Bickford in wooden fuse *Right:* Bickford in cardboard disc

The firing tubes or mortars for the shells consist of strong steel, fiberglass, aluminum, polythene, or paper tubes. The length of the tube varies with the type of shell, but mortars normally are 30 cm to 50 cm for round shells or short cylinders, and 75 cm to 125 cm for long cylinder shells. Paper tubes are adequate up to about 125 mm in diameter and should be stout, straight rolled tubes about 12 mm thick, fitted with a strong wooden base firmly nailed in position. Spirally wound tubes can be used, but they are less satisfactory. Cylindrical shells should always be fired from steel tubes, but maroons or salutes should be fired from paper tubes in case they explode in the tube. Steel mortars are made from ordinary steel tube fitted with a firm steel base that is carefully welded in position. It is a good idea to countersink the mortar base and fill the space with weld. Cast iron previously was not recommended, but it would be difficult to find now. In the late 1960s, the author pioneered the use of polythene (polyethylene) mortars for firing single-break shells and is still using some of the very same tubes after 30 years! They have several commendable characteristics.

1. They are lighter than metal and do not corrode.
2. They are easily washed out and are unaffected by rain.
3. They do not shatter.

A 75-mm salute in one of these tubes blew a hole in the side of the tube after "belling" the center of the tube, but it did not shatter. PVC, of course, is entirely different and must not be used because of the sharp fragments it can produce.

Mortars for small shells frequently are placed in long wooden boxes that must be very securely held with stakes to prevent them from falling over during the firing operation. Larger tubes are buried in the ground or secured in drums of sand. A thick wooden board should be placed under mortars because they tend to dig themselves in the ground and can be very difficult to remove from the wet earth or clay. Beware also of the dangers of using such a board. If you have a particularly heavy kick from a shell bursting in the tube, this can cause the wood under the mortars to act like a "springboard" and throw all of them out of the trench. It is sensible to drill a hole through the top of mortars that are subject to considerable recoil. A steel rod pushed through the holes makes the mortar much easier to lift. Mortars also become held in wet clay. A steel bar to push down the outside and get air under the tube also facilitates removal, particularly in heavy clay.

Sensitive compositions should have been eliminated by now, but when they are used, it is a wise precaution to place sandbags between the mortars and the operator or crowd because detonations have caused serious accidents. Flash charges containing chlorates and sulfur or sulfides in large cylinder shells are particularly liable to detonate and (in the opinion of the author) should be illegal in all countries.

Articles in *Pyrotechnica* (29) and Hardt (78) have described the manufacture of cylinder shells, and it must be said that some of the best effects are produced by them. Unfortunately, cylinder shells also are the most expensive items because of the tremendous amount of handwork that goes into making the more effective ones.

A former and some cardboard discs are required to make a cylinder shell, the diameter being about 12 mm less than that of the mortar. A sheet of medium unglazed kraft paper is then cut so that it is about 65 mm to 80 mm longer than the length of the completed shell and sufficiently long to go approximately three times around the former. Sometimes a thicker sheet of paper the length of the finished shell also is wrapped inside. The paper is rolled around the former and pasted just on the outer edge to form a bag. A disc is placed in the bottom of the bag, and the bottom 40 mm of paper is cut into 18-mm strips, which then are turned over as in Figure 16.6. The bag is then filled to within 40 mm of the top with stars or other effects, carefully consolidating the stars but not breaking them. Gunpowder is then poured into the stars to fill up the spaces so that the bag is quite full. This is a necessary operation because the shell wall is flimsy and gains its strength only from the filled interior. In special shells, the powder is placed only in the center. In this case, the stars are filled around a hollow brass or copper tube in the center. Powder is then placed in the center of the tube, which is slowly withdrawn, leaving the powder in the middle (Fig. 16.7).

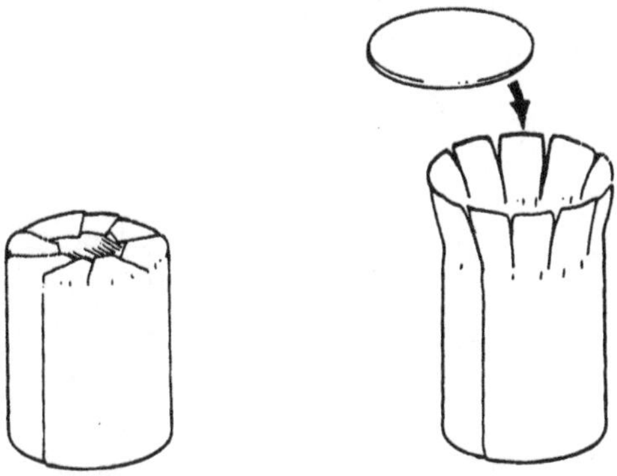

Figure 16.6 Shell bag

Figure 16.7 Filling the stars and powder

The next operation is to add the fuse that has been previously glued into the center of a disc. This disc is placed in the bag on top of the stars and gently tapped down. The shell wall is then cut into strips as at the beginning, turned down, and held in position with another disc (Fig. 16.8).

Stringing takes place next. The thin flax string is soaked in paste or used dry and wound around the shell, first down the length of the shell and then around the sides so that the whole shell is covered in squares with a side of anything from 6 mm to 24 mm, depending on the type of shell. The stringing must be even and quite taut to produce an even burst (Fig. 16.9).

The final operation consists of covering the outside of the shell with two or three layers of pasted kraft paper. Small shells require only a small quantity of paper, but larger shells take more. The shells then are completed by adding the lifting charge and leaders.

Cylinder shells are the best for carrying smaller tubes such as whistles, hummers, tourbillions, and flash charges because they are more economical of space. It is common practice to place the special effects in the bottom of the shell with a charge of stars on top. Shell tourbillions usually are about 12 mm in diameter, 100 mm long, and drilled with two holes, but, of course, they require no stick. Flash charges usually are thick tubes with wooden end plugs, one plug being furnished with a length of Bickford fuse. Sometimes a number of small shells are

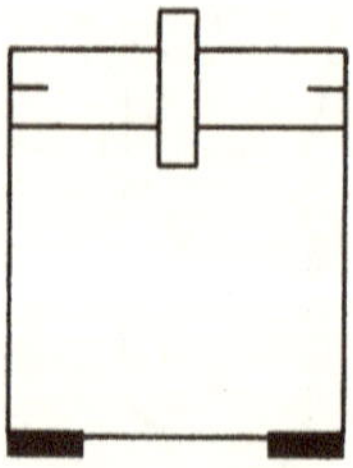

Figure 16.8 Filled bag

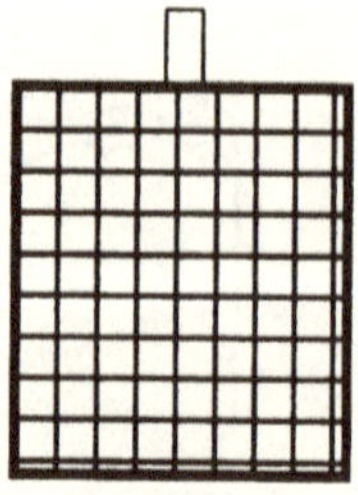

Figure 16.9 Shell after stringing

Figure 16.10 Shell manufacture

placed inside a large shell so that after the initial burst of stars, the smaller ones burst in a ring below.

The small bombettes that are used as shell inserts are many and varied, but basically they are small paper containers bound with string. The most important factor in making them is to be sure that the hot gas produced by the exploding shell cannot penetrate the paper and preignite the bombette. Paper wrapping serves this purpose very well and often is necessary even when plastic casings are used.

Small salutes often are called "saettine" (small shot) or "lampi" after the Italian tradition but not in the United Kingdom.

The bursting charge of cylindrical shells, particularly the smaller sizes, tends to use some form of flash powder almost universally. Invariably this is a small paper bag tied to the end of the main fuse (spolette). Shell makers have their own preferences, but there can be no doubt that some breaks are much more powerful than is necessary. The main requirement is that, apart from being powerful enough for the necessary spread, it also should be hot for good ignition.

The normal mixtures of potassium perchlorate and pyroaluminum are used. The author has tended to use either a mixture of two parts potassium perchlorate mixed with one part pyroaluminum and one part bright aluminum, but in recent times tends to substitute the bright aluminum with fine magnesium. In some parts of the United States, loose potassium benzoate whistle composition has been added to shells to assist the break, but it may attract moisture and may well be an added danger.

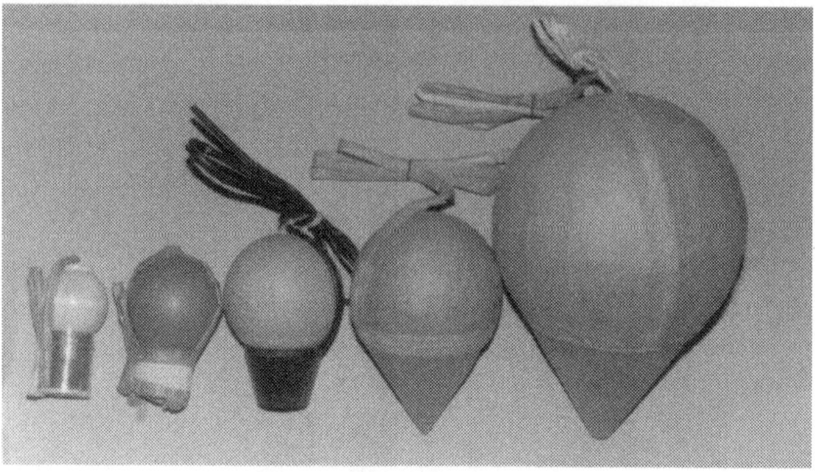

Figure 16.11 Round shells

Figure 16.12 Cylinder shells

The larger shells frequently are made without any flash burst at all. This certainly would apply to the multibreak 100-mm and 125-mm colored shells that are a feature of some well-known U.K. regattas. Henley Regatta and Cowes Regatta certainly expect these items, which are fired one after the other up to six breaks as the audience counts. A six-break 125-mm shell may not be excessive in Malta, but in the United Kingdom we regard it as approaching the limit, and when the sponsors requested seven breaks, we were a little anxious. Many years ago at Cowes, Ted Heath was the Prime Minister and I recall him waving to us as his yacht passed by. Instead of breaking the first shot at the top of the flight, we decided to shorten the first fuse and start the breaks on the way up, but it was not quite right and the well-made pill box stars (those were the days!) were seen bouncing on the deck of a yacht. We expected a nasty insurance claim, but we were lucky.

The old 200-mm comet shells contained only a single stacked ring of 35-mm comets or crossettes, and the whole of the empty space in the center was filled with lovely 2 mm to 8 mm polished blasting powder. The string was quite thick on the outside, but the outer paper was a double crown kraft that had been soaked in paste and rubbed up like chamois leather (see also Chapter 9). This produces a very tough outer layer with strong resistance. A less precise burst for round shells can be obtained with just a few layers of pasted kraft paper outside a round shell, but a central flash burst is required.

The reader is referred to *Pyrotecnica* (29) and Hardt (78) for more precise details of shell manufacture.

Inexpensive cylinder shells sometimes ar manufactured out of canisters fitted with cap ends and two discs (Fig. 16.13). The effect is not as good as the more complex cylinder shells, but is quite useful for ordinary work.

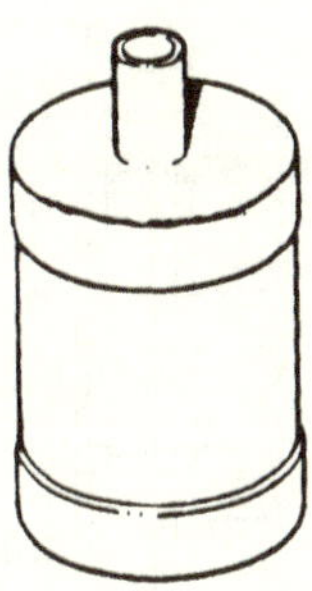

Figure 16.13 Cylinder shell with cap ends

The older methods employing tough cylinders with wooden ends are not commonly used now because they are too complex and expensive.

Repeating shells are somewhat more complicated but are made in basically the same manner as ordinary cylinder shells, although two techniques can be used. One technique involves making short cylinders all of the same diameter and height and marrying them to each other so that the fuse of one projects into the section above. The sections are all strung together, this being a vital operation for success.

The sections of repeaters usually are fitted with two second fuses. The first fuse may also be of two seconds if the shells are to burst on the way up; otherwise it is four of five seconds where the shell reaches its fullest height before the initial burst. Needless to say, the lifting charge increases with the number of repetitions, and shells bursting on the way up require larger lifting charges than the others (Fig. 16.14). The second technique for repeating shells is the one more commonly used in Europe and involves the placing of one shell inside the other. The last section to burst frequently is a flash maroon that is carefully made and strung before being placed in the bottom of the bag of the

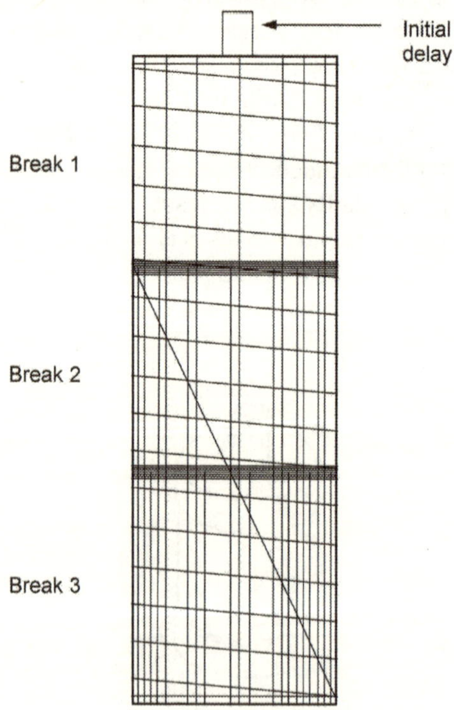

Fig. 16.14 Repeating shell

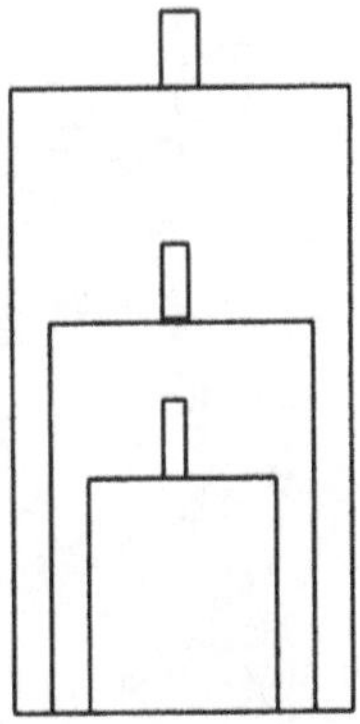

Fig. 16.15 Repeating shell

second section. Section two, when complete, then goes into the bottom of the bag of the first burst (Fig. 16.15). Careful packing and stringing is essential or the inside shell is destroyed by the main shell burst. It also means the use of many different sizes of discs and formers. Figure 16.16 shows a set of repeating shells. Further detail on shell manufacture is pointless because every shell maker adjusts his or her technique to fit the materials at hand, and the unreliability of fireworks makes her or him adhere to those methods he or she has tested in personal experience. Again, a list of types of shells is pointless because the numbers of permutations are limited and obvious.

Fig. 16.16 Set of repeating shells (Courtesy St. Michael's Firework Factory, Lija, Malta)

Fig. 16.17 Loading a 915-mm shell (Courtesy of Mr. Seiji Kase, Nagaoka-Shi, Japan)

Perhaps the 915-mm Japanese shell is the most unusual of shells. We are grateful to Dr. Shimizu for this photograph of such a shell being loaded for the Nagaoka festival by the Kashiwazaki Fireworks Co. (Fig. 16.17).

The comet shell is an interesting variation. A large pressed comet is fastened onto the end of the shell so that it will produce a spiral of fire as it rises in the air. The Japanese glue a small comet onto the outside of some of their shells, but they also sometimes coat the whole of the outside of a shell with a streamer composition.

A smaller version of the comet bomb is to press a star into the bottom of a tube about 35 mm in diameter and then place stars on the top of it (Fig. 16.18). Small units like these also are fired out of Roman candles.

Parachute shells are made in much the same manner as other shells, although the bursting charge must be cut down and the parachute must be protected with sawdust. A good way is to tie the flare into the small bursting charge, which is in turn tied onto the inside of the fuse. The rest of the shell is filled with the parachute and sawdust (Fig. 16.19).

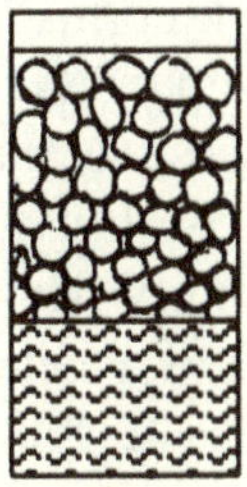

Fig. 16.18 Small comet bomb

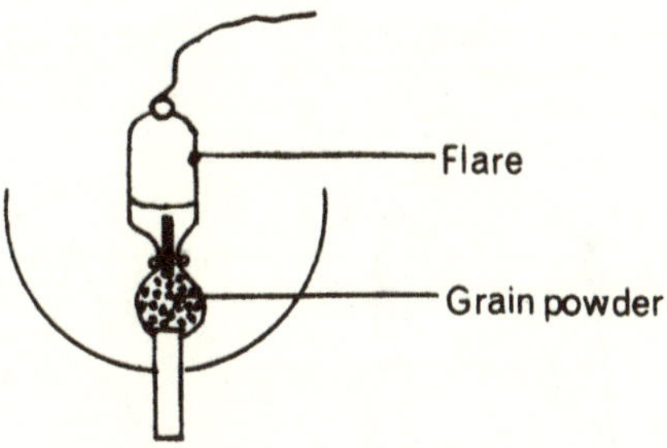

Fig. 16.19 Parachute shell

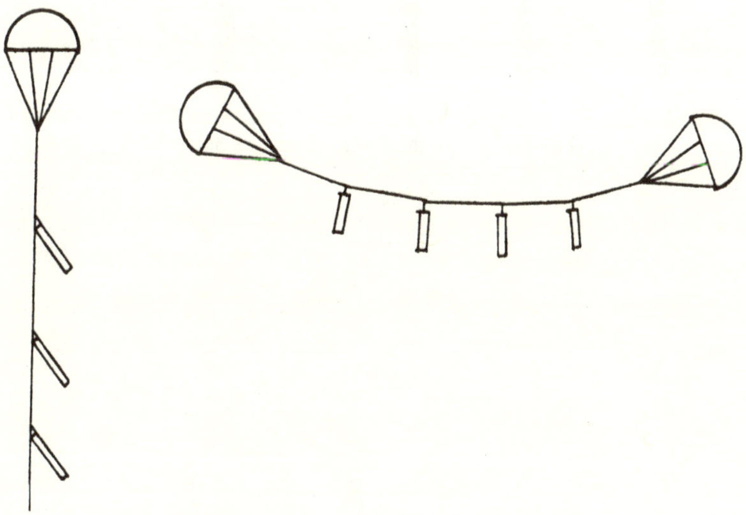

Fig. 16.20 Parachute chains or festoons

Parachutes for ordinary work normally are made of cotton cambric or good-quality tissue paper with 0.38 cm of twine tied onto each corner. These corners are tied together at their ends and fixed to about 30 cm of wire or fireproof string. This last 30 mm must be fireproof so that it will not be burned by the flare. When it is required to make a string of lights, there must be as much as a meter of string between each light; otherwise they appear to be close together in the sky. Sometimes festoons are made by attaching parachutes to both ends of a string of lights (Fig. 16.20).

Parachutes must be specially folded so that they open correctly. The following diagrams illustrate the procedure, and it is wise to sprinkle the chute with talcum powder before folding it. The folding procedure applies equally well to rockets and shells (Fig. 16.21).

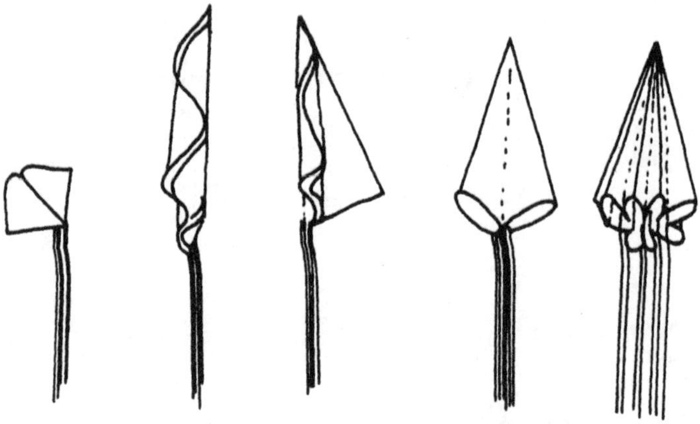

Fig. 16.21 Method of folding parachutes

17

GERBS, FOUNTAINS, RAINS, SQUIBS, AND CONES

GERBS

The French word *gerbe*, meaning a spray or sheaf of corn, is still used for firework fountains, particularly for those that produce force in choked tubes. A wide variety of fountains can be found within the firework range, but generally there are two basic types: (1) those of the long, narrow variety, and (2) the short, squat fountains. The mixtures naturally vary immensely with the type of tube and method of filling. The long narrow "fix" used for "feu croisée" in set pieces, for example, requires fierce compositions that must be carefully charged, while at the other extreme, some short, squat fountains are so slow-burning that the composition requires little more than being poured into the tube and shaken down.

FOUNTAINS

Tubes for fountains usually are about 25 cm long and from 12 mm to 25 mm in diameter. There also are other sizes that are smaller or larger than those mentioned. The wall thickness is sufficient to prevent the fire from bursting the side of the tube. Long, narrow fountains are either choked with a clay washer to one-half or one-third of a diameter or just "pulled in," as described in Chapter 9. Pulled in tubes sometimes are essential; some glitter fountains, for example, will not function correctly in clayed tubes, mainly because the solid burning products do not seem to be able to get past a clay choke, with the consequent "coking up" of the tube. Machine-choked tubes also present problems with very fierce compositions. It seems that the tube relaxes when the pressure is released after pressing, and sometimes loose powder causes an explosion at this point. This is overcome by charging the first increment with a slow composition.

The so-called display "fix" are 12 mm to 16 mm inside diameter and 25 cm long. They are choked to half a diameter and charged by hand or pressed in about eight increments. The compositions are quite fierce because they are required to produce one or two meters of concentrated fire for set piece pattern work. The following compositions are used for this purpose:

	A Weight (%)	B Weight (%)	C Weight (%)
Meal gunpowder	84	73	64
Charcoal 150 mesh	16	—	—
Potassium nitrate	—	—	8
Sulfur	—	—	8
Iron 60 mesh	—	27	20

Fountains of 18 mm to 75 mm inside diameter require less fierce mixtures, but if they are not strong enough, they will "deteriorate" lower down the tube because of blockage from dross. These compositions usually are less expensive because they contain less gunpowder, but they invariably are dirty because finely powdered materials are required. The following mixtures are fairly uncomplicated

	A Weight (%)	B Weight (%)	C Weight (%)	D Weight (%)
Potassium nitrate 150 mesh	72	40	44	27
Sulfur	8	8	9	14
Charcoal 40/100 mesh	16	—	—	9
Charcoal 28 mesh	4	—	—	—
Charcoal 150 mesh	—	8	8	—
Meal gunpowder	—	16	9	40
Coated iron 20 mesh	—	24	30	—
Titanium 40 mesh	—	—	—	10
Aluminum flitter 10/30 mesh	—	4	—	—

The old English "flower pot" is produced from a combination of red orpiment and lampblack. The burning is very drossy, with the result that bits of dross are forced into the cool air, where they sparkle with their characteristic golden coruscations. The following compositions are typical, but it should be emphasized that only certain types of gas black will function correctly and the material must be chosen by trial and error.

An excellent article in *Pyrotechnica* (XVII) well illustrates the problems of the flower pot effect, and the reader also is referred to Dr Shimizu's comments in Chapter 6.

The traditional English flower pot gerb was a long tube of about 16 mm inside diameter and 200 mm long. It was fitted with a clay choke with a hole of about 9 mm. The composition was not rammed, but charged with a funnel and rod. A good composition was:

	Weight (%)
Potassium nitrate	54
Sulfur	24
Mealed gunpowder	8
Red orpiment	8
Gas black	4
Charcoal + 150 mesh	2

A smaller fountain used to be sold in the shops under different names such as Zodiac or Orion. The tube was about 35 mm inside diameter and only about 40 mm high. It was choked with a stout cardboard washer that was forced into the top. The washer had a central hole about 7 mm or 8 mm and was closed at the base with another disc. Again, the composition was similar to the preceding and was something like:

	Weight (%)
Potassium nitrate fine crystals	55
Sulfur	25
Gas black	14
Red orpiment	6

The mixture frequently was adjusted by adding fine potassium nitrate and reducing the gas black.

A bubbling cone also was also made (lava cone). It actually looked like a volcano pouring out red lava and producing the characteristic gold coruscations. For some the effect is not exciting, but for those of discernment it is great.

	Weight (%)
Potassium nitrate	57
Sulfur	25
Gas black	9
Iron 20–40 mesh coated	9

But everything depends on the gas black or lampblack.

Silver fountains are made with titanium or aluminum. Titanium is uncomplicated, but aluminum powder varies tremendously from one manufacturer to another. Fountains made with aluminum often start off well and then become blocked with molten dross lower down the tube. As a general rule, it is necessary to use two or more grades of aluminum in the one composition and keep the quantity as low as possible.

	Weight (%)
Meal gunpowder	72
Potassium nitrate 150 mesh	7
Charcoal 40/100 mesh	7
Aluminum dark pyro	7
Aluminum 80/120	7

A rather more difficult fountain is a silver fountain using barium nitrate as the oxidizer. The tube is about 18 mm in diameter and 15 cm long. The choke must be of clay (about half a diameter), and the wall of the tube must be very strong because the temperature is very high indeed. Mixtures of barium nitrate and aluminum are, of course, sensitive, so they should not be charged with a mallet and drift, but should be pressed. Again, everything depends on the aluminum for success. The following mixture represents only an approximate guide:

	Weight (%)
Barium nitrate	45
Potassium nitrate	5

Meal gunpowder	5
Aluminum	45

The aluminum here is a mixture of dark pyro, bright polished, and flitter grades.

Glitter and flitter gerbs are modifications of the corresponding star mixes. Yellow glitter is made with sodium oxalate, and white glitter is made with barium carbonate or strontium oxalate. As a general rule, they perform well in unchoked tubes, but it is not always straightforward in long, narrow choked tubes where clay chokes should be avoided. The heat opens up a paper choke, thus preventing "coking up" toward the bottom of the tube. Manufactured gunpowder should be used for the best results, and there is an advantage in using a mixture of mealed gunpowder and soft grain powder. The following is typical:

	Weight (%)
Gunpowder	68
Antimony sulfide	14
Sodium oxalate	10
Aluminum	7
Boric acid	1

The aluminum can be flake "bright" powder + 120 mesh or fine atomized aluminum.

Compositions for unchoked tubes about 125 mm long are:

	A Weight (%)	B Weight (%)	C Weight (%)	D Weight (%)
Mealed gunpowder	67	38	66	76
Potassium nitrate	—	28	—	—
Strontium carbonate	—	—	—	10
Antimony sulfide	—	—	—	8
Charcoal 40/100 mesh	—	—	20	—
Charcoal +150 mesh	8	—	7	—
Aluminum dark pyro	8	—	—	—
Aluminum atomized +120 mesh	17	—	—	—
Aluminum atomized +350 mesh	—	—	—	6
Sulfur 40/80 mesh	—	27	—	—
Titanium 20/40 mesh	—	7	—	—
Iron 60 mesh coated	—	—	7	—

	A Weight (%)	B Weight (%)	C Weight (%)
Meal gunpowder	42	17	15
Potassium nitrate 150 mesh	—	46	45
Charcoal 40/100 mesh	—	17	—
Sulfur	7	3	6
Charcoal 150 mesh	10	—	9
Iron coated 60 mesh	—	—	8
Antimony sulfide	20	—	—
Sodium oxalate	14	—	—
Aluminum "bright"	7	—	—
Charcoal 28 mesh	—	17	—
Lampblack	—	—	12
Aluminum flitter 10/30 mesh	—	—	5

RAINS

Gold and silver rains consist of long, narrow fountains of small bore that are fierce enough to produce a good show of sparks. The tubes are about 6 mm to 9 mm in diameter and from 7.5 cm to 12.5 cm long. As they are inexpensive items, they invariably are dry-rolled, being merely pasted at the edge of the paper. Tubes are charged with a funnel and wire or other means, but it should be pointed out that dry compositions containing more than a small percentage of aluminum should not be used in a funnel and wire apparatus. The following compositions are useful:

	A Weight (%)	B Weight (%)	C Weight (%)
Meal gunpowder	75	75	80
Charcoal 40/100 mesh	25	23	5
Aluminum dark pyro	—	—	5
Aluminum "bright"	—	1	5
Aluminum 80/120 mesh	—	1	5

Short rains for rockets and shells occasionally are charged as short lances.

Squibs are no longer manufactured in Britain, but they were, in effect, golden rains with a small gunpowder "bounce" at the end. They

Figure 17.1 Flying squib

also had a practical use because they were sold to householders, who placed them inside coal-burning oven "flues" to discharge collections of soot from places difficult to reach by other means.

FLYING SQUIBS

In effect, flying squibs were fierce little drivers, but they are no longer sold to the public because of their erratic and consequently dangerous behavior. Tough little tubes about 8 mm in diameter with a "pulled in" choke were firmly charged with a fierce driver composition of gunpowder and charcoal (Fig. 17.1). When ignited, they scurried along the ground and could be both amusing and irritating. They are, however, very effective in bags, mines, and rockets, and are still used for this purpose.

Compositions:

	A Weight (%)	B Weight (%)	C Weight (%)
Meal gunpowder	64	91	—
Barium nitrate	—	—	60
Potassium nitrate	8	—	5
Sulfur	4	—	—
Charcoal 150 mesh	24	6	—
Titanium	—	3	—
Aluminum dark pyro	—	—	25
Aluminum 30/80 flitter	—	—	10

Composition C requires a starter fire such as composition B.

CONES

Cones have been included in this chapter because they really are fountains, although they usually are not charged very solidly.

The cases are not too easy to roll and normally are wet-rolled on a cone-shaped brass or aluminum former from a number of semicircular pieces of paper cut out as shown in Figure 17.2.

The narrow end of the cone is then covered with touch paper or a plain paper circle pasted over the end, and they are placed narrow end downward in a type of egg rack with holes large enough to support each cone in the center. The composition is poured into the cone, lightly consolidated by hand pressure, and then filled with sawdust to the top. Finally, a disc is firmly glued in place or, in some cases, the bottom of the cone is spun over onto the disc.

Some cones are fitted with a washer inside the cone itself to give a secondary effect, but this is not common (Fig. 17.3).

Tough textile cones are quite effective for fireworks, although the author used to be somewhat unimpressed by the huge cones seen on some markets because close examination revealed only about 50 mm of composition at the narrow end! Things have changed in recent years, however. Large cones from China with as much as 1 kilo of composition are now commonplace.

Large cones with a high composition content must be carefully pressed and quite dry at the time of pressing. Any loose composition or shrinkage through drying can cause a buildup of pressure due to burning along the composition surface inside the cone. The very shape of the composition also adds to the problem because it can be pushed downward from the narrow top to the wide base. Poorly manufactured cones can blow out the base or even explode.

Cone varieties are many and various, although, as a general rule, one is limited to the usual fountain effects—gold, silver, iron, or gold and silver. Some cones commence their performance with colored fire, but

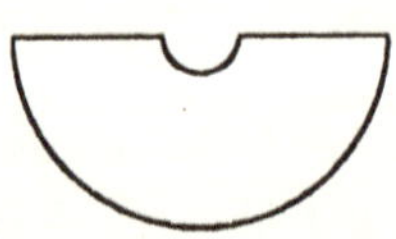

Figure 17.2 Paper cut for rolling cones

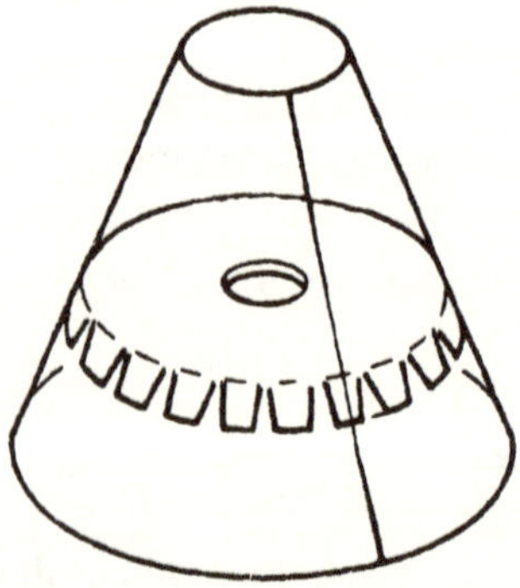

Figure 17.3 Cone with secondary effect

this should always be perchlorate if it is to be followed by a mixture containing sulfur. In fact, it is not even wise to mix perchlorate and sulfur; the articles can be knocked about on the market, and it is preferable to leave the cones as simply colored fire in this case.

Cone compositions are as follows:

	A Weight (%)	B Weight (%)	C Weight (%)	D Weight (%)
Potassium nitrate	55	54	55	—
Sulfur	11	9	8	—
Charcoal 40/100 mesh	—	13	—	—
Charcoal 20/40 mesh	19	—	11	—
Sawdust 20/40 mesh	6	—	4	—
Titanium 20/40 mesh	9	—	—	—
Iron 60 mesh coated	—	24	10	—
Mealed gunpowder	—	—	10	—
Aluminum flitter 30–80 mesh	—	—	2	10
Potassium perchlorate	—	—	—	63
Aluminum flake "bright"	—	—	—	18
Shellac 60 mesh	—	—	—	9

B is taken from Weingart and works very well. D is charged after damping with alcohol and requires a starting fire of red fire or something similar without sulfur.

CHIP GERBS

Fountains and cones have always been made with micro stars, which were made from either small pieces of colored star composition or silver granules. These were mixed with a gold charcoal mix; needless to say, stars made with chlorate were hazardous. There also also a second problem in that the stars often burned out inside the tube before they reached the outside.

The silver granules were made with barium nitrate, aluminum, and sulfur, but these have become more or less obsolete because the Chinese system using nitrates and magnalium works so much better. The nature of these micro stars is such that they light after a short delay, and this gives them time to get out of the tube. However, the compositions must be lightly charged to avoid breaking the stars and to get fast, clean burning.

The reader is referred to Dr. Shimizu's notes in Chapter 6.

Figure 17.4 Old cam shaft, weighted table charging machine (Courtesy Kimbolton Fireworks)

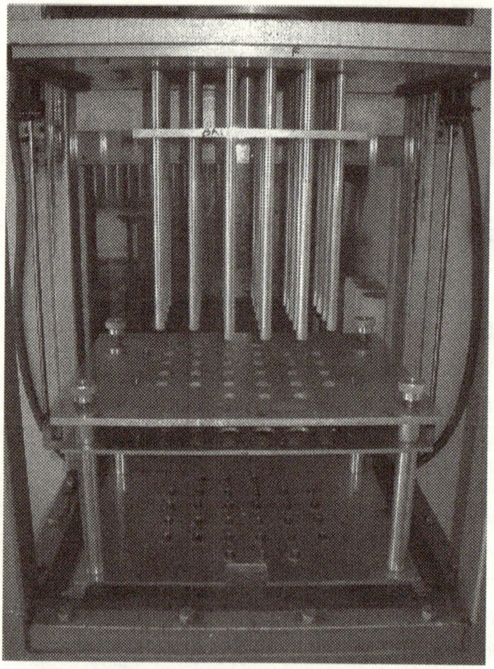

Figure 17.5 Large modern erb press (Courtesy Kimbolton Fireworks)

18

PINWHEELS AND CRACKERS

PINWHEELS

There are several ways to make pinwheels, but good pinwheels are not easy to manufacture. The usual British method employs a long, narrow pipe about 5 mm in diameter, dry-rolled against the grain, and pasted only at the edge. The length of the tube depends on the size of the pinwheel, of course, but it could be 30 cm to 50 cm long. The narrow-bore tubes are not easy to fill, but they can be charged in bundles by funnel and wire or the composition can be shaken down the tubes. In England the wheels are filled on automatic funnel and wire machines operating on a bouncing principle.

If the composition is filled by hand-bouncing, a sheet of paper is wrapped around a bundle of tubes so that it extends the open ends (one end of the tube having been previously closed with touch paper or sealed off in some other way). Figure 18.1 will make this clear.

The composition is then poured onto the open ends of the tubes until they appear to be full. The whole bundle is then lifted and dropped sharply onto the bench several times, so that the composition is consolidated in the pipes. The whole operation is then repeated until the pipes are full. Some manufacturers conduct this operation inside a container with a close-fitting lid to reduce the dust hazard.

After charging, the pipes are placed inside damp cloths so that the paper will become sufficiently damp to avoid breakage during the coiling operation. When the tube is sufficiently damp, it is neatly coiled around a cardboard or plastic disc, fixed with sealing wax or tape, and set aside to dry (Fig. 18.2).

Figure 18.1 Bundle of tubes ready for charging

Figure 18.2 Pinwheel

Compositions are a little more complex than might be expected. In fact, the mixtures must run down very narrow pipes and therefore contain a high percentage of grain gunpowder, coarsely ground sulfur, and potassium nitrate crystals. They also must be fast-burning, and they do contain a higher percentage of gunpowder than most printed formulations would seem to indicate. The following basic formulation is typical, but manufacturers naturally adjust the particle sizes of the ingredients according to their technique and granulate the composition.

	Weight (%)
Meal and grain gunpowder	74
Potassium nitrate	13
Sulfur	13

Silver wheels can be made with gunpowder mixed with about 12% of bright aluminum powder.

Some silver spinning wheels of this type employ a mixture of barium nitrate, special kinds of dark pyro aluminum, potassium nitrate, and aluminum flitter, similar to the silver gerb composition, but there are many snags to the manufacture of this firework. In the first place, the composition is very hot, of course, so it is essential to have a thicker and larger bore tube than the smaller pinwheels or it will burn through all the coils of tube on ignition. In the second place, such a mixture also is virtually a flash composition, so it will explode if there are air locks in the filled tube; this can be partially overcome by rolling the tube between heavy rollers after charging. A third drawback is that it would be dangerous to funnel and wire such a mixture, but it is very dirty to bounce it. The fourth problem is the type of aluminum to use. If the aluminum is too fast-burning, explosion is likely, but if the aluminum is too slow-burning, the wheel will not turn. These wheels are a problem, but the Germans make good ones.

English crackers are made in exactly the same way as pinwheels in that the grain gunpowder is poured over a bundle of pipes and then shaken down. The pipes are made of brown, dry-rolled kraft paper. Manufacturers vary in their tastes as to which type of grain to employ but often use FFF or F or mixtures of these and other grades.

When the pipes are full of grain, they are rolled several times between heavy rollers so that the grain is crushed and the pipes are flattened.

The extent to which the pipes are rolled is a matter of experience and can be determined only by trial and error.

After charging and rolling, the pipes are damped in the same manner as pinwheels so that they can be easily bent backward and forward into their characteristic form. Weingart and others have described the method by which the pipes are bent backwardsand forward as in Figure 18.4. Finally, the tubes are held in place by a tie of twine or adhesive tape (Fig. 18.5). Some manufacturers use machines to bend the pipes, but others think that the time taken to load the machine is almost as long as it takes to bend the pipes manually.

Figure 18.3 Finishing pinwheels. Early twentieth century (Brock's Fireworks)

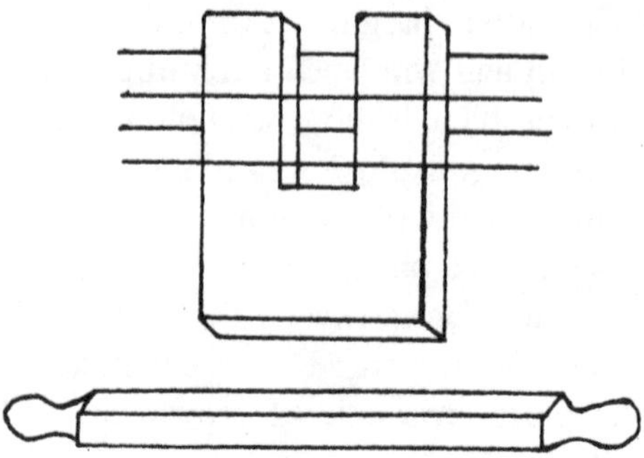

Figure 18.4 Cracker bending tools

Figure 18.5 Cracker

Large-bore English crackers were rolled and charged on a manually operated machine that is fed with a continuous strip of paper and a hopper of gunpowder. A continuous length of charged cracker piping can be made in this way.

Some European manufacturers used an entirely different method for making crackers. The method is basically as follows: The long narrow strip of paper for the tube is laid on the bench and pasted down one of the long sides; in fact, several sheets usually are fanned out and pasted at the same time. The sheet of paper is then dry-rolled around a cylindrical brass former, but only half the paper is rolled onto the former. At this stage, a specially long, narrow scoopful of meal gunpowder is carefully placed onto the paper alongside the paper-covered former and the tube finally is rolled up and stuck down at the edge. The special gunpowder scoop naturally must be the same length as the cracker pipe itself. Finally, the former is removed from the pipe, which is then rolled and completed in the usual manner. The tools are shown in Figure 18.6.

Narrow-bore English crackers and the other wider-bore crackers are somewhat different in performance. The former type tends to fizz in

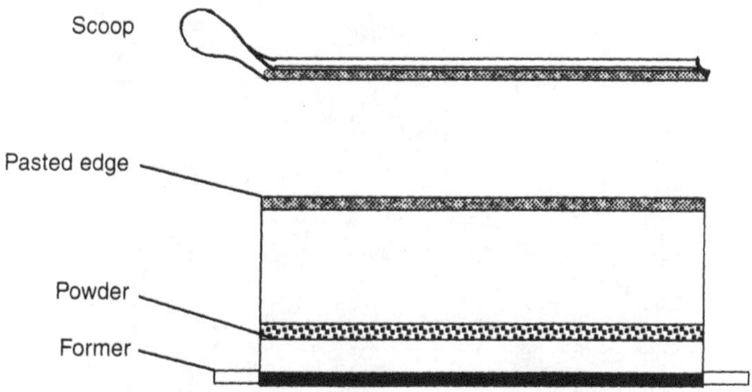

Figure 18.6 German method of cracker manufacture

straight portions of the tube and then crack at the bends in a rather leisurely manner, but the wide-bore crackers usually produce a quick succession of cracks rather like a burst of machine gun fire.

A rather inexpensive and less successful method of making crackers is to enclose a length of thick, good-quality match in a close-fitting length of match pipe and complete it in the usual way.

Figure 18.7 Filling pinwheels. Early twentieth century (Brock's Fireworks)

Figure 18.8 Finishing crackers. Early twentieth century (Brock's Fireworks)

TORPEDOES, THROW DOWN CRACKERS, AND AMORCES

These specialty items are outside the normal range of firework manufacture because they employ impact-sensitive materials that are very dangerous to handle. Silver fulminate is much too dangerous to handle commercially, and mixtures of potassium chlorate and red phosphorus are quite lethal. Although the quantities used in amorces and torpedoes are quite minute and safe to handle, the actual process of manufacture is frighteningly dangerous and should be attempted only by specialists.

19

INDOOR FIREWORKS

The manufacture of indoor fireworks tends to be somewhat specialized and frequently is undertaken by manufacturers who restrict themselves to these alone. They also have the advantage of being able to operate under less restrictive legal provisions.

The following is a brief summary of the more usual types of indoor fireworks, although this sphere of activity is outside the direct experience of the author, who can do little more than comment in certain cases on confidential industrial information. However, modern attitudes to some of these chemicals prohibits their sale, and many indoor fireworks have disappeared.

FERN PAPER

This unusual effect is obtained by soaking brown ribbed kraft paper in hot water containing a mixture consisting of 5% potassium bichromate and 2.5% potassium nitrate. When the paper is dry, a sheet about 7.5 cm × 10 cm is folded backward and forward like the bellows of a concertina and fastened at one end to a cork or a block of wood. When the paper is touched at the top with a burning cigarette, it smolders away to leave a beautiful and unusual fernlike ash (Fig. 19.1). Today it would be impossible to distribute such paper soaked in a chromate.

FIRE PELLETS

A compressed tablet of metaldehyde or hexamine will burn with a steady smokeless flame. The addition of lithium, or copper in some

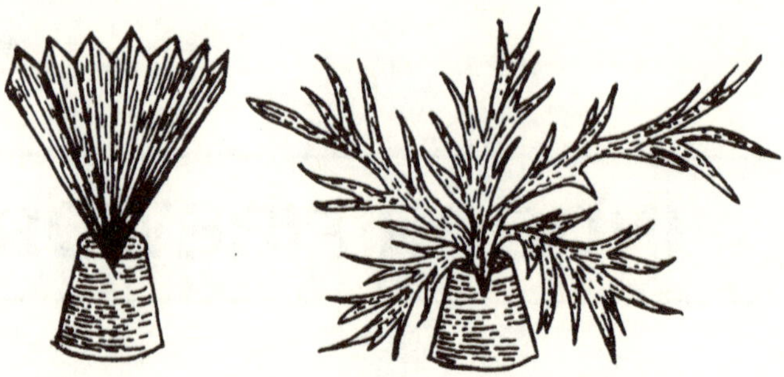

Figure 19.1 Fern paper

form, and magnesium powder will create quite attractive variations in color and effect. It should be noted, nevertheless, that the fumes produced by burning metaldehyde are objectionable in quantity and, in fact, are poisonous. Patents have been taken out for these fireworks in the past, but they have since disappeared from the market.

FIRE PICTURES

Dr. Shimizu also mentions fire pictures in Chapter 23. A sheet of semi-absorbent paper can be used for the picture, which is drawn with a solution of potassium nitrate. The result is invisible when dry, but if any part of the picture is touched with a burning cigarette, the rest of the picture gradually is burned out. Sometimes the papers also are printed in ink with additional detail. For example, a man may be depicted firing a gun. The trajectory of the bullet is invisible until it is burned on the saltpeter trail, and it culminates in the ignition of a paper amorce fixed at the end.

FLASH PAPER

Sometimes unsized paper is nitrated with the usual nitrating mixture of concentrated nitric and sulfuric acids. When it is thoroughly washed and dried, it burns with a brilliant flash. It is well known, however, that such nitrations require careful control, and the products can be very unstable unless they are carefully washed and expertly made.

FOUNTAINS

Indoor fountains are almost odorless and smokeless, being made principally from nitrocellulose. The addition of iron, magnesium, and aluminum, along with salts of strontium and copper, produces some very pleasant effects. The main problem is knowing where to obtain powdered nitrocellulose with the correct percentage of nitrogen and the correct—and very necessary—stabilizer. When conditions are not right, the fountains sometimes produce gas but no flame.

MATCHES

The manufacture of match heads has been well covered elsewhere (5), but the so-called Bengal matches do come within the scope of fireworks even though they normally are made by match manufacturers. It is well known, of course, that the choice of adhesives as well as the viscosity and the temperature of the slurry are all-important when the splints are dipped because this affects the way in which the matches dry. Some prewar German formulations set out in the B.I.O.S. Reports (12) were:

Red	Strontium nitrate	2,500
	Shellac	500
	Potassium chlorate	300
	Fine charcoal	250
Green	Barium nitrate	3,000
	Shellac	500
	Potassium chlorate	300
White	Potassium nitrate	3,500
	Sulfur	1,125
	Shellac	1,000
Blue	Potassium chlorate	900
	Paris green	300
	Resin	135

The author has never seen any blue ones, but red and green are still available in the United Kingdom.

Storm matches are simply Bengal matches that produce flame but not color. They were used extensively at sea for the older signals, which had to be ignited by ordinary methods, and are still used for lighting portfires at firework displays.

Exploding matches usually are book matches that have a tiny dab of fulminate and adhesive at the base of the match head so that they produce a sharp report soon after ignition.

SMOKES

Firework smokes frequently are used in the theater, but naturally they must be as innocuous as possible. Sometimes a smoke puff is produced by the combustion of a flash powder made with aluminum, magnesium, or zinc; larger quantities of smoke can be made with the insecticidal types of smoke (see Chapter 21).

Colored smokes are made with the usual combination of dye, potassium chlorate, and sucrose or lactose. An interesting colored smoke stick can be made by forming the composition into a slurry with adhesive and dipping the sticks in the same way as for sparklers.

An amusing little non-pyrotechnic smoke seen on the market uses celluloid to give the illusion of a toy monkey smoking a cigarette.

SNAKES

The best snakes are the black ones that are made from pitch. This curious German discovery is said to have been made by G. Vorbringer in 1867 and since then has given an immense amount of pleasure. Recent research by Adrian Shell of the United Kingdom has confirmed that there are two main types of snakes but that many substitutes have been tried throughout the ages, including one recipe using cigar ash and a cough sweet. The snakes are quite remarkable, and those made of naphthol pitch will produce an ash 25 mm in diameter and over a meter long. This pitch is the residue after the distillation of naphthol and appears to be a mixture of many compounds, including naphthol and its dimers. The material is not readily available, and its producers naturally make no guarantees for what is a waste product; samples must be obtained and tested. Weingart and Davis (6,7) have written at some length about the manufacturing procedure, so there is no need to repeat this except to add a few caveats omitted by them. Davis maintains that a mixture of roofing pitch and Syrian asphalt is a good substitute. Some manage to use ordinary coal tar pitch.

The actual nitration is an unpleasant procedure that is best carried out in an aluminum or glazed earthenware bowl. Pitch, linseed oil, and nitrocellulose powder are mixed together and the acid is slowly added. The reaction is very vigorous and copious clouds of nitrogen dioxide are evolved. Ordinary concentrated nitric acid will not work and fuming nitric acid is too strong; a small amount of water therefore is added to the fuming nitric acid, and this enables the reaction to proceed smoothly.

Weingart and Davis do not make it clear that the oxidation process is incomplete. On the one hand, the addition of too much fuming nitric acid frequently causes the mass to catch fire, destroying the whole experiment. On the other hand, insufficient acid for the completion of the reaction causes the mass to be sticky because of the presence of too much oil. The correct quantity of acid leaves a hot molten mass with no excess acid present, presumably due to the usual breakdown of nitric acid in the presence of hot carbon. Washing thus is unnecessary, although as a precaution a tiny percentage of barium carbonate sometimes is incorporated into the final operation.

After the mass has cooled and solidified, it is ground and pressed into pellets with the addition of a small amount of a lubricant such as graphite. Picric acid has been used to burn the pellets to produce the ash, but it is quite useless commercially because it stains the fingers and clothing and has a bitter taste. Nitrocellulose is best and can be incorporated into the nitration. It appears that the Japanese have used ammonium perchlorate.

When the snake pellet is ignited, it burns with a sooty orange flame, and a smooth snake with a metallic luster winds out of the flame. The shiny outer skin has small carbon particles on it; inside is a mass of bubbles (vesicular). The snake is composed of an aromatic polymer system swollen to many times its original size by gaseous decomposition products.

The snake-in-the-grass is an interesting variation that consists of green Cr_2O_3 from the thermal decomposition of ammonium dichromate $(NH_4)_2Cr_2O_7$. A small cone made of aluminum foil is filled with a mixture of ammonium dichromate mixed with a small amount of fuel and oxidizer. At the base of this, a pinch of loose black snake composition is added before the cone is sealed off. On ignition, it produces a spray of green chromic oxide followed by the black snake (Fig. 19.2).

Smaller white snakes called Pharaoh's serpents used to be made with mercury (II) thiocyanate (sulfocyanide) and apparently were discovered serendipitously and independently by both Berzelius and Wöhler in 1821. This obscure salt is insoluble in water and can be made by adding potassium thiocyanate solution to mercury (11) chloride solution to which a small amount of iron (III) chloride has been added, until

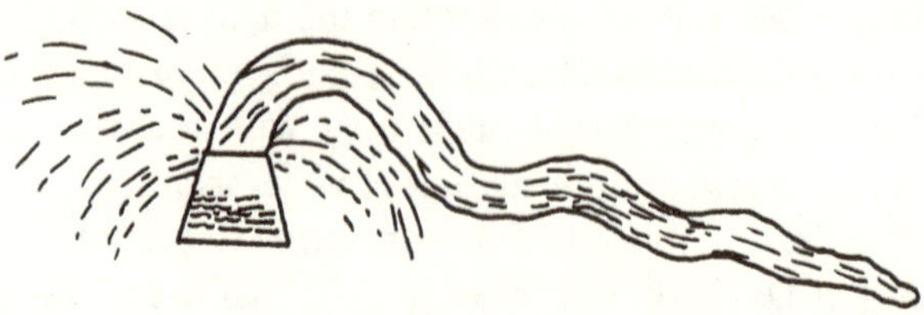

Figure 19.2 Snake-in-the-grass

the first permanent red color remains on stirring. When the precipitate has been washed and dried, it is made up into tiny pellets with a solution of gum arabic and water; a little potassium nitrate is added. This snake is less exciting, is more brittle, and has the disadvantage of being poisonous, as a certain young Prince O. discovered in 1865 when he ate a snake pellet mistaking it for a sweet; he survived the poisoning, but it killed off his tapeworm!

The pellet burns with a blue flame, giving off mercury vapor, and expands (intumesces) to approximately 50 times its original length. The composition of the snake is quoted as $C_6H_3N_9$ and is orange-white on the outside and gray-black on the inside. It is said to consist of a mixture of melam, mellone, mercury sulfide, and mercury. The pellets at one time were placed in wine corks and into porcelain caricatures so that pigtails and curly tails emerged once lit!

Outside the pyrotechnic world it is well known among the chemical fraternity that a snake can be made from 4-nitroacetanilide. To achieve the snake, heat a 1:1 mole ratio of concentrated sulfuric acid with (3 g at most) 4-nitroacetanilide in a beaker. This gives an intumescence of 100 times the original size. The snake consists of a polyaniline-like polymer swollen by gas. It is black and almost metallic in color but more wartlike than the smooth naphtha snake.

For obvious reasons, only the black snakes are available today. Germany made the best ones, but they are no longer being manufactured.

SNOW CONES

These little cones are made in exactly the same manner as the snake-in-the grass. Circles of aluminum foil are cut out with a radial cut as shown in Figure 19.3, and after forming into cones, they are placed in a board with a series of cone-shaped holes of the same size. When the cones are filled in this way, they retain their shape and are easily closed up at the base by simply bending the foil over on top of the composition.

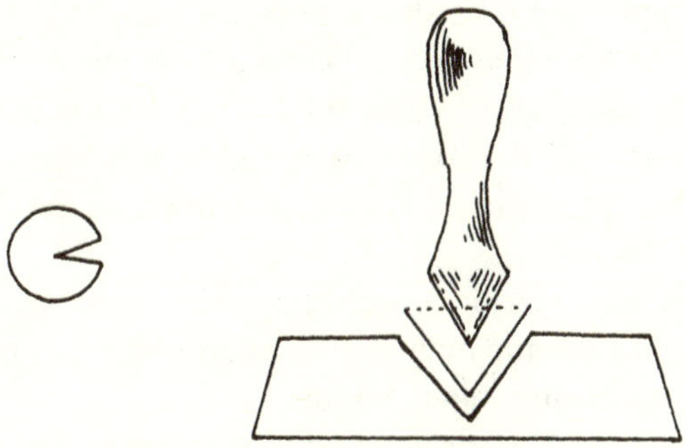

Figure 19.3 Tools for snow cones

The snow cone merely burns magnesium powder with the production of large flocks of white magnesium oxide, although why any person should wish to allow such a thing to happen indoors is anybody's guess. The basic composition is:

	(%)
Fine magnesium powder	40–50
Potassium nitrate	5–10
Flour or woodmeal	about 40

Small tablets of metaldehyde also produce a snowlike effect when placed on a cigarette due to the sublimation of the metaldehyde.

SPARKLERS AND FIRE STICKS

The well-known sparkler is basically made of barium nitrate, aluminum, and steel:

	Weight (%)
Barium nitrate	50
Dextrin	10
Steel	30
Aluminum powder	8
Charcoal 150 mesh	1/2
Neutralizer	1.5

The materials are ground with suitable adhesives and water to make a thick slurry. Wires are fixed into wooden frames to space them suitably before dipping into the slurry. Sometimes the sparklers have to be dipped twice. As might be expected, the drying process is quite important and usually takes quite a long time to prevent the mass from cracking. The iron usually is coated in good-quality sparklers, although judging by the appearance of many commercial products, this frequently is not the case.

The best quality sparklers are dipped like matches at an elevated temperature, adding some hide glue as an adhesive. Dipping techniques require some skill using this method.

Other fire sticks also are made, but, of course, they should not be burned indoors as the compositions suggest. The following composition is from an earlier American Patent No. 1,936,221 (1933):

	Weight (%)
Potassium chlorate	46
Barium nitrate	17
Strontium carbonate	12
Shellac	11
Cryolite	8
Dextrin	6

The outside of the composition is coated with magnesium-aluminum alloy grit. Weingart (6) gives other compositions, but these have not been tested by the author.

TABLE BOMBS

Table bombs are, in effect, little mines filled with toys and other items. The lid is fixed quite lightly so that it will not blow off violently, and instead of blackpowder, a nitrated cotton is used to eject the toys. The cotton is ignited with Bickford fuse (Fig. 19.4).

THEATER FIRES

Smokeless fires sometimes are required, but it is not possible to produce good colors without smoke or smell. The following compositions sometimes are used, but they must be kept dry:

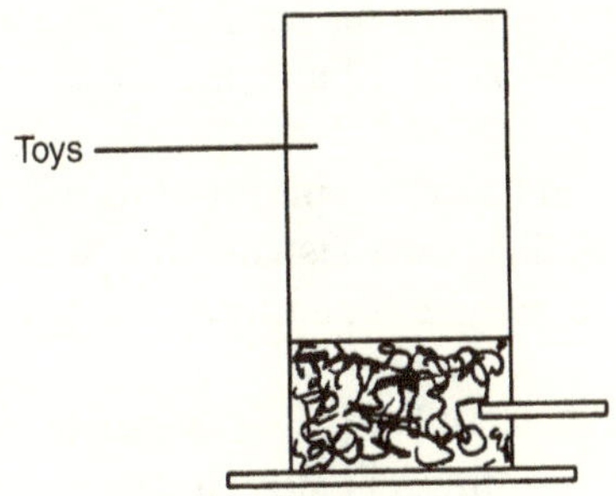

Figure 19.4 Table bomb

	Weight (%)		Weight (%)
Strontium nitrate	68	Barium nitrate	56
Potassium chlorate	24	Potassium chlorate	26
Shellac 30/120 mesh	8	Shellac 30/120 mesh	18

THROW DOWN CRACKERS

Thrown down crackers consist of small twists of tissue paper containing sand and a small amount of silver fulminate. They crack sharply when thrown onto the pavement. Large quantities are manufactured in Brazil. Packaging of these items is quite an important manufacturing procedure.

PROXIMATE SPECIAL EFFECTS

A number of new words have entered the English language in recent years, and although they use the same format, they have tended to be used differently. The word *proximate* is quite a new word in the United Kingdom, although, clearly, in the sense of "close by" the word is being used for theatrical and special effects. Strange as it may seem, pyrotechnics tended to fall into disuse with the advent of modern cinema and lighting effects, but now they are back in a big way.

The smell of firework smoke, particularly that of gunpowder or similar mixtures with a low sulfur content, was not particularly unpleasant to most people, and it had a very evocative association for many people. Needless to say, there were exceptions where the smell from blue flames was particularly unpleasant, and it was somewhat surprising to discover that blue sparklers were being manufactured with copper compounds.

It was inevitable that the development of theatrical pyrotechnics and the constant use of well-known display sites would lead to questions about the problems of pollution and the dangers that might be caused to operators.

In the past questions would not be asked about either toxic or irritating emissions, but it is a different story today, and there is much ignorance and speculation about the real or anticipated dangers. Years ago the smoke from the annual pantomime came from magnesium flash-powder and there was not too much of it, but today the audience can be bathed many times with great clouds of oil smoke, and some people are more likely to worry about it than others.

A number of articles have been written about the nature of the gases and dusts that are spread by pyrotechnics, and the form they take tends to reflect the point of view of the authors. An article such as the one written in the *Journal of Pyrotechnics* (81) tends to take the broadest of outlooks and takes the widest view of materials that could be used.

It also found that the industry itself tended to be somewhat secretive, which has the potential for observers to take the worst view, even to the point where regular operators at big events had to wear respirators. We all know that Health and Safety people often are humorless, but respirators and entertainment do not exactly go hand in hand.

The burning products of materials such as gunpowder and nitrocellulose are mainly producing oxides of nitrogen and carbon, and there may be traces of hydrogen sulfide and dust products. It also is clear that there must be traces of strontium, barium, antimony perhaps, and some chlorine compounds, but it would be an irresponsible manufacturer who used mercury or lead these days.

An article in the United Kingdom condemning the pollution from fireworks on November 5th or New Year's Eve stated that there would be tons of mercury, cadmium, and chromium in the atmosphere over Stockholm at New Year. We had to point out that mercury had not been used in more than 50 years, but that there would be a regular supply from the local crematoria from silver amalgam tooth fillings and interesting products from cremated pacemakers; cadmium is not used in fireworks, and small amounts of chromium might come only from magnalium passifyers used in wet processing. Oxides of aluminum and magnesium could well be irritating, but an occasional exposure probably would be less than that consumed with an antacid tablet.

Accepting that responsible manufacturers of proximate materials are constantly seeking to minimize risks to audiences, and accepting that basic formulations are at this stage similar to traditional ways, questions have been raised about the use of antimony, barium, and lead. The former can be easily avoided for indoor use, and the latter may be in only small amounts in electric fuses, but barium is not as easy to avoid for green colors.

At Lake Buena Vista at Orlando, Disney tested mud in the lake after fireworks over 10 years. The levels of antimony and barium were no greater than normal (81).

Also in Germany, tests on barium in the soil found that it was converted to insoluble sulfate and carbonate and had no significant effect on plants and animals. Only concentrations of 1,500 mg per kilo of soil started to have any toxicity.

Modern concerns are understandable, and there must be a sensible balance between the extremes. Knowing how people feel, industry must do its best. It might be lovely to live on an unpolluted island for a short time until we became bored, but it might not necessarily prevent a heart attack.

It might well be that the greatest dangers come simply from the use of theatrical effects where a number of well-known events have taken place through lack of care and judgment. The height of the sparks and flames, low ceilings, and flammable materials that can take fire have all contributed to the experience of the industry. However, it is not only a case of unsuitable situations, but also the lack of thought in setting up electric circuits where operators have been unexpectedly injured or even the alteration or use of manufactured products that were not designed for the work at hand. A famous case in the past in the United Kingdom involved a large flash pot that was placed inside a piano. The explosion wrecked the piano and seriously injured the player. To make things worse, however, it seems that wheat flour had been placed on top of the flash powder, supposedly to enhance the visible effect.

Approved Codes of Practice along with regular training schemes have done much to improve the industry, and this along with the detailed instructions for use has done much to improve Health and Safety at Work. Too many instructions can sometimes in themselves be an additional hazard, but in the United Kingdom "The Labelling of Dangerous Substances Acts" along with labeling requirements under the "British Standard for Fireworks" were one of the more useful aspects of legislation in recent years.

Further discussion of compositions in proximate pyrotechnics also is pointless because the principles are discussed elsewhere in the book. The fact that a silver gerb is made with a gunpowder-type mixture and titanium is one thing, but the height of the sparks and the actual, precise burning time may be of greater significance. Similarly, the amount of flash powder and the number of decibels produced by a salute may require careful precision, depending on the application.

The United States has a BATF "Mixing Binary Explosives" license. This allows the mixing of small quantities for use in special effects. Packed in two containers, one being a fuel and the other an oxidizer, there are clear advantages for classification and transport. There also are inherent dangers, and many countries do not accept such material.

A useful small book, *Proximate Special Effects. Familiarization and Safety*, by Mattingly, Opperman, and Pinkerton is published by American Firework News (89).

20

FUSES AND QUICKMATCH

SAFETY FUSE

In the early days of mining, many accidents were associated with the blasting operations that used mainly blackpowder. These powder charges were ignited with quills that were filled with gunpowder. It was an extremely hazardous and unreliable operation due to the fact that some quill fuses flashed off irregularly and in other cases breaks in the powder train led to a dangerous misfire. The matter was further complicated by moisture, and there were many instances in which miners returned to misfires too quickly only to be blasted by the explosion that followed.

William Bickford was born in Devonshire in England in 1774 and was baptized in Bickington. Earl (26) tells us that Bickford later moved to Tuckingmill near Camborne in Cornwall, where he became a leather merchant in the heart of the mining area. Earl makes the point that Bickford was a humane man who was particularly distressed by the mining accidents and set about trying to make the ignition systems more reliable. While visiting a friend who was a ropemaker, he conceived the idea that it might be possible to pour gunpowder through a funnel into the center of a rope as it was being spun. A Methodist, Bickford further discussed the matter with another Methodist friend who was a worker miner called Thomas Davey, and after some trials they were delighted to find that their method of making fuses worked.

The first fuse to be made was simply a continuous core of powder that was surrounded by a number of flax yarns. The yarns were twisted as they were pulled down around the continuous core. The next stage was to place a second layer of yarns around the first stage but in the opposite direction when twisted. The final stage was to pass the fuse through a bath of tar to hold the fuse together, promote even burning, and keep out moisture.

The first product was called "The Patent Safety Rod" and was 10 mm to 13 mm in diameter. British Patent No. 6159 was taken out in 1831. Unfortunately, Bickford was taken ill in 1832 and died in 1834. Bickford made an enormous contribution to safety with his invention and set up a business at Tuckingmill that remained with many changes until July 1961.

The Bickford family had some problems protecting their valuable invention, but many changes were made and in time the manufacturing method was turned into a continuous process. Earl tells us that in the beginning they were making about 72 km a year. In 1930, however, they were making 167,272 km a year, and that did not include the material made by competitors.

The biggest change in the process came in 1840, and after that the method has remained pretty well unchanged until the present time. Quoting Earl (26), machines were made up from cast iron framework, a brass powder funnel being fixed at the top. The powder from this fell into the middle of the flax, later jute yarns, that were brought together to form a tube in a cone-shaped hole "die" machined to the diameter needed for the first spin cord. The die was arranged vertically and coaxially in the center of the circular rotating "plate" that carried the bobbins of flax yarns. Each thread was guided into place in the die by small holes drilled down at an angle through the plate into its central axial opening, ensuring that they were evenly spaced as they entered the die. From here the cord, with its gunpowder core, was drawn down and away by a pulley that was geared to turn at the right speed in relation to the spinning plate so that the desired pitch-lay was given to the spinning yarns. The cord so made then was countered by a second plate—the "countering plate"—charged with bobbins of yarn, turning in the opposite sense, rotating faster for a given "draw down" (length of fuse) so as to give a steeper pitch to the threads. Finally, the semifuse rod so made was "struck in" to a varnish mixture as in the original method. An outline of this process is given in the French patent taken out by Bickford, Smith, and Davey during 1842, when Davey's factory at Rouen was going into production. The fuse was world famous by this time.

Two kinds of fuses in were in production: the common fuse and the sump fuse. The latter was for use in wet places or under water and was much the same as the common fuse except that it had extra cover and varnish. A rather larger diameter fuse of 10 mm diameter has been made in the United States even in recent times. This has been particularly useful in the firework trade for making shells. However, the fuse did reduce in diameter over the years and normally was about 4.5 mm in diameter. Earl makes the point that in the early days, the fuse cost three pence per fathom. Post decimalization of the U.K. currency, this would be 1.25 pence per 1.83 meters. In 1990 it could cost 30 pence per meter.

In the older Bickford, the powder charge was 11 to 12 g per meter, and Earl quoted two compositions:

	Common Fuse Weight (%)	Sump Fuse Weight (%)
Potassium nitrate	73.0	77.0
Charcoal	15.5	13.5
Sulfur	11.5	9.5

Inevitably, the main problem with manufacture was the possibility of having gaps in the powder core that would cause the fuse to "hang fire," This problem was overcome by introducing two central identification cotton strands into the center of the powder stream. By making these cotton strands slightly damp, the powder actually stuck to them.

The burning time of the fuse powder was determined by charging the powder into a lead pipe about 300 mm long and about 6 mm in diameter. This was then repeatedly rolled out until it was just over three times its length, and finally a length of 0.9144 m was timed by burning it on a sand tray.

Even though the fast-burning or "running" fuse had its problems for the makers, the time came when there was a demand for an instant fuse and this was available in the United Kingdom until World War II. Large quantities of old stock of this material were used up in the postwar years as the first fuse for lighting the lifting charge on firework shells. This fuse was developed in the Davey plant at Rouen, and it was colored orange to distinguish it from the other fuses. This fuse was made by spinning three strands of quickmatch into an impervious cover. This technique provided an airspace between the quickmatch and the outer cover, with the result that an increase in pressure caused the fuse to explode. The manufacture of this type of fuse was discontinued after the early 1950s because of the new detonating cords.

The firework maker primarily requires fairly short lengths of fuse that provide delay times up to about 10 seconds at the most. Bickford fuse serves this purpose well and has the additional advantages that it is reasonably waterproof, the fire is not communicated laterally, it is easily cut and pierced, and it is tough enough to block up narrow bore tubes in addition to acting as a delay.

Bickford fuse normally is about 4.5 mm in diameter. This narrow bore is commonly used in small explosive fireworks and ignites quite easily from a gunpowder charge. To ensure ignition, the fuse often is cut at 45° and then primed with gunpowder paste.

The best way to be certain of ignition is to punch a hole through the center of the Bickford from one side to the other and then thread this with a length of quickmatch. The small diameter of the British product does not readily lend itself to this technique, but the larger 9 mm sizes made in other parts of the world are excellent for this purpose (Fig. 20.1). An excellent plastic-coated Bickford fuse is made by Martins and Martins of Portugal.

In addition to making the larger size of Bickford fuse, Americans also make a self-consuming type of Bickford fuse that is useful for firework purposes where it is intended to ignite a succession of items. This Red Visco, as it is called, has only a small amount of yarn on the outside and is finally coated with a layer of red nitrocellulose dope. These fuses are now extensively made in China.

More recent times have seen the introduction of plastic igniter cords. Again, according to Fordham (13), the fast cord burns at a rate of about 1 second per 30 cm and consists of paper or textile yarns coated with blackpowder at the center. This is then covered with a plastic incendiary composition beneath an external coat of waterproof polyethylene. The slow cord is made in a similar manner except that a copper wire replaces the blackpowder center of the fast cord. The copper wire conducts heat from the burning front into the unburned composition, thereby controlling and speeding up the rate of burning. The slow cord burns at the rate of about 10 seconds per 30 cm. British plastic incendiary

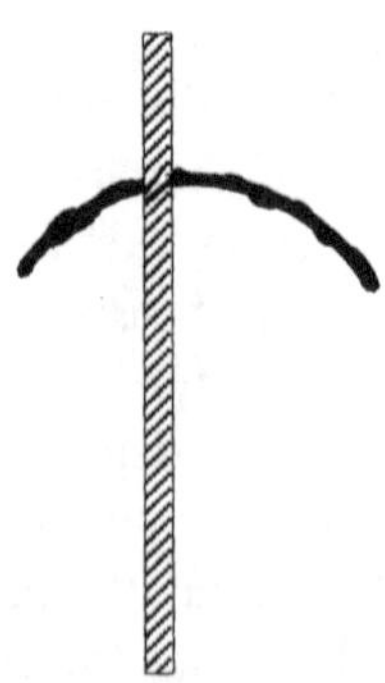

Figure 20.1 Bickford fuse threaded with quickmatch

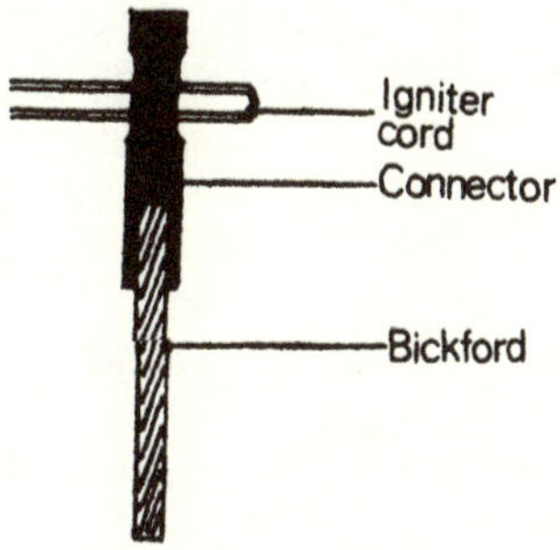

Figure 20.2 Method of connecting several pieces of Bickford fuse

compositions usually are made from potassium nitrate or potassium perchlorate, red lead, and silicon with a nitrocellulose binder. Igniter cords are now much more expensive than piped match, but they have the advantage of being waterproof. On the other hand, they are much slower than good piped match and burn irregularly in long lengths, appearing to hang fire momentarily in an unusual manner. A slight disadvantage of the igniter cord is that when it is used for lancework, it is more likely to be accidentally ignited from dross than paper match pipe. However, igniter cord is very useful for igniting a number of lengths of Bickford fuse by using special bean hole connectors, which also are made by I.C.I. The technique is shown in Figure 20.2. Igniter cords also are clean to handle and do not drop bits of blackpowder when cut. Fast and slow cords based on blackpowder also are made by Wano of Germany.

Firework makers sometimes make their own time fuses with blackpowder. Narrow-bore tubes with strong walls are charged with fine-grained powder. Each increment is rammed with about 12 blows of a mallet, and there are about 8 increments per 24 mm. Much, of course, depends on the powder that is used, but such fuses usually burn at the rate of about 3 seconds per 24 mm. These fuses also can be pressed.

Occasionally fuses are made with modified gunpowder mixtures. For example:

	Weight (%)
Potassium nitrate	25
Sulfur	10
Meal gunpowder	65

Fuses for small mines and cannon crackers are long, narrow-bore tubes charged with a reasonably fast composition of the golden rain type. For example:

	Weight (%)
Meal gunpowder	75
Potassium nitrate	20
Sulfur	5

The composition is charged with a funnel and wire.

QUICKMATCH

One of the most essential items to the firework maker, quickmatch also is one of the most unpleasant to make. Cotton strands are passed through a slurry of gunpowder, adhesive, and water and then are wound onto a drying frame and allowed to dry.

The match usually is made up from 4 to 10 strands of cotton that are run off reels into a bath of alcohol and water, which thoroughly wets the cotton. As the cotton leaves the bath, it passes between rollers, which remove the alcohol solution before passing into the gunpowder slurry. As the cotton leaves the slurry, it passes through a small funnel that removes the excess.

The match is wound onto large wooden frames, about 180 cm long and 120 cm wide, where it stays until it is dry. Frequently, the box of slurry moves along a threaded rod, which in turn is geared to the winding frame so that the match is carefully spaced on the frame as it leaves the slurry container.

When the frame is full of match, it is dusted with fine-grained powder while it is still wet. Manufacturers use a good-quality powder for match, (e.g., FFF), but they vary in their choice of adhesives. European firework makers tend to favor gum arabic or dextrin, but there is a preference for boiled starch solution in Britain. Perhaps the damp British climate is the reason for this preference since there is no doubt that a match made with dextrin becomes quite limp in damp weather.

A typical match formulation would consist of approximately 100 parts of grain powder mixed with 100 parts of water that have been boiled with 6 parts of starch. Another formulation used 1 part of gum arabic in 4 parts of water to which is added 8 parts of sulfurless gunpowder.

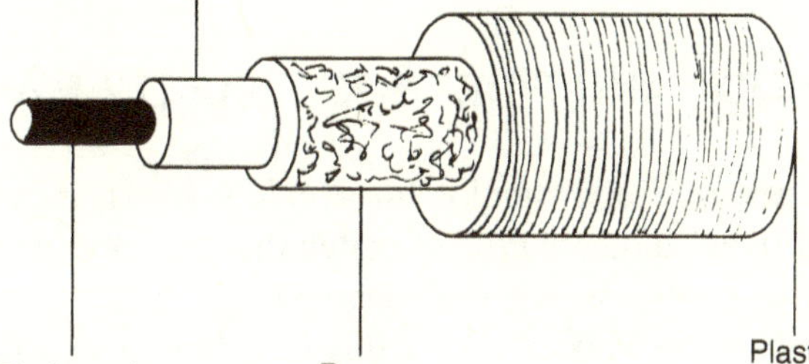

Figure 20.3 Mexican protected match

Good match burns steadily and fiercely at the rate of about 5 seconds per 30 cm in the open. When placed in a match pipe, however, it is almost instantaneous.

Match pipes are made from brown kraft or white sulfite paper. The tubes are dry-rolled on a former about 6 mm in diameter; they must be dry-rolled because they are bent so much when in use, but the diameter of the pipe is not especially important. Some manufacturers make their match pipe in continuous lengths from rolls of strip paper, but there is still a tendency in Britain to roll the pipes in 0.5-meter lengths and join them together as required. For this purpose, the former is slightly tapered so that the narrow end of one pipe will fit into the wide end of the next (Fig. 20.3). The newer continuous processes for making match pipe use plastic-covered papers (waterproof) and aluminum-covered papers (sparkproof).

TAPE MATCH

It is not clear whether the U.K. firm of Standard Fireworks was the first to use this type of fuse, but Edward Greenhalgh certainly first made tape match commercially in the late 1950s. It consisted of a train of grain gunpowder sealed in the center of a tape, which consisted of two pieces of paper that were stuck together. It was extremely useful for conveying fire around units in Roman candles, for example.

A less sophisticated tape also is marketed in the form of a train of grainpowder down the center of wide "scotch" tape. Many people use this tape for connecting lances on set pieces, combining both speed of burning and ease of application. It also has some drawbacks in that it dribbles gunpowder, is easily ignited by stray sparks, and does not look very elegant.

WATERPROOF AND FIREPROOF SAFETY MATCH

Dr. Pablo Hernandez of Articulos Pirotécnicos Mexicanos has kindly provided information on the type of match that they use in their special firework pieces known as castillos (page 67).

Firework shows in Mexico are mainly based on this *castillo*, or castle. It is a structure of 15 to 20 m in height on which a great variety of wheels, logos, and frames are set up and burned in sequence. It can burn as many as 15 times.

The show must be synchronized and needs to keep a burning sequence that allows gerbs, drivers, lances, and special effects to be burned as they have been programmed.

However, in varying conditions of rain, humidity, and the close proximity of the fireworks, it has been necessary to look for other ignition options in order to guarantee harmony and continuity.

It meant that the lances, drivers, and quickmatch had to be protected from fire residue, hot ashes from nearby fireworks that would alter the burning sequence. There also must be protection from humidity and rain, both of which are enemies of quickmatch (Fig. 20.3).

Plastics, aluminum foil, different waxes, and paper nosings to protect primes have all been investigated so that quickmatch can maintain its necessary characteristics:

- Flexibility
- Waterproof
- Fireproof
- Minimum ash residue
- Burning speed

Flexibility

It is essential that blackmatch have the correct amount of gunpowder and binder so that the cotton strings are completely impregnated. If there is insufficient binder, the blackpowder will become detached and spoil the fire transmission. Conversely, too much binder will make the match brittle and burn slower.

The paper and plastic pipe is also important because it protects the match. The paper should have a long fiber so that it is strong under tension, this being further reinforced by the plastic cover, which helps to prevent rupture.

This flexibility allows the work to be done quickly, but more importantly, there is no need to fix the quickmatch since the tube does not break easily.

Impermeability

The plastic cover affords perfect protection from the weather, but it also enables work to be done more quickly (e.g., using plastic tapes to join up pieces of quickmatch).

Fire Protection

The aluminum foil pipe around the quickmatch is vital to protect the match from fire residues and hot ashes from other performing fireworks. It also aids the continuity of the show and prevents preignition of other devices.

In a recent test, a length of blackmatch was wrapped around this safety match and ignited. The safety match did not ignite after ignition and even subsequent cooling.

Residual Ash

After the safety match has burned, there is very little residue of any size and very few ashes. This is because the protective aluminum foil is in intimate contact with the blackmatch, thus preventing the propagating flame from igniting the paper and plastic layers. Since they do not burn, there is little ash.

Transmission Speed

Accurate control is required becauses it is necessary to have a definite burning speed in terms of meters per second. It is related to the size of the explosion, and consequently the explosion increases with the burning speed.

All drivers and gerbs are fitted with a paper nosing. This protects the prime and allows the fuse to be tied in to give a seal and avoid accidental fire and moisture. Good safety fuse is vital because if it burns too violently, the explosion may be so great that it prevents ignition.

ELECTRIC IGNITERS

Modern ways of presenting firework shows require the use of huge numbers of electric igniters. These fuses are extremely useful for a number of reasons. In the first place, they allow the operator to maintain a reasonable distance from the object to be ignited; more importantly, they allow many items to be fired instantaneously over a large area of ground.

Electric igniters basically consist of a blob of pyrotechnic composition on a bridgewire across copper foil that is fitted with two leg wires. The pyrotechnic composition usually contains lead thiocyanate or lead mononitroresorcinate, which is then covered with a couple of layers of protective lacquer coating. Ideally, this matchhead is covered with a protective shroud in the form of a plastic tube.

As with most match compositions, these items are sensitive to impact and friction and most certainly have been the cause of a number of accidents. The most dangerous types are the bare match heads, which undoubtedly are convenient for slipping into piped match or inside certain manufactured items, but they are easily ignited with a sharp knock. Even though most workshops have rules about the storage and movement of match heads, it is quite perverse that a bent piece of leg wire will contrive to get the bead between the lid and the top of a container. One accident in the United Kingdom probably was caused by cutting a bead with a pair of scissors; the bead was buried in the tied match of a group of shells returned from a display.

Experiments conducted by K. Kosanke and B. Kosanke (*Journal of Pyrotechnics*, No. 15, 2002) revealed that match heads with shrouds were less sensitive to impact by a factor as much as 5. This study also considered, among other things, the well-known anxiety about the suspected sensitivity of these items to electrostatic discharge. Not only is it a fact that humans vary somewhat in this respect, but also conditions of humidity, radio frequency, and other factors mean that there is an obligation to treat these items with great respect. Lest this sounds rather trite, it must be said that these are items that frequently are not treated with respect. Many have been the occasions when the author has seen bundles of igniters thrown into boxes of tackle or loose on the floors of vehicles before and after displays.

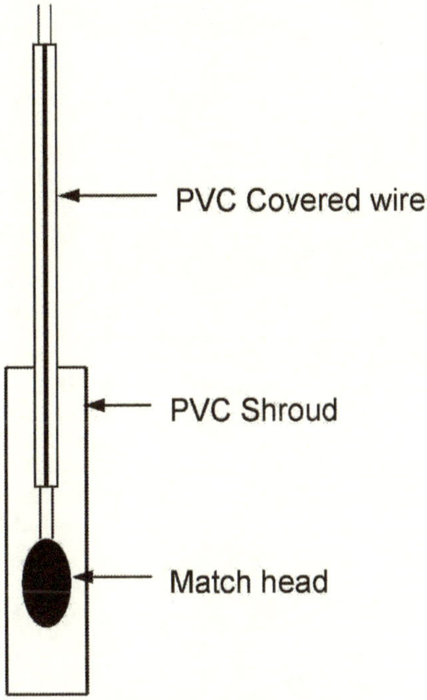

Figure 20.4 Electric match

One practical matter that should be borne in mind is that different makes of igniters do, in fact, perform slightly differently. It would be very unwise to use more than one make of igniter in any particular situation, particularly where several items were wired in series; a marginally different resistance from another make of igniter could cause a misfire and break in the line.

21

SMOKE

The extensive use of smoke for screening and signaling purposes during World War I and World War II has produced a very specialized branch of pyrotechnics that is not directly concerned with fireworks. As there is an abundance of specialized literature on the subject, the reader who is looking for greater detail is referred to the following references: (4,5,9,12,22, and 34).

Colored smokes usually are made by vaporizing a suitable dye by mixing it with a heating mixture of potassium chlorate and a fuel. In order to make the mixture as cool-burning as possible, lactose or a mixture of lactose and sucrose is used as the fuel. Thiourea also is sometimes used in certain types of white smokes.

As a general rule with colored smoke formulations, it is necessary to use dyes with a melting point that is as consistent and as low as possible. The dye also should be free from organic salts and should have a uniform particle size. Smoke compositions are mixed for long periods to produce uniformity and consistency because this, along with the particle size, affects the burning rate, which is further controlled by other additives.

A simple smoke formulation would consist of approximately 60% dye, 20% potassium chlorate, and 20% sucrose or lactose or a mixture of the two. The material is charged into a metal container with hydraulic pressure, leaving a space down the center of the composition rather like the spindle hole in a rocket. The composition must not be too dense or voluminous or else the carbonaceous residue will decompose the smoke before it leaves the container. A metal gauze frequently is placed on top of the smoke composition, which must not completely fill the cannister. The cannister is then closed with a suitable lid that contains holes for the smoke to issue forth and a suitable ignition. The empty space at the top of the container is essential in order to allow cooling to take place. In the

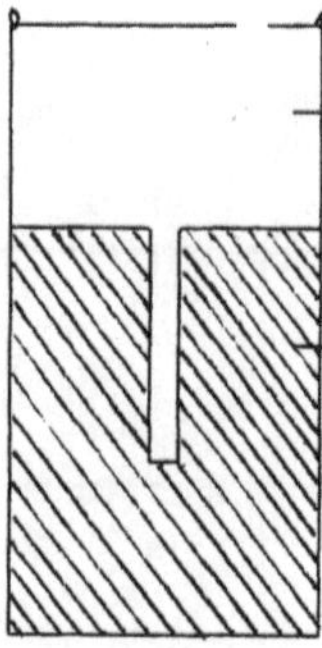

Figure 21.1 Smoke cannister

absence of this, the smoke will catch fire and merely burn. These smoke containers and ignitions have become highly complex and efficient. A smoke cannister is shown in Figure 21.1.

Black smokes gain their effect from the production of tiny particles of carbon when oxygen-negative mixtures containing naphthalene or anthracene are burned. These mixtures consist of potassium perchlorate and anthracene or potassium perchlorate, hexachloroethane, and naphthalene. During World War II, the Germans used a somewhat explosive mixture of hexachloroethane, anthracene, and magnesium powder. For example:

	Weight (%)
Hexachloroethane	60
Anthracene	20
Magnesium-fine	20

Brown smokes frequently have been made with pitch or tar. Hard pitch is ground to a powder and mixed with potassium nitrate and sulfur. Glue frequently is added to control the reaction. The mixture is placed in a cannister similar to colored smoke, but it tends to become very hot. The following composition given by Faber (9) is typical:

	Weight (%)
Pitch	29.2
Potassium nitrate	47.4
Borax	10.6

Calcium carbonate	4.9
Sand	4.0
Sulfur	3.9

A better method is to use liquid tar that is absorbed into sawdust. The mixture has better burning characteristics and can be used in thick-walled paper tubes with a washer at the exit to prevent the smoke from burning.

Gray/Brown Smoke	Weight (%)
Potassium nitrate	60
Sulfur	3
Sawdust	26
Liquid tar (coal)	7
Accaroid resin	4

Gray smokes invariably are of the so-called HC type involving the use of hexachloroethane (HC) and zinc in some form. The gray color is formed because of the production of zinc chloride and carbon particles, but the zinc chloride also forms zinc hydroxide and hydrochloric acid in the presence of moisture. Mixtures containing zinc powder are liable to react with water, and it is usual practice to make sure that moisture is excluded before the cannister is sealed. One magazine was destroyed when smoke spontaneously ignited because a worker allowed beads of perspiration to fall into the composition during processing. In view of this hazard, HC smokes more frequently are made with a mixture of hexachloroethane and zinc oxide, with smaller percentages of calcium silicide and potassium nitrate. The high vapor pressure of hexachloroethane also means that all HC smokes must be sealed in air-tight cannisters.

		Weight (%)
A.	Hexachloroethane	50
	Zinc powder	25
	Zinc oxide	10
	Potassium nitrate	10
	Colophony resin	5

(continued on next page)

B.	Hexachloroethane	45.5
	Zinc oxide	45.5
	Calcium silicide	9

A is from Izzo (8), and B is from Ellern (5).

White smokes invariably were made by the old firework makers with ammonium chloride. Strictly speaking, a mixture of potassium chlorate and ammonium chloride is highly unstable as a result of the formation of ammonium chlorate. Nevertheless, there seems to be good evidence of the stability of such mixtures (Chapter 5, page 147).

The following is typical:

	Weight (%)
Potassium chlorate	33
Ammonium chloride	14
Colophony resin	53

Dr. Becher (10) quotes the following:

	Weight (%)
Potassium chlorate	40
Ammonium chloride	45
Montan wax	12
Kieselguhr	3

Formulations of the latter type occasionally employ a little paraffin oil, presumably to help exclude moisture and reduce sensitivity.

More useful white smokes are made with HC and zinc oxide or mixtures of potassium chlorate, thiourea, lactose, and colophony resin. Colophony resin is said to exacerbate asthma. Mixtures of potassium chlorate and thiourea can be somewhat explosive, and they now are somewhat frowned upon. In addition, there now are anxieties about the toxic nature of HC smokes, and many of them have been replaced with cinnamic acid. The composition percentages can be changed to vary the burning time. However, these can be explosive.

	Weight (%)
Potassium chlorate	29

Cinnamic acid	27
Lactose	29
Kaolin	15

In more recent times, many commercial and military smokes have returned to the potassium chlorate and ammonium chloride mixture, with the addition of lactose and colophony resin. These formulations naturally vary with the requirements, such as the burning time and the type of container. These smokes, however, would be based on percentage of 40% to 50% of the pine resin and about 10% ammonium chloride, the remainder being the heating mixture of potassium chlorate and lactose.

As a general rule, black smokes are still made with anthracene and naphthalene. Mixtures of potassium perchlorate and anthracene undoubtedly are the easiest to handle, but this material is increasingly difficult to find in commercial grades. The alternative is to use naphthalene, but much depends on the type of material, the particle size, and the circumstances in which it is burned. The following is such an example

Potassium perchlorate	51%
Potassium nitrate	9%
Charcoal	1%
Dextrin	7%
Naphthalene	32%

Naphthalene flame and smoke bursts are commonly used in outdoor special effects, and it must be remembered that this chemical is extremely toxic to fish. It also is inevitable that both naphthalene and anthracene produce carcinogenic burning products. Reference to these materials also is made in Chapter 23.

Many of the old smoke mixtures were formulated to simply burn in the open and often were used on the stage, although they would be regarded with some caution today.

The following would be typical:

	Weight (%)
Potassium chlorate	50
Potassium nitrate	33
Magnesium powder	8
Gum copal	9

An interesting communication from Dr. Yurí Frolov of the Moscow Academy reports of investigations at the Institute of Chemical Physics into the use of smoke for fire suppression. It was found that smokes containing alkali metal ions (potassium and sodium ions) were particularly effective in buildings such as libraries and museums. Examples are as follows:

1.	Weight (%)	2.	Weight (%)
Potassium nitrate	60	Sodium azide	63.5
Epoxy	25	Sodium nitrate	21.8
Iditol	15	Urotropine	13.7

Iditol is phenol formaldehyde resin $C_{13}H_{12}O_2$. Sodium azide is extremely toxic.

The earliest types of insecticidal smoke generators were used in World War II in the form of hand grenades for fumigating dugouts in the jungle and the Far East. Later these were developed, under patent, for use in greenhouses.

Insecticidal smoke formulations are much the same as those for colored smokes that utilize dyestuffs, although in this case the pyrotechnic mixture must be carefully regulated to give the right amount of heat to vaporize the maximum quantity of active ingredient. The earliest ingredients in these smokes were DDT and BHC.

DDT 1,1,1-Trichloro 2,2-D1-(4-chlorophenyl) ethane.

Lindane (gamma BHC γ-1,2,3,4,5,6-Hexachlorocyclohexane, the pure gamma isomer, later replaced BHC. Later developments used azobenzene and tecnazene (1,2,4,5,-tetrachloro-3-nitrobenzene); development since has been extensive.

After ignition of the smoke composition, a dense smoke is produced with particles less than 3μ in diameter consisting of super-cooled droplets of insecticide. As the insects fly around (particularly at optimum temperatures of 21°C), they collect the insecticide particles on the hairs of their bodies and antennae and soon die. Once again, environmental and "green" attitudes have played their part in the revival of natural substances such as pyrethrum.

Other firework smokes are by Dr. Shimizu in Chapter 23.

Smoke ordinarily has no direct application in fireworks; on the contrary, it can be a nuisance at a firework display on a still night. However, apart from daylight fireworks, smoke is used commercially because of its penetrating capacity. It is used to detect leaks in drains and tanks because it not only penetrates all available space, but also is visible when it emerges.

Drain testers are simply thick-walled tubes clayed like fountains, but with a smaller hole, and charged with a simple mixture of potassium nitrate, sulfur, and a small amount of additional fuel. These smokes contain large amounts of sulfur dioxide and also are frequently used to discourage mice and moles. Mole smokes usually are made of roughly equal quantities of potassium nitrate, sulfur, and sawdust. They do not often kill the animals, but instead drive them to the next property!

22

EXHIBITION FIREWORKS

Exhibition fireworks do not vary very much from one manufacturer to another in their general arrangements because the number of possibilities are limited. However, the finer details vary enormously from one country to another, although we can see that each individual country tends to follow its own general pattern.

In England displays usually are fairly leisurely affairs, striking a balance between set ground pieces, Roman candles, rockets, and shells. The Italians and Germans, however, prefer fast-firing mass aerial effects of shorter but more spectacular duration. Many English people also prefer the latter type of display, but the public usually thinks that a show should last half an hour if they are to get good value for money. Apparently the English also are more reluctant to spend large sums of money on displays, preferring to buy a few smaller fireworks of their own. Large displays are still very much a feature of European fetes, festivals, and carnivals, but they are becoming more common in England. Very large quantities of small fireworks are still sold in England for the traditional festivals on November 5th (Guy Fawkes Day).

SHELLS

Shells for display work have been quite adequately covered in Chapter 16. Reloading is virtually impossible in fast-firing displays, making it quite imperative to have one mortar per shell. For the smaller sizes, paper mortars in wooden boxes surrounded by sand are quite adequate, but care must be taken not to allow the boxes to fall over. The shells frequently are fitted with automatic delays of the type mentioned earlier, but if these are used, the match pipe must be tied to the mortar or a shell leaving a mortar can sometimes drag unfired shells after it (Fig. 22.1). It is an even greater problem with wires attached to electric fuses.

Figure 22.1 Mortar boxes

Figure 22.2 Roman candle batteries

There has been great interest over the past years in the types of mortars that should be used, the way they are used in displays, the distances from the public, and the heights to which shells might climb and drift.

Traditionally, mortars have been of either paper or steel, but one or two nasty accidents with cylinder shells put the use of steel into sharp decline. More than 30 years ago Kimbolton started to use polythene water/gas tubing from a local firm; they did not shatter easily and, amazingly, the originals are mainly still available for use. They have the advantage of being waterproof, and explosions of round shells often cause only bulging with gas escaping from small melted holes. On the other hand, they are not as suitable for reloading because they become hot and are heavy to carry about. This in turn has led to the use of expensive aluminum and more recently to the widely used fiberglass.

An engineer friend, L. Jackson, makes the point that the forces that are being applied to the bung at the point of "lift off" are entirely dependent on the screw or nail fasteners, their size and intimate contacts. In order to avoid failure, it also is necessary to have enough material in that part of the bung above the nails and in the tube below the nails to prevent shearout (i.e., the nails trying to rip through either material). Failure through shear across a nail cross section at the interface of the bung and tube should not be a problem because the metal is much stronger than either the bung or the HDPE tube.

The light, thin-walled, resin-bonded, glass fiber tubes can be quite strong and are able to withstand the repeated explosive forces acting around their circumference, but problems can develop when the bungs are fitted as described previously. Direction of ply and quality of bonding can be factors, in addition to the "crazing" of the matrix around the holes if they are not drilled carefully. Careful drilling and hammering is essential when nails are used. When minute breakdowns occur in the matrix, the load-carrying capacity is reduced and repeated firings will cause failures.

A survey of factors related to the use of mortars in the United Kingdom has been published, and research recently has been undertaken to quantify the hazards posed by the high-velocity projectiles that can be produced on the rupture of spirally wound and straight wound mortar tubes (84). This work also has examined the effectiveness of rubber tire collars, sand bags, and sand barrels in mitigating the effects resulting from a shell misfire in the mortar tube (85).

Empirical relationships have been published that enable estimates to be made of the height attained by spherical shells from knowledge of either their mass or diameter.

The most useful of these, suggested by Dr. R.Wharton, probably is:

$$\text{Mean burst height (m)} = 1.4 \times \text{(shell diameter in mm) (ref. 83)}$$

Safety distances for the general public at displays where shells are fired represent a constant problem. The idea of having a meter for every millimeter of a shell caliber is a nice idea, but the majority of firework displays simply do not have this luxury. There might be difficulty in finding a space of even 75 meters.

A number of articles have been published in the *Journal of Pyrotechnics* on these matters, and the reader may well refer to Issue 17 (2003) to an article by Marc Speer, in which details on shells, mortar lengths, shell heights, and distances found to be safe in Germany could be of interest. (86).

ROMAN CANDLES

Display candles usually are fired in batteries of five or more. The best effect is gained by arranging the candles in a fan-shaped board or spline. A piece of timber drilled with a few holes to take the tubes is all that is required (Fig. 22.2). Large candles should be wired to wooden frames. Candles sometimes are fired in bundles, but these are less effective.

At this stage, it would be appropriate to mention the manner in which candles and other fireworks are joined together with piped match. The candles and gerbs usually are capped with about three turns of a tough thin paper (e.g., ribbed kraft). The paper is pasted only at the edge and where it comes into contact with the tube. It should overlap the end of the tube by about 40 mm.

When it is necessary to join several tubes together, a length of piped match is used to connect them. The end of the match pipe with about 12 mm of bare match protruding is pushed inside the end of the tube and the capping paper is gathered around it. The capping paper is then tied with a clove hitch. The match pipe is then cut further along the pipe near to the next tube with a V-shaped snip to bare the match again, and after being inserted into the capping, it is tied off as before. The method is illustrated in Figure 22.3.

Clove hitches are almost always used by fireworkers and are tied in two ways (Fig. 22.4). Both methods must be mastered because the one method, involving the formation of two loops on top of each other, can be used only in those situations in which it is possible to slip the loops over the end of the object to be tied. It also is important to note that when the cappings are tied, they must be neither too tight nor too slack; in the former situation the fire does not always get through, and in the latter case the match can be pulled or blown out of the cap without causing ignition.

Gradually, since the end of World War II, the use of Roman candles in public displays has become increasingly important. In earlier times candle batteries alternated between rockets, set pieces, and shells, but now they are vital elements in aerial shows for both continuity and precise timing.

Chinese multishot candles can be wonderfully effective, at an economic price, but it would be unwise to use a number of them together if precise timing is required. Modern candles made with precise black-

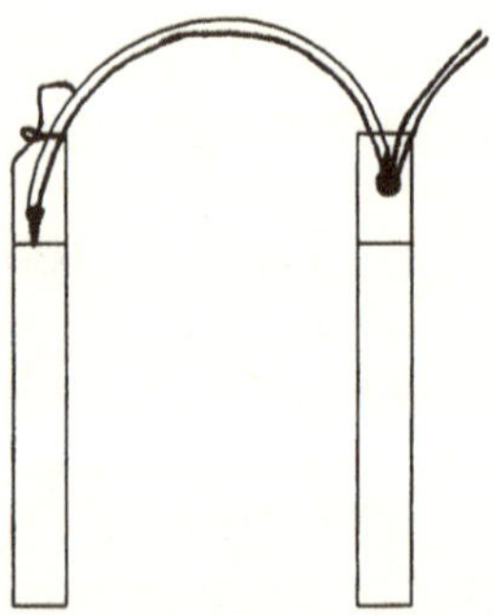

Figure 22.3 Matching tubes

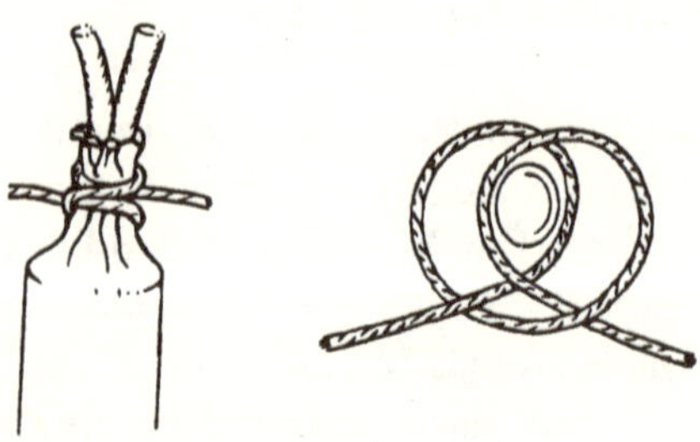

Figure 22.4 Clove hitch

powder delays or reliable Bickford type fuses can be manufactured for precision and are a valuable part of modern digitally fired shows. It also has increased the demand for electrically fired single-shot candles and mines, which have been a feature of the sequential firing seen in huge shows fired for the millennium celebrations or the many competitions around the world.

Where such candles are used regularly, it is tempting to make metal frames where candles can just be slotted in and connected up. It also is a bit of a snare and delusion because unless they are fastened down, they occasionally do jump out of such holders and the comets can travel long distances in undesired directions. It is still wise to fasten candles to wooden frames with wire or cable ties.

Candles with felt delays have cut down the smoke from the old-fashioned candle fuse and gained in precision, but one sometimes longs to see the old-fashioned batteries with showers of golden sparks peppering the air with their colored jewels.

ROCKETS

Large display rockets are fired from frames with a double row of screw-eye staples about 75 mm apart. The frame usually is collapsible for ease of transportation and is arranged so that the rockets can be fired at any angle (Fig. 22.5).

For this type of firing, the base of the rocket is covered with a paper drumhead so that the sparks do not ignite other rockets in the frame. Touch-papered rockets cannot, of course, be fixed in these frames. Sometimes the rockets are furnished with thin cappings and matched together for rapid firing.

Mass firing of small rockets is obtained by placing many rockets onto a board drilled with holes, or chicken wire covered with a sheet of paper, or simply in long rows. Each rocket has a piece of match in the vent, and in each case ignition is obtained from one of the rockets, which produces enough sparks to light the rest. Batteries of rockets of this type require some kind of cover lest they ignite prematurely as a result of stray sparks from other items in the display (Fig. 22.6).

Falling rocket sticks are a potential hazard, of course, but it is remarkable how few seem to cause any damage, particularly to people. It also is necessary to exercise some caution in firing parachute rockets, since chutes occasionally fail and drop burning flares to the ground. In Spain impressive volleys of rockets are fired from metal cones.

Figure 22.5 Collapsible rocket frame

Figure 22.6 Rockets arranged to fire in quick succession

FIRE PICTURES

Before World War II, lancework was very much a feature of firework displays. Huge portraits of kings and queens 6 × 9 m, Indian palaces, and triumphal arches were to be found, involving the use of thousands of lances and many more feet of match. The assembly of these items involved large quantities of timber, posts, ropes, and pulleys, as well as many hours of manual labor. A whole day or more would be spent setting up these displays with rain clouds hovering menacingly above. Needless to say, such displays are now most uncommon; four or five frames 3 × 1.5 m is the most one would expect to see.

Lance-frames usually are a standard size of 3 × 1.5 m in Britain, being made in 30 cm squares from light lathes. The portrait or design is first drawn out on squared paper or graph paper and then reproduced exactly on the 3 × 1.5 m frame using thick cane that is fastened to the frame with wire nails.

The next stage is to fix double-pointed nails into the canework at 75-mm intervals. These nails are about 12 mm long and are allowed to project about 6 mm into the cane (Fig. 22.7).

The unprimed ends of the lances are dipped in glue and then pushed onto the nails, where the glue sets sufficiently to hold the lances temporarily.

Techniques for connecting the lances with piped match vary somewhat, but all are more or less similar. The most laborious method uses a large pin to secure raw quickmatch to the lance. This is then covered between each lance with short lengths of piping that are slid along the match. Pasted paper strip is then used to cover the top of each lance.

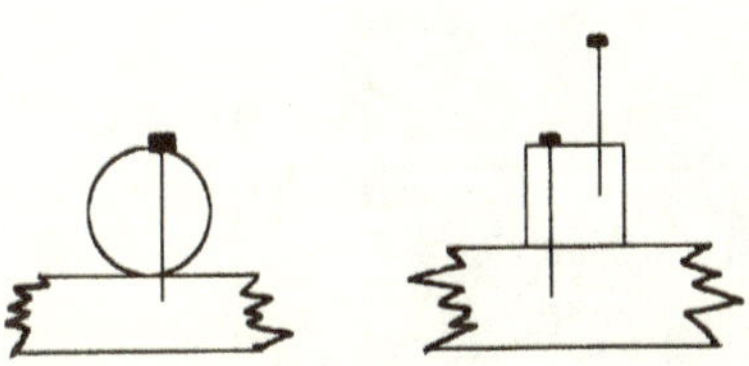

Figure 22.7 Making lancework frames

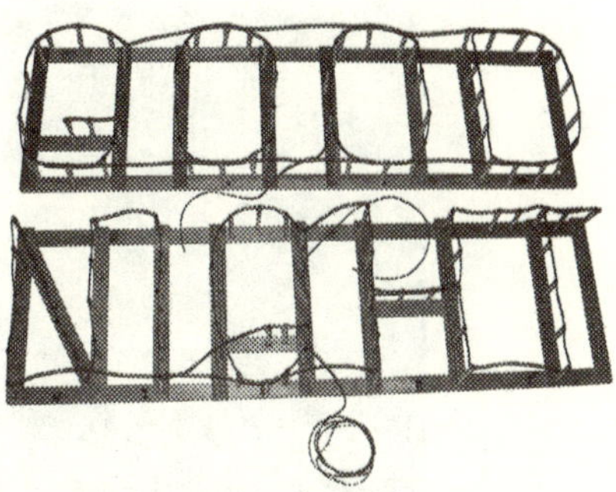

Figure 22.8 Lancework device

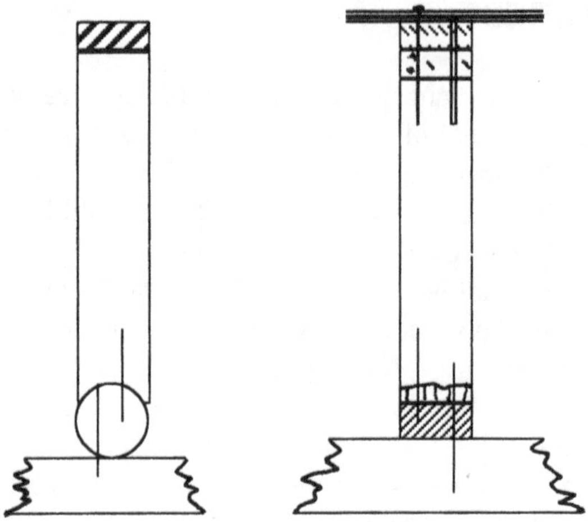

Figure 22.9 Matching lances

This method is ridiculously time-consuming, but it does ensure that all the lances ignite (Fig. 22.9).

An easier method is merely to lay piped match across the tops of the lances and drive a large pin through the pipe into the lance. A narrow staple also is used for this purpose, but many ignitions have been caused by stapling guns. It also is fairly common practice to make an additional hole through the match pipe into the lance prime to ensure the fire transfer from the match. The final operation is to cover the top of the lance and the match pipe with either pasted paper or adhesive tape. This operation helps to prevent the pins from being pulled out

Figure 22.10 Novelty set pieces. Early twentieth century (Brock's Fireworks)

and provides protection from rain and stray sparks. The lancework also can be protected from rain with a solution of shellac or paraffin wax. Plastic strips that hold a number of lances have now appeared on the market.

FOUNTAINS

Fierce burning fountains, or "fix," are arranged on light timber to produce geometric designs or "feux croisées." In situations in which it is desirable for fires to touch or cross, the spread of the fire must be measured so that distances can be worked out. Saxons and small wheels invariably are added to produce variety and movement. It is here that the greatest variations are to be found, each fireworker being able to devise his own pieces. The suggestions in Figures 22.11, 22.12, and 22.13 are typical.

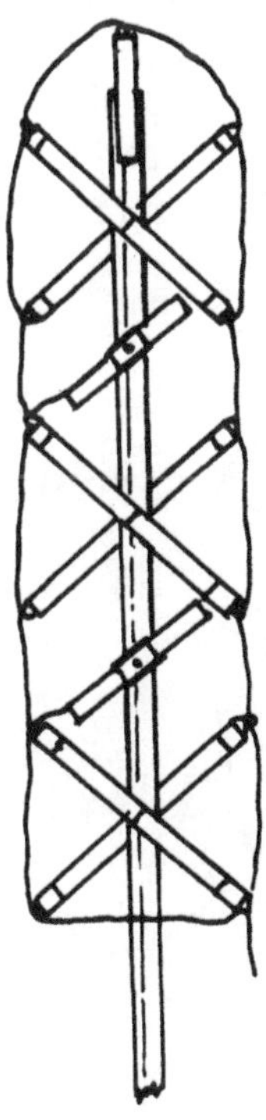

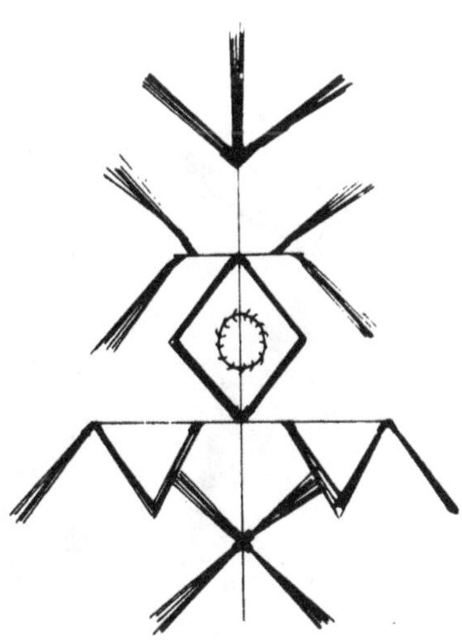

Figure 22.11 Lattice pole before ignition **Figure 22.12** Gerbs forming a tree piece

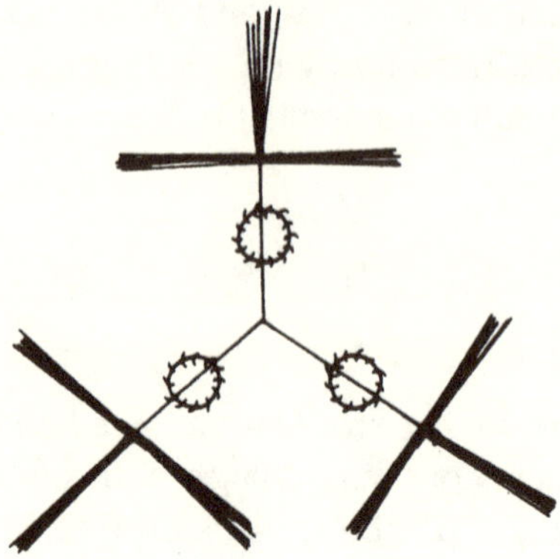

Figure 22.13 Gerbs and saxons in geometric design

WHEELS

Display wheels vary in size from 0.5 m up to 5 m in diameter. The smaller wheels (up to about 1.5 m) frequently are made in the old-fashioned manner with a heavy central boss, six spokes, and a circular beech or ash hoop around the circumference.

Larger wheels are made so that they can be assembled on-site. In this case, the spokes slot into the boss in tapered sockets and are held in place by a smaller steel hoop and a number of bolts (Fig. 22.14).

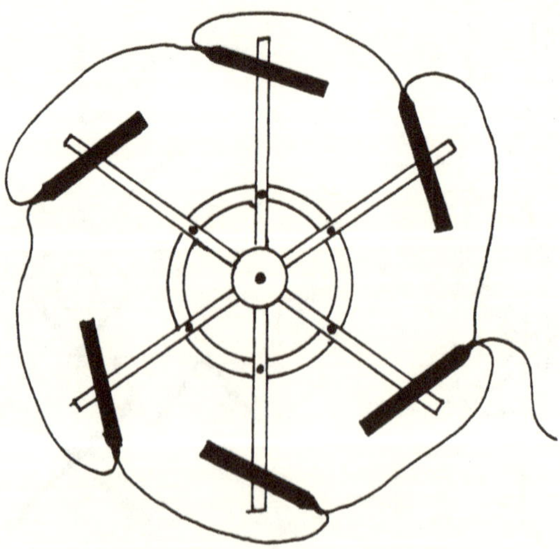

Figure 22.14 Revolving sun

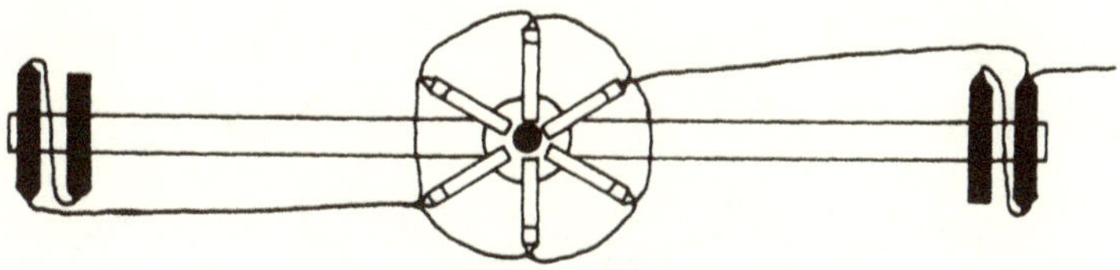

Figure 22.15 Large vertical wheel

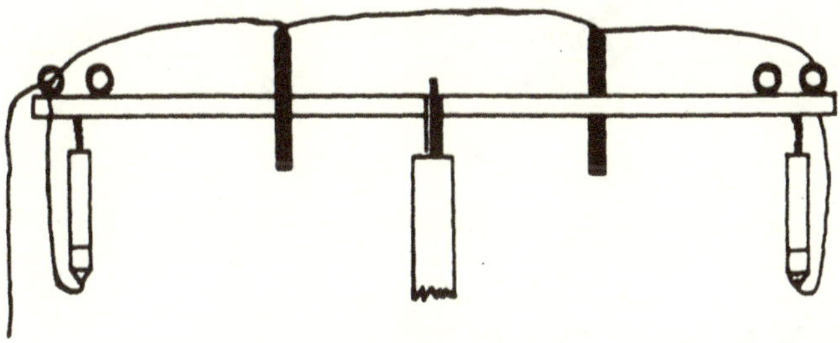

Figure 22.16 Horizontal wheel

Many horizontal and vertical wheels are made from a long piece of timber with a hole drilled through the center to take the spindle upon which the timber revolves (Fig. 22.15).

Vertical wheels often are fitted with color pots, whistles, gerbs, and even waterfall cases, strings of lights, or flash charges (Fig. 22.16).

Drivers for wheels vary greatly with the size and speed of turning. Small wheels about 1 m in diameter use 18-mm drivers; up to 3 m take 24-mm drivers; 5-m wheels take large gerbs about 35 mm in internal diameter and 25 cm long.

The old wheels were very cumbersome and inefficient, and drivers were not powerful enough, but new aluminum materials combined with a bit of intelligent engineering can produce some fine effects.

COMBINATION FRONTS

Very large spectacular effects are obtained by combining several identical pieces or a mixture of set pieces, wheels, and so on (Fig. 22.17).

Battle scenes also use various combinations. Two ships in lancework, for example, can be made to fire streamer Roman candle stars at each other to the accompaniment of Roman candle whistles, flash charges, thunderflashes, and crackers. The possibilities are quite endless.

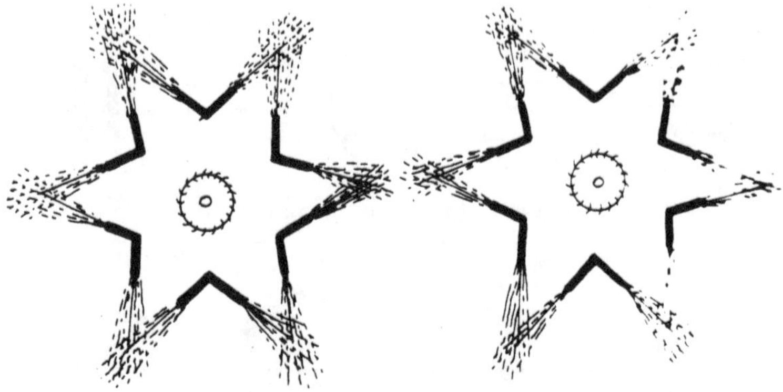

Figure 22.17 Combination fronts

FIREWORK DISPLAYS IN THE TWENTY-FIRST CENTURY

Darryl Fleming

Over the past decade, we have seen some significant and far-reaching changes to how we now operate and fire large-scale firework displays. In the 1980s and early 1990s, hand-firing a small show was very much the norm. Even on large-scale displays such as the 800th anniversary of the City of London in 1988, the display was manually electrically fired using a shot firer and a nail board, while the shell sequences were still hand-fired. There was widespread use of pyrotechnic delays, which allowed a significant number of shells to be matched together to start their firing initiation from the hand-fired fuse.

In the 1990s, with the advent of a large number of displays fired to music, greater synchronization and control were desired. Many companies invested in relatively simple manual electrically fired firing consoles. They were limited in that they were still manual and there was little scope for any kind of automation. Semiautomation was achievable with the use of preprogrammable sequencers, which could be initiated by firing a cue from the main manual console. They could be preset to fire a number of firing cues over a set time to create a chase sequence by firing up to, say, 16 cues over a predetermined time duration.

The main disadvantage with all these manual systems was that they were time-consuming to rig, required a large amount of bell wire, and still did not offer the control that can now be achieved with computer technology. The only advantage was that they offered a higher degree of safety for the operator because for the first time there was no need to be very close to the fireworks.

Many companies have over recent years designed and manufactured manual and semiautomated firing systems. They have been very limited and require a great deal of skill and experience to use them, as the design tends to reflect the particular way that company designs and fires displays. The systems have been specific and do not offer universal use. Their reliability and durability also have been questionable.

Reliability and durability are the most important factors, especially for a U.K.-based company that must contend with adverse weather conditions most of the time. The busiest time of the year in the United Kingdom is November 5th, when more than 60 large public displays may be fired over 12 days. The likelihood of experiencing some rain is very high. The thought of trying to fire a large show from a laptop in a cold, wet, and muddy field has always offered some reservations and lack of confidence in the show firing.

Over the past five years, the benefits and the scope of computerized firing systems has increased. In this age of technical advancement and mass global coverage of large-scale firework displays such as the Olympic games and the millennium celebrations, it is clear what can be achieved by using a fully computerized firing system.

Probably the most famous and instantly recognizable display might be the Sydney Harbour Bridge. The Australians know how to celebrate New Year's Eve and have created an event that is watched by millions around the world. Similarly, the display fired for the Sydney Olympic Games in 2000 has almost set the benchmark for stunning and spectacular displays, which would be impossible to fire were it not for the computerized firing system.

The main bridge effect is made up of 21 separate firing positions equidistant along the span of the bridge. From each position a range of single-shot and multishot effects are fired by either a chase from one end to the other and back, from the center out, or a hit from every position at the same time. In addition to the bridge, three large barges provide the platform to fire the aerial shells. In the 20-minute program, more than 6,000 cues will be fired to initiate the sequences seen on television, with each cue being fired up to 100th of a second of each other.

Clearly, the environment and the backdrop lend themselves to providing a spectacular display and the Harbour Bridge is an amazing structure to use as a firing platform, but the effects and sequences fired from the bridge are instantly recognizable to anyone with a passing interest in fireworks and firework displays.

Initially, the advent of computerized systems was met with skepticism and hostility by much of the industry. The benefits were not immediately understood and, more importantly, electrical engineers and not pyrotechnitions were manufacturing the technology. This meant that while the ability to fire a display with a computer was available, it was not much good without the ability to design and create firing sequences. This required pyrotechnition-friendly software and the ability to synchronize firing sequences to music.

The main advantage of computer control is the ability to truly synchronize a display to music. In addition, the ability to fire from a great number of locations or multiple sites offers tremendous applications when designing a display. Computer firing control has now matured significantly and proven itself as a viable and, in some cases, essential technology providing a new tool and resource for advancing the art of firing firework displays.

With the advent of the millennium displays, particularly in Sydney and Paris, the public at large has been able to witness truly amazing effects. While these displays set a benchmark, organizers of firework events, and indeed the public who come to watch them, come to expect something more than just aerial star shells being fired one after the other. They want a performance and desire to be amazed with stunning sequences, which only the computer can achieve. Five years ago, there probably were two companies in the United Kingdom that had embraced computer technology; now there are more than 15 companies that have invested in computer technology in one form or another. In fact, it has become a selling tool to potential clients to gain the advantage over another company in order to win new business.

The two most widely used and international renowned systems in the market are Pyrodigital and Fireone, both of which are made in the United States. Pyrodigital was the first purpose-built firework computerized firing system formed in 1980. Fireone was born not many years later but started life as a firing system for explosives in quarries and was adapted to be used for fireworks. Each system has unique and individual advantages and disadvantages.

After three years of using Fireone, the style and contrast of the displays that Kimbolton now fire compared to that of 10 years ago is significant. The benefits and scope of the system can allow the creation of effects that were impossible to achieve.

This has been illustrated by our success at recent international firework competitions, particularly Cannes, where in 2001 we won the Silver award, which is first place out of six other international companies. Every seventh year the previous silver award winners then compete for the gold award. It just so happened that for us the next gold award was in 2002, so we found ourselves having to compete in Cannes for two consecutive years. The gold award is a cup winner's cup, and we were up against some of the best display companies from around the world. Kimbolton Fireworks rose to the challenge, and we were delighted to win the gold award as well.

Fireone proved itself to be the main reason we won. Not only because we could fire a large number of cues in quick succession of each other, or that the whole display was choreographed and synchronized perfectly to music, but also because of the advantages it offers when firing from multiple sites.

The main technical problem in Cannes is that not only are you firing a 25-minute display synchronized to music, but also you have to fire the display from three barges moored 150 meters apart and 500 meters out from the promenade in the bay. With any traditional electrical firing system, to fire from three remote barges in synchronization would be virtually impossible. The benefit of the computer system really does come into its own when one is firing from more than one location.

The design of the display was split into three separate events, so each barge was independent of the others. We may have had a single 6-inch red and blue shell fired from the main barge and then the same shell firing 2.1 seconds later from left and right, or the same shell from all three barges at exactly the same time. Each barge was fired using its own Fireone main module where we had up to 25 firing modules on each barge, each module firing 32 cues. A laptop was connected to each main module on each barge with its own specific firing cues for that particular barge loaded on it; this is known as a fire file, a sequence of cues at a given moment in time.

In order to achieve the synchronization, a radio receiver was connected to the time code input on the main module, where on the shore a time code signal was broadcast so that each computer picked up the same time code at exactly the same time. The time code was burned onto a CD with the time code on one channel and the music in mono on the second channel. The CD could then be played from either the PA system with the time code being broadcast from a radio, or played from a desktop computer with the time code sent through a radio via a modem. The sound card from the computer linked to the PA system, to broadcast the music to the audience.

The result was stunning, with all three barges firing in absolute synchronization with each other and the burst of the shells in time with the beat and crescendo of the music. In our second year for the gold award, we further complicated issues by adding nine floating pontoons in-between the three barges to fire chaser effects from right to left or from the center out. This created much more depth and width to the display

Show Design

When designing a pyromusical display, the first place to start is the music. The music must be instantly recognizable and lively, and, if possible, it should vary in tempo and volume so that one can pick out the differences in the score with the various fireworks and effects. The musical score is just as important as the quality and effect of the fireworks and how you use them when synchronizing to the music.

The most difficult aspect about choosing the music is to make sure that you appeal to as wide a range of the audience as possible and not to concentrate to much on any one particular genre. Once the final music selection has been made, a CD or Wave file is produced by editing the various titles together, and the process of designing the display starts.

With the 25-minute music track for Cannes, we ended up with 42 separate sequences. The majority of our Roman candles have a duration of 30 seconds, but in some instances we required shorter duration candles. The advantage we have as manufacturers is that if we need a Roman candle to last for 18 seconds, we will make one that lasts for 18 seconds.

With the Roman candles in place, it is relatively easy to select the shells within each section. The long process of selecting a shell for each of the cues created begins; the caliber is very much dictated by the pace and volume of the music, and slowly the display starts to take shape. The great thing about Fireone is the Shot Display feature, which allows you to select the different positions and play back the music with the computer displaying the type of shell and the time at which it fires. This is a very easy tool to visualize the display and make changes where required.

Once the design is complete, the various reports created by Fireone are printed, and the process of manufacturing and collecting the fireworks begins. After this, each firework must be sequenced and boxed into module numbers to simplify the loading process on-site.

Kimbolton Fireworks recently fired the St. Patrick's Day celebrations in Dublin, which is one of the biggest annual displays in the firework calendar. This display was a real challenge. The remit was to fire a large aerial display from a main central barge and from the North and South Quay of the River Liffey, synchronized to a 20-minute soundtrack.

Unfortunately, the river is not wide enough to fire large-caliber candles or shells of any size, so this meant that the viewing area was very linear in nature. The audience, totaling nearly half a million people, would be spread along both sides for nearly a kilometer. The festival organizers were keen to keep the display as inclusive to as many people as possible. The idea was to bring the display as close and as safely as we could down the river. In order to achieve this, we devised 30 small floating pontoons from which we fired a variety of Bengal's, fountains, small Roman candles, mines, and a range of close-proximity single-shot zero-debris effects. The pontoons were moored 15 meters apart along the river from the main barge toward the audience. Each pontoon had about 65 fireworks and effects on it fired from one module on each pontoon. In addition to the three aerial locations, we also fired a range of low-debris effects from the top of 150-ft cranes, where we also suspended a waterfall across the width of the river.

Effectively, the display would be fired from 38 separate locations over nearly 1 kilometer distance of the river using 3,000 cues synchronized to music. In addition, we had to provide audio in mono to 12 PA stacks spread along both sides of the river together with a stereo feed for a live broadcast on both radio and television. This is achieved by using an 8-track digital hard drive; channel one contained the time code, channel two contained the music in mono, and channels three and four contained the stereo left and right version of the music. In addition, a sixth channel was required for audio vocal cues to three firers located around the firing site to light the lancework and act as backup to manually fire the waterfall should it fail.

THE WAY FORWARD

With the advent of digital computer technology, one may ask "What is the way forward?" Advances already are in the pipeline at the moment, where the software is refined and designed in a way that is more user-friendly. The major new technology is the wireless modules. These are the small site modules, which are fitted with an aerial and will receive time code and firing commands. At present, the digital modules require a two-core wire connection in which each module is daisy chained in series to the main data cable coming from the main firing module. The wireless modules will eliminate the need for hard wiring, and the main advantage will be the flexibility to position the modules wherever you want around a site. The cost of having to physically connect each module will be eliminated, and this also will make less accessible firing locations such as rooftops accessible.

Computer chips used in the manufacture of shells as opposed to pressed delays now take technology to the extreme, but as people become accustomed to seeing choreographed displays using computers, they will demand more. It is always the case that in order to stay competitive and remain the best, one will always have to move forward and push technology to the limits.

Computers have embraced most industries and proved themselves indispensable. Computers have the distinct advantage of having perfect memories and perfect timing. It is for this reason that they have become so widespread in the use of firing firework displays. With the advent of multimedia experiences such as water screens, lights, lasers, and flame projectors, computer technology will continue to play an important role in the design and firing of firework displays.

Figure 22.18 Displays fired from different positions in London

23

MANUFACTURING PROCESSES FOR FIREWORK COMPOSITIONS: JAPANESE FIREWORKS

Takeo Shimizu

These processes generally are very simple and once the manufacture of the most dangerous compositions have been mastered, it is easy to manufacture others that are safer. Compositions that contain potassium chlorate and red phosphorus, realgar, antimony trisulfide, or metal powder are the most dangerous. The best way is to use none of them, but we cannot always find safer compositions of the same character, and we often have no alternative but to use them.

We have two processes: the wet process and the dry process.

The *wet process* applies to compositions of high sensitivity, which do not explode in the wet state but explode when the moisture is removed after drying. An example of the wet process is as follows: A quantity of water is added to an oxidizer to make it wet. Then fuel is added to it and the whole is well mixed into a paste, keeping the mass wet. Compositions of potassium chlorate and realgar, or potassium chlorate and red phosphorus, usually are produced by this process. When the process is finished, it must be used quickly by wrapping it in small pieces of paper, or some other method of separating the composition into small amounts, before it dries. The composition that is enveloped in each piece is left in the air so that it may dry gradually and gain its sensitive character.

It must be noted that the wet process does not apply to ingredients of high solubility or to ingredients that cause some chemical reaction or create acid in the water. We have had experiences of spontaneous explosions that were thought to have been brought about because the acid concentration became high enough to cause the explosion. Each component has a different solubility in water, and when the wet composition is left to stand for a while, some particles of the ingredients that are heavier than others sink to the bottom and create a lack of uniformity. An adequate amount of water and, if necessary, gelatin can be used to moderate the viscosity of the composition.

The minimum amount of water that ensures safety during mixing depends on the composition. For example, in a composition that contains red phosphorus and potassium chlorate, 30 g of water per 100 g of the composition will not completely suppress its flammability and explosive nature; in fact, the amount of water should not be less than 50 g. A water content as high as this makes the composition muddy. A composition that contains potassium chlorate and realgar, sulfur, or antimony trisulfide explodes even when mixed with 20 g of water per 100 g of the composition, but becomes nonexplosive with 30 g of water per 100 g, so in this case the operation is permissible with not less than 40 g of water. Care must be taken during the operation that neither the rim of the container nor the operator's clothes are stained with the slurry, which often can dry and ignite without one's being aware of it.

The *dry process* applies to compositions that are relatively safe to handle. Sieves are used to mix dried powdered ingredients, and this does not change the component ratios like the wet process does. (The wet process causes the oxidizer to ooze out of products, thus changing the ratios of the components.) First, the ingredients are ground to a fine powder with a chemist's mortar, wooden roller, or ball mill. It is one of the common rules to sieve each ingredient before mixing so that the particle size is uniform and foreign substances are eliminated. A sieve made of hair generally is used for mixtures of relatively high danger such as a potassium chlorate or barium chlorate composition, and a sieve made of copper wires may be used for less dangerous compositions. Apertures of 800μ to 1,000μ are usual for mixing sieves. Ingredients that agglomerate easily (e.g., potassium chlorate, barium nitrate, potassium nitrate, and colophony) should be sieved just before the mixing process. (Potassium chlorate can now be purchased as a free-flowing powder.)

As an example, we can cite here the mixing process of the composition H3, a name that derives from the percentage of hemp coal it contains (i.e., the ratio of hemp coal to potassium chlorate is 3:10). H3 sometimes is used as a bursting charge for shells in Japan. This process also applies in the same way to other less dangerous compositions. A sheet of kraft paper is laid on a worktable. (A sheet of synthetic resin is not recommended for this purpose because it is too soft to handle.) The paper should be used only once and should not be used for other compositions, so it is best to use something like inexpensive cement bag paper. The prescribed amounts of potassium chlorate, hemp, coal, and glutinous rice powder are weighed onto the paper. The ratio of the three components generally is 10:3:0.3. A part of the hemp coal and all the rice powder are added to the potassium chlorate; it is well mixed by hand and then is passed through a hair sieve once. The potassium chlorate then becomes free-flowing and can be mixed easily. The residue in the sieve should be left untouched and not rubbed with the hands. The rest of the hemp coal is mixed into the premixed potassium chlorate and allowed to pass through the hair sieve two to three times. In this case, the residue in the sieve also is left untouched. The final residue that remains in the sieve after the mixing process is thrown away in a safe place. The finished composition is packed into a suitable container, with a label showing the kind of composition, the formula, the date of manufacture, and the name of the worker, and is taken to an expense magazine. It is forbidden to leave explosive or inflammable products in the workroom. The quantity of material that can be mixed at one time depends on the composition. In the case of sensitive compositions, it should be less than 5 kg. To correctly mix more than this amount by hand takes longer and is more dangerous. In place of hand mixing, it is possible to use a V-type or cylinder-type mixer, but special care must be taken in charging or discharging its content.

In the case of blackpowder, even when the three components (potassium nitrate, sulfur, and charcoal) are mixed together by the preceding process, the force of the powder is very small. In blackpowder production as an industrial explosive on a large scale, the three components are brought together by pressing or milling or by beating in a mortar. The manufacture of blackpowder in small amounts for fireworks is very dangerous, and the semi-wet process is much easier to handle. A mixture of 7.5 kg of potassium nitrate and 2.5 kg of water is charged into an iron pan and heated to 100°C, while stirring constantly. Hot water also may be used to heat the pan. Next the contents are taken out of the pan and charged into a mixer; 1.5 kg of hemp charcoal is then added, and the two are mixed for an hour to soak the hemp coal into the potassium nitrate. This process is safe because there is still some moisture in the mixture. Then the mixture is dried and charged into a ball mill, which is made of wood; an eight-angle drum of 33 cm diameter and 35 cm in length. This is placed in a workroom surrounded by defense walls. First, 1.0 kg of sulfur is added to the mixture, and then about 100 glassy porcelain balls, each one being 33 mm in diameter and 30 g in weight; these are charged into the drum, which is driven for 24 hours. This three-component mixing process serves two purposes: mixing and milling. It is a dangerous operation and is not always to be recommended, but it has been used for a long time in my factory without an accident. The "three-component" mixing room should be capable of bursting in one direction, and the operator should not enter the room while the drum is revolving. The force of the blackpowder produced by this process was compared with that of ordinary hand-mixed blackpowder by comparing their burning ratios in the atmosphere. If we take the value of the hand-mixed powder as 1, then the value of the powder made by the above method in the wet state becomes 12, and "three-component" dry mixing gives a value of 6. This proves the superiority of the semi-wet process described above. Never use a porcelain mortar or drum for mixing the three components because it is very dangerous.

BURSTING CHARGE

The bursting charge is the powder that when charged into a shell causes it to break by means of an explosion. The purpose is to throw out and spread the contents of the shell, to ignite the contents, and to give velocity to the stars. We have two kinds of shells, the so-called "Warimono" and the "Poka." Warimono is a chrysanthemum shell that requires a fairly large amount of strong bursting charge. It is capable of giving proper velocity to stars so that they form a round flower in the sky. Poka is an ordinary shell with a flag, willow, or flare; it does not require such a strong charge as the warimono.

The force of the charge for poka should be only enough to break the shell, and the commercial blackpowder or powder made in the factory generally is used for this purpose. The powder is mixed with water to a paste and chaff is covered with the slurry.

The bursting charge for Warimono should have a large explosive force and proper burning rate. When the value for the burning rate is too large, the stars in the shell are destroyed or not ignited. On the contrary, when the value of the burning rate is too low, it is not possible to have a good radius of fire in the chrysanthemum.

Potassium Perchlorate Bursting Charge I	Weight (%)
Potassium perchlorate	70
Hemp coal	18
Sulfur	12
Soluble glutinous rice starch	2 (additional %)

This formula has already been used for more than 20 years and no accident has happened. It is useful in practice and is thought to be safer to handle than the potassium chlorate bursting charge. A 2 kg drop hammer test showed that the nonexplosive point is 37 cm (in the case of picric acid, 23 cm) By Yamada's friction test apparatus, we obtained a nonexplosive weight of 75 kg (in the case of picric acid, 56 kg). The burning rate of this composition in the atmosphere when it is pasted on cottonseeds is 1.0 mm/sec and is rather small. A calculated value of the force of explosion shows 0.71×106 dm. This value is twice that of blackpowder. The values of the heat of explosion, the specific volume (= the volume of the gas created by explosion measured at 0°C, 1 atm) and the explosion temperature are 690 kcal/kg, 4,801/kg, and 3000°C, respectively. The author proposes the use of the following formula, which is thought to be safer than the former because of the absence of sulfur:

Potassium Perchlorate Bursting Charge II	Weight (%)
Potassium perchlorate	70
Hemp coal	25
Lampblack (Pine black)	5
Potassium bichromate	5 (additional %)
Soluble glutinous rice starch	5 (additional %)

Generally, it is better to use very fine charcoal to get a high burning rate, (i.e., the finer the better). The force of explosion of the preceding formula seems to be somewhat smaller than that of the former. The values of the nonexplosive point and the nonexplosive weight, by the same test apparatus, are 60 cm and 75 kg, respectively.

Potassium Chlorate Bursting Charge H3	Weight (%)
Potassium chlorate	77
Hemp coal	23
Soluble glutinous rice starch	2 (additional %)

This is the famous old formula for the bursting charge H3, which is still widely used in Japan. The ratio of hemp coal to potassium chlorate is about 3:10, from which the name H3 comes. Compositions that come into contact with this formula increase its sensitivity when they contain sulfur, realgar, red phosphorus, or antimony trisulfide, and we must pay close attention to this in production management. Compositions of this kind are thought to be the ones that most frequently cause accidents. Many fireworkers replace part of the potassium chlorate with potassium nitrate, but it is not always certain to be safe enough. The value of the nonexplosive point obtained by a 2 kg drop hammer test was 50 cm. By Yamada's friction test apparatus, we had the value of the nonexplosive weight of 75 kg. The latter is the same as that of the potassium perchlorate charge. The value of the burning rate of this composition pasted on cottonseeds was 3 mm/sec at 1 atm. This value is three times as large as that of the potassium perchlorate composition. The calculated values of the force of explosion, the heat of explosion, the specific volume, and explosion temperature are 0.74 × 106 dm, 640 kcal/kg, 5,70l/kg, and 3100°C, respectively. This composition has been widely used because of its large explosive force and the simplicity of the manufacturing process. It must not be mixed with blackpowder, but since both compositions have the same black color, care must be taken to take the right one.

POTASSIUM-NITRATE BURSTING CHARGE

The formula is no different from the normal formula for blackpowder (i.e., the weight ratio of potassium nitrate, charcoal, and sulfur is 75:15:10). As an adhesive, 2% of glutinous rice starch is added to the total percentage. It is thought to be the safest composition with regard to friction and shock. The 2 kg drop hammer test showed the nonexplosion point of 85 cm. The burning rate of grains made of cottonseeds, pasted with this composition, is about 2 mm/sec at 1 atm. This value is twice as large as that of the potassium perchlorate composition. In practice, in place of charcoal we use hemp coal, which contains 10% moisture and 10% ash, and the pure coal remains as 80%. The counted value of the force of explosion is 0.3 × 106 dm, which is smaller than ordinary rifle powder. This is because a smaller volume of explosive gas is created. The values of the heat of explosion, the specific volume of the explosion gas, and the explosion temperature are counted as 400 kcal/kg, 3,70l/kg, and 1800°C, respectively. This composition is extremely ignitable and has a large value for the burning rate, but its disadvantage is that the force of explosion is too small. It is used for relatively large shells.

To make the fire spread rapidly when the shell is ignited, the composition is pasted onto a nucleus. Cottonseeds, chaff, or immature ears of rice most frequently are used as nuclei. With cottonseeds the ratio of the composition is 1:1.3–1.6, and in the case of chaff its ratio to the composition increases. In general, the charge made of cottonseeds is used for relatively large shells, which require only a small loading density, and that made of chaff is used for small shells, which require a relatively high loading density. Sometimes these are mixed together. For example:

The Grains of Bursting Charge	Weight (%)
For Small Shells	
(less than 85 mm in diameter)	
Composition	80
Rice chaff	20
For Intermediate Shells (125–175 mm)	
Composition	
(Potassium Perchlorate Bursting Charge)	52
Cottonseeds	48
For Large Shells (200–300 mm in diameter)	
Composition	
(Potassium Perchlorate or Nitrate Bursting Charge)	52
Cottonseeds	48

The composition is poured, with the nuclei and a suitable amount of water, into a vessel and stirred until the nuclei are covered with the composition. The grains produced in this manner are dried in the sun. The thickness of the pasted composition on the nuclei is irregular in the case of chaff but is 0.47 to 0.49 mm on average in the case of cottonseeds. More than this is not recommended because of its performance character as the bursting charge. Bursting charges that follow these formulas show an apparent specific gravity of 0.48 to 0.50. The loading density of the bursting charge is counted, rejecting the nuclei that have values of 0.38 to 0.40 and 0.25 to 0.26, in the case of cottonseeds and chaff, respectively. For example, the amounts of the bursting charge used in shells without nuclei according to the preceding specification are:

The Amounts of Bursting Charge for Chrysanthemum Shells

Diameter of the Shell (mm)	Amount of Bursting Charge (g)
89	44
125	70
150	140
178	290
215	400
240	940
315	1,250

When cottonseeds and chaff are used as the nuclei, they should be well dried beforehand.

COLORED FLAME COMPOSITIONS

In general, a colored flame composition consists of an oxidizer, a fuel, and a color-creating material. Some metal salts play the role of both oxidizer and color-creating material (e.g., barium nitrate, barium chlorate, and strontium nitrate).

OXIDIZERS

When ammonium perchlorate is used as an oxidizer, it produces a relatively small amount of smoke in air with a low moisture content because it creates very little solid material during burning. In moist air, the burning gas produces small water particles and creates a dense smoke cloud. Ammonium perchlorate gives the most clear and beautiful colored flame when it is used with all the color-creating materials. For frame fireworks, ammonium perchlorate composition is recommended most if the cost permits. However, the contact of ammonium perchlorate composition with blackpowder or potassium nitrate creates ammonium nitrate, which is very hygroscopic and damages the fireworks. Ammonium perchlorate composition is very ignitable and generally requires no igniting composition.

Potassium perchlorate composition creates more smoke and less depth of color than ammonium perchlorate composition, but it is more practicable. The fuel that is used to the best advantage in ammonium perchlorate compositions will not always give the same effect in potassium perchlorate compositions. When potassium chlorate is used as an oxidizer, the effect is not so different from that of potassium perchlorate, but the amount of smoke seems to be greater than the former and around the base of the flame, potassium chlorate creates a tubular ash, which disturbs the flame projection. For example, in a lance, the flame does not come out of the end of the lance but branches out as it is diverted by the ash. The tubular ash also diminishes the light intensity of the flame.

The color of the flame depends on the kind of oxidizer, even when we use the same fuel and color-creating material. For example, an ammonium perchlorate composition with strontium carbonate as the color-creating material can give a very clear and deep flame, but with potassium perchlorate in place of ammonium perchlorate, the flame looks reddish pink.

As for the fuel, it is advisable to use a material that gives no color to the flame, but such material cannot be obtained in practice, and we must be content with a material that gives as weak a color to the flame as possible.

Using ammonium perchlorate as the oxidizer, fuels were tested to discover which were least likely to disturb the flame color. The order of preference was: shellac, wood meal, pine root pitch, amber powder, colophony, and charcoal. Samples 115 mm in length and 7 mm in diameter, each of which consists of a brown paper tube and 6 g of a composition pressed in it to an apparent specific gravity of about 1.35, are ignited one by one at the end and their burning state is observed. The composition consists of 17% of a fuel and 83% of ammonium perchlorate, and contains no color-creating material so that the color-disturbing effect of the fuels can be observed.

The result is summarized as follows:

Find of Fuel	Flame Condiion	Burning Rate (mm/sec)	Flame Length (mm)
Shellac	Weak reddish orange	1.51	90
Wood meal	Weak reddish orange lines	1.27	60
Pine root pitch	Bright white	2.09	110
Amber powder	Brilliant white at base	1.44	70
Colophony	Brilliant showing lines	2.80	100
Hemp coal	Yellow and brilliant	2.80	90

For potassium perchlorate, the most suitable fuel is pine root pitch, and then it should be arranged in order of merit as follows: colophony, amber powder, hemp coal, and wood meal. Shellac and potassium perchlorate burn with some difficulty. In the same manner as described, with composition of 17% of a fuel and 73% of potassium perchlorate, we obtained the following results:

Find of Fuel	Flame Condiion	Burning Rate (mm/sec)	Flame Length (mm)
Pine root pitch	Weak violet at the top and brilliant at the base	1.83	80
Colophony	Weak violet at the top and the rest white	1.24	50
Amber powder	Accompanied by much white smoke	1.51	60
Hemp coal	Yellow at the center and the rest reddish	4.80	—
Wood meal	Violet white, burning with difficulty	0.69	20
Shellac	White, burning with difficulty	0.84	30

These compositions create much smoke. When we use wood meal or shellac with potassium perchlorate, we also must add other more combustible fuels.

When potassium chlorate is the oxidizer, pine root pitch is the most suitable fuel and shellac follows next. Colophony burns with difficulty. Amber powder gives a white flame and much smoke. The mixture of charcoal and potassium chlorate burns very rapidly and dangerously, and charcoal alone is seldom used as the fuel. Charcoal generally is used to adjust the burning rate of composition, to deepen the color of the flame, to increase its brilliancy, or to be a supplementary fuel for some other purpose. The result of a test for compositions, in which 83% of potassium chlorate and 17% of a fuel are used as described previously, is as follows:

Find of Fuel	Flame Condiion	Burning Rate (mm/sec)	Flame Length (mm)
Pine root pitch	Much smoke and violet flame	2.13	—
Shellac	Light violet, ash remains	1.20	80
Amber powder	Much smoke, white flame	1.60	—
Colophony	No burning	2.26	—
Wood meal	Violet flame, vibrational burning	17.4	—

Coal tar pitch disturbs the color of the flame remarkably and cannot be used as a fuel for colored flame compositions.

COLOR

Color-creating materials have the following characteristics:

Red color–creating materials: Strontium carbonate is one of the red color –creating materials, and it gives the most beautiful red flame, but a high percentage in a composition causes the generation of a greal deal of smoke and creates burning difficulties. The effect of strontium oxalate is the same as that of strontium carbonate. Strontium nitrate gives a very beautiful red flame, but it is a weakness of strontium nitrate that it is rather hygroscopic.

Yellow color–creating materials: Sodium oxalate gives a clear yellow color to the flame even in small amounts, but the color is somewhat reddish and not pure yellow. Borax gives almost pure yellow. Ultramarine leaves a residue on burning and is not as commendable.

Green color–creating materials: Barium chlorate gives the deepest green, but it creates much smoke when the amount is increased. This material resolves easily and is not recommended. Barium nitrate cannot give the depth of color, but when we have a good formula and a high temperature, it can give a green flame of practical use.

Blue color–creating materials: Copper sulfate gives a good blue color to the flame in an ammonium perchlorate composition, and its effect is not inferior to that of emerald green (Paris green). Copper sulfate powder is neither as fine nor as good as that of Paris green. As a safety precaution, it is forbidden to mix copper sulfate with potassium chlorate or other chlorates. Paris green will give a good flame color with any kind of oxidizer. The only defect is that Paris green is easily scattered in the air and it is poisonous when workers inhale it. Copper arsenite shows almost the same character as Paris green in handling and in coloring the flame.

As an inexpensive ingredient, calcium carbonate sometimes is used in place of other red color–creating materials. In this case, the flame is reddish orange and not as beautiful.

The colored flames for fireworks are founded on the useful band or line spectra, which are inherent in metal or metal compounds. But the color that appears to the eyes depends not only on the strength of the spectra caused by the color-creating material, but also on the following conditions: Every flame has background spectra, which depend on the kinds of metal salts and fuels used in the composition. The flame temperature also has an influence on the flame color. The paper tube that contains the composition sometimes disturbs the coloring of the flames. The physiological condition of eyes, moisture, dust, or mist in the air change the color of the flame. Note that an electric light that is white close-up looks reddish yellow at a distance of 1 to 2 km. A green star, which looks light yellowish green at a short distance and appears to be useless, gives a deep, clear, and beautiful green when observed at a distance of 50 m. It is one of the general rules when designing the formula of a star composition that we must observe the burning star at a long distance.

The principle of designing colored flame compositions is as follows: First, the oxidizer is selected according to the use (e.g., for stars, frame fireworks, flares) and then the color-creating material and fuel that are suitable for the oxidizer. The flame color should be carefully chosen so that it is best for the purpose. The fuel should be carefully arranged to give the most adequate burning rate. To obtain a light of high intensity, the flame temperature must be as high as possible, and the composition should be rich in oxygen or should contain a metal fuel such as magnesium powder. Chlorine gas in the flame deepens the color of the flame, especially when barium and copper salts (in the case of green or blue flame) are used as the color-creating material. Potassium nitrate generally is not used as the oxidizer in colored flame composition because it gives a low burning temperature, which is insufficient to excite the coloring molecules or atoms in the flame. Ammonium perchlorate is not used in practice for shell stars because of its hygroscopic nature and difficulties in ignition at high velocity in the air. In the case of shell stars, the composition may be allowed to create smoke to some extent and the flame is somewhat inferior in color, but it should be well-dried and nonhygroscopic. The compositions for frame fireworks should not create so much smoke, and their tendency to absorb some water should not disturb their ignitability because they are not exposed to strong winds, as shells are. Moreover, they are loaded as powder, which is very ignitable.

When we solidify and form the composition into a certain shape for use, we must add some adequate solidifying material, a so-called "binder" to the composition. When water is used as a solvent, soluble glutinous rice starch is recommended, and shellac is used when alcohol is the solvent. If water is forbidden, it may be possible to use linseed oil or another drying oil, and in this case, care must be taken to disperse the heat of oxidation of the oil. Celluloid dissolved in amyl acetate also can be used as a binder. The amount of these binders should be as small as possible.

Colored flame compositions of practical use are:

Colored Flame Compositions for Frame Fireworks (Lances)	(%)
Red Flame:	
Ammonium perchlorate	70
Strontium carbonate	10
Wood meal	20
Yellow Flame:	
Ammonium perchlorate	75
Sodium oxalate	5
Wood meal	15
Colophony	5
Green Flame:	
Ammonium perchlorate	50
Barium nitrate	34
Wood meal	8
Shellac	8

(continued on next page)

Blue Flame:	
Ammonium perchlorate	70
Copper sulfate	10
Wood meal	10
Shellac	10
White Flame:	
Ammonium perchlorate	40
Potassium perchlorate	30
Antimony trisulfide	14
Starch	11
Wood meal	5

Each of these compositions is loaded in a thin brown paper tube of 9 mm in diameter with a density of 1.3 to 1.4 and designed so that the rate of burning should be 1.0 to 1.1 mm/sec. Generally, when a powdered composition is pressed by hand into a paper tube and ignited at one end, the relationship between the burning rate and the diameter of the tube differs with the type of composition, and the greater the diameter, the greater the burning rate below a diameter of 20 mm. For larger sizes than this limit, the burning surface becomes unsteady and causes flake dropping. When we take the burning rate at the diameter of 10 mm as the unit 1, the burning rate at a diameter of 20 mm reaches 1.7 to 3.0.

Color Flame Compositions for Stars	A Weight (%)	B Weight (%)
Red Flame:		
Potassium perchlorate	67	54
Pine root pitch	13.5	—
Accaroid resin	—	4
Magnesium/aluminum alloy (50/50)	—	14
Strontium carbonate	13.5	10
Parlon	—	13
Soluble glutinous rice starch	6	5
Green Flame:		
Potassium perchlorate	46	10
Barium nitrate	32	50
Pine root pitch	16	—
Accaroid resin	—	7
Magnesium/aluminum alloy (50/50)	—	13
Parlon	—	15

Soluble glutinous rice starch	6	5
Yellow Flame:		
Potassium perchlorate	72	
Sodium oxalate	7	
Pine root pitch	12	
Colophony	3	
Soluble glutinous rice starch	6	
Blue Flame:		
Potassium perchlorate	64	
Pine root pitch	13	
Paris green	17	
Soluble glutinous rice starch	6	

Red Flame A and Green Flame B produce a rather high temperature and a brilliant flame. These compositions can be consolidated using water, but dry the stars without heat to avoid cracks due to the reaction of the magnesium/aluminum alloy and water. In other compositions, when pine root pitch is not available, accaroid resin can be used in place of it. In this case, the stars burn rather slowly.

For stars used in chrysanthemum shells, the design conditions depend especially on the burning rate of the stars. Pine root pitch is used here to make the burning as fast as possible, and if there is another fuel that gives the same character, it is used in place of pine root pitch because the pitch is difficult to obtain. In order to obtain a good chrysanthemum, fuels must be chosen so that the compositions burn quickly without disturbing the flame color. The color of the green flame is not as good at a short distance, but it is practicable from a long distance because of the physiological nature of the eye. If the relationship of the burning rates between the solidified composition, which is made by the wet method, and that of pressed powder composition of the same formula were previously studied, it would be better to design the stars from the data of the pressed composition, which is easier to make. Insufficient study has been made hitherto, but a few examples are shown in Table 23.1.

Table 23.1 Comparison Between the Burning Rates of Solidified and Pressed Compositions

	Solidified Star (mm/sec)		Pressed Star (mm/sec)	
	S.G.	B.R.	L.D.	B.R.
A Blue Star	1.6	1.79	1.47	1.70
A Green Star	1.64	1.18	1.60	1.52

(S.G. = specific gravity; B.R. = burning rate; L.D. = loading density)

Each of the solidified stars is a sphere of 12 mm in diameter, and the pressed stars are cylindrical using a paper tube 9 mm in diameter, into which the powder is pressed. The table shows that the relationship between the burning rates of the solid and powdered states depends on the kind of composition; in practice, the influence of the solidifying material (binder) must be considered. Studies of the burning rates of stars of different diameter but the same composition have not yet been sufficiently made, but we know that the larger the diameter, the larger the burning rate. A few examples are shown in Table 23.2.

In the case of the composition for shell stars (e.g., for color-changing stars), there are many successive ignitions from one composition to another. For ignition without failure, the amount of heat of burning, and the burning rate of the former layer and the specific heat of the latter layer, must be adequate. The star compositions for shells described here do not have very good ignition because of the potassium perchlorate, and the construction of stars requires some device to make them ignite easily.

Table 23.2 The Relationship Between the Diameter and Burning Rate of Spherical and Cylindrical Stars

	Spherical Stars (solidified)	
Diameter	18 mm	32 mm
A Blue Star	1.79 mm/sec	1.92 mm/sec
	Cylindrical Stars (pressed)	
Diameter	9 mm	21 mm
White Star	1.95 mm/sec	2.10 mm/sec
Yellow Star	1.15 mm/sec	3.00 mm/sec
Red Star	0.98 mm/sec	1.80 mm/sec
Blue Star	1.00 mm/sec	1.28 mm/sec
Green Star	1.04 mm/sec	1.60 mm/sec

Generally, each of the colored flame compositions has a sufficient quantity of oxygen to oxidize the fuel it contains, and it burns progressively in the solidified or pressed state when it is ignited, but causes detonation when it is initiated with a percussion cap. We must pay close attention to this characteristic of color composition.

FIRE DUST SPARK COMPOSITION

Colored flame is appreciated at both short and long distances, but the sparks described here, which are like fire dust, are admired at a long distance. The fire dust sparks are the phenomena caused by charcoal, metal, and some ash projected at high temperature from a burning composition as unreacted matter, or newly formed matter, and make a second burning with oxygen in the air. This can be verified by burning the fire dust over a vessel of water when it will be possible to detect the unreacted or yet reactable charcoal, metal, or dust sparks that did not contain enough oxygen to oxidize the fuel it contained. (Viz. the composition for fire dust sparks consists of a smaller amount of oxidizer and a larger amount of fuel than colored flame composition.)

The colored flame is admired for its spectra, which is peculiar to the metal or the metal compound in the vapor phase. But in the case of fire dust sparks, it is concerned with the temperature of solid or liquid particles. The color of the light emitted by a radiator changes according to its temperature, and in ordinary cases it follows thus:

500°C	dark red
700°C	red
850°C	orange
1000°C	yellow
1100°C	white

To obtain beautiful fire dust sparks, it is necessary to adjust the combustion temperature of the composition.

Other problems arise with fire dust stars. Ignition characteristics and the separation of the fire must be good, and generally the appearance, the particle size of the dust, the density of the fire particles, and the lifetime of the particles are the most important factors. When planning fire dust spark compositions, we must have a clear knowledge of these conditions, but this has not yet been studied fully and systematically, and here we can describe only a few fragments that have been made clear. In compositions of this kind, we often see rather complicated formulas, but it seems that they depend on the conditions described previously and especially on the regulation of the burning temperature of the compositions. Now we shall study the influence of the components on the temperature of some simple compositions.

The character of the fire dust sparks caused by charcoal depends on the formula, the character of the charcoal, and its particle size. It is impossible to adjust the color of the fire dust very much, for the color at best lies only between reddish orange and orange red. The composition of potassium nitrate, sulfur, and charcoal can produce the most beautiful fire dust sparks, but composition that does not contain potassium nitrate cannot produce such sparks. Charcoal, for example, with ammonium perchlorate, burns totally in the flame without discharging sparks. In the case of potassium chlorate, it is the same as above, except that the burning reaction is more violent than the former. It creates a few sparks with potassium perchlorate, but most of the reaction finishes in the flame. Only in the case of potassium nitrate, where the flame is relatively small, can a large amount of fire dust sparks be produced from the flame. It seems that the potassium sulfide produced in the flame surrounds the particles of charcoal and disturbs the strong reaction in the flame. The smaller the particle size of the charcoal, the shorter will be the life of the fire dust sparks, and the phenomenon also depends on the amount of sulfur and the kind of charcoal contained in the composition. Some examples of the formula are as follows (these compositions are called "Tail" or "Chrysanthemum"):

	Weight (%)
Chrysanthemum 6	
Potassium nitrate	58
Sulfur	7
Pine charcoal	35
Chrysanthemum 8	
Potassium nitrate	52
Sulfur	6
Pine charcoal	42

The number six or eight means that the weight of pine charcoal is 6 or 8 relative to 10 parts of potassium nitrate. When a high velocity of burning or a large quantity of heat is required, Chrysanthemum 6 is used, and for the opposite requirement Chrysanthemum 8 is used. An intermediate composition or compositions outside this range also are used. To solidify the stars, soluble glutinous rice starch is added as a binder (normally about 6%). To make a composition of very fine fire dust, which can produce sparks of high concentration, the preceding formula is prepared and charged into a wooden ball mill with porcelain balls and it is then driven for 24 hours. The life of the sparks is short in this case, but they are very beautiful and elegant. When the composition has no sulfur, it is difficult to burn well, and the branching out of the fire dust is not good, for the stars keep the fire to themselves for a long time. This phenomenon produced by moving stars looks like the dawning of a weak fire dust band or tail in the sky, and it appears peculiarly elegant.

Many fragments of the fire are seen to be falling down. This star is called "Chrysanthemum of Mystery":

	Weight (%)
Pine charcoal	50
Potassium nitrate	45
Soluble glutinous rice starch	5

Sometimes 3% to 4% of red lead oxide (Pb_3O_4) is added to this composition. The purpose may be to use the special after-reaction that is peculiar to minium, or to increase the weight of stars, but it seems that adding minium has little influence on the fire dust sparks phenomenon. Other kinds of charcoal and other formulas also are used. A composition that contains black lampblack produces sparks with a reddish orange flame.

The defect of composition made of potassium nitrate, sulfur, and charcoal is that it has less ability to ignite the next layer of the star than that of the colored flame compositions because of the small calorific value and low burning temperature. This would be expected because of the lack of oxygen in the composition. Therefore, when it is hard to ignite the second layer from the first layer (which consists of a blackpowder-type composition), it is necessary to put a new layer of a more easily ignitable and hot composition between the first layer and the second layer.

As a source of metal fire dust sparks, we usually use aluminum powder and mix it with potassium chlorate, potassium perchlorate, potassium nitrate, or barium nitrate, but from the standpoint of ignition and safety, chlorate is not recommended. The sparks from these sources have a peculiar beauty. Sulfur, realgar, antimony trisulfide, or charcoal are added to regulate the temperature of the sparks or to give good ignition to the stars.

Potassium nitrate and aluminum spark composition produce beautiful fire dust sparks that are reddish orange, gold, or yellow white.

The so-called "Fire Dust Branching" phenomenon occurs when a melted mass of high temperature branches out into fine fire dust by the action of air currents. To see this phenomenon, it is important to test the burning star by projecting it in the sky. A star that burns on the ground and generates very good sparks does not give such a good effect in the sky because the duration of the sparks looks too short. In the ground test, it is seen that the melted ash after the burning reaction becomes a mass and has brightness without producing many sparks, although it creates good sparks in the air. On the other hand, it should be noted that the overmelted ash does not branch into much fire dust; in fact, the duration of the sparks becomes too long and the beauty diminishes. It is the same when potassium perchlorate is the oxidizer.

Falls	Weight (%)
Potassium nitrate	41
Aluminum	49
Sulfur	4
Soluble glutinous rice starch	6

A small amount of water is added to the composition and it is kneaded well; it is then pressed into a thin paper tube with a wooden or brass pounder. It generally is said that thorough kneading gives a long life to the sparks, but that does not always seem to be true. Coarse aluminum flake powder is good for making the sparks burn long enough, and sometimes we use aluminum foil flakes (flitter) in the falls composition. When the tube has been loaded with the composition, it is dried in the sun. If it is noticed that the tubes are generating heat, they must be separated from each other in the shade to cool them. If a long time is taken to dry the tubes, the fire dust sparks will become somewhat reddish. The burning rate of a waterfall of 15 mm in diameter was 0.52 mm/sec. This falls composition can produce fire dust sparks of silver color that reach to the ground, hanging at a height of 3 m. It is used for set fireworks "falls."

Composition of this kind, which contains nitrate and aluminum as well as a small amount of water, sometimes generates enough heat to catch fire when it is left in the wet state for a long time. The heat comes from the reaction between nitrate and aluminum, which creates H_2, NH_3, NO, or NO_2. To avoid accidents, never leave the composition in the wet state. The addition of 0.5% boric acid usually prevents the composition from heating up.

In general, it is difficult to ignite stars that consist of the two components (i.e., potassium nitrate and aluminum). When the amount of aluminum increases to 50%, it is impossible to ignite it by ordinary methods, but the larger the amount of aluminum, the more beautiful the sparks are, and we add other ingredients to the composition to obtain good ignition even if it contains a large amount of aluminum. Sulfur is the most suitable ignition ingredient, but it seems to create trouble when stored for long periods. Next, in order of preference, antimony trisulfide or charcoal is recommended, but realgar is not as good for this purpose. The color of the fire dust sparks depends on the burning temperature, and accordingly it is peculiar to the kind of fuel used. The following compositions are all of practical use, but note that they are difficult to ignite when they are made into stars, and it is necessary to cover the stars with an easily ignitable composition that is specially selected.

	Weight (%)
Golden Wave A	
Potassium nitrate	40
Aluminum	50
Antimony trisulfide	10
Soluble glutinous rice starch	5 (additional %)
Golden Wave B	
Potassium nitrate	40
Aluminum	50
Sulfur	10
Soluble glutinous rice starch	5 (additional %)
Golden Wave C	
Potassium nitrate	40
Aluminum	50
Realgar	10
Soluble glutinous rice starch	5 (additional %)

Compositions A and B produce very fine golden fire dust sparks. Composition C produces reddish golden sparks, which are of special elegance and are different from the sparks of A and B.

Potassium Perchlorate and Aluminum Spark Composition: A composition of potassium perchlorate and aluminum is more ignitable than the aforementioned potassium nitrate compositions, and it can be used without adding other fuel. We can increase the amount of aluminum to 75% without losing its ignitability, but when the amount of aluminum is over 65%, the fire dust branching is not as good. The practical compositions are as follows:

Silver Waves	A Weight (%)	B Weight (%)	C Weight (%)	D Weight (%)
Potassium perchlorate	50	45	40	35
Aluminum	50	55	60	65
Soluble glutinous rice starch	5	5	5	5 (additional %)

The amount of soluble glutinous rice starch in the preceding formulas also is suitable for pressed stars, and in round stars it should be 8% to 10% for ease of manufacture, but in this instance they are more difficult to ignite. The compositions that contain metal powder, such as golden waves or silver waves, are less easily ignited than ordinary compositions for stars and must be formed by means of a special process described later (see page 417). Composition A appears with a short silver stream that has a short life in the air. As the composition changes, its constitution from B to D, the brilliancy of the stars gradually diminishes and the stream of the sparks becomes larger and longer.

Magnesium cannot produce such good fire dust sparks as aluminum when it is used with potassium nitrate or potassium perchlorate because magnesium is vaporized at the relatively low temperature of about 1100°C and burns completely without producing hot liquid or solid matter that can be projected out of the flame.

The phenomenon of the fire dust sparks is somewhat complicated, but it may be possible to plan various compositions for practical use by the application of the fundamentals described previously.

FIRE-BRANCHING SPARKS COMPOSITION

The so-called "fire-branching sparks" is a phenomenon produced when liquid or solid matter of high temperature is projected into the air and explosively separates into many fire branches like a pine needle. It therefore is different from the fire dust spark described previously, which does not branch explosively. This kind of spark is seen at short distances. There are two kinds of composition for the sparks: One contains charcoal, and the other contains iron or magnesium powder. The former belongs to the toy fireworks "senko hanabi" and the latter to "fountains," "sparklers," and so forth.

The spark composition using charcoal consists of potassium nitrate, sulfur, and charcoal and belongs to the blackpowder-type of composition, but the peculiar distinction from blackpowder is that the amount of charcoal is less than the amount of sulfur. The composition that produces good sparks contains 10% to 15% of charcoal. It seems that any kind of charcoal or soot, except crystalline carbon such as graphite, can produce the pine needle–like sparks, but the finer the particles of carbon, the better the sparks. Accordingly, pine soot is best, followed by paulownia charcoal or hemp coal. Pine charcoal also can produce sparks, which are not as large as those of pine soot. A test showed that carbon black produces almost no sparks, although it does create, on rare occasions, very large ones. The examples of the formulas used in old times are:

Senko Hanabi (Japanese Sparkler):

	A (%)	B (%)	C (%)	D (%)	E (%)
Potassium nitrate	50	61	36	48	59
Sulfur	34	24	22	34	24
Pine soot	—	15	7	12	14
Paulownia charcoal	8	—	35	—	3
Pine charcoal	8	—	—	6	—

In the author's laboratory, the compositions were studied by observing the sparks that are projected from the hot liquid drops produced by burning compositions (of pine soot, potassium nitrate, and sulfur) that are pressed in thin paper tubes. The results are shown with a triangular graph as in Figure 23.1. The area with the single shadow shows the spark-producing zone, while the double shadow shows the zone with excellent sparks.

When we use other oxidizers in place of potassium nitrate, we do not get these pine needle–like sparks, which are peculiar to senko hanabi. Sulfur is replaceable by realgar, but other materials are unable to produce such sparks. Charcoal is not replaceable by other materials.

To manufacture senko hanabi (Japanese sparkler), about 0.1 g of the composition is twisted in a Japanese paper tape (20 × 2.5 cm) or cong-

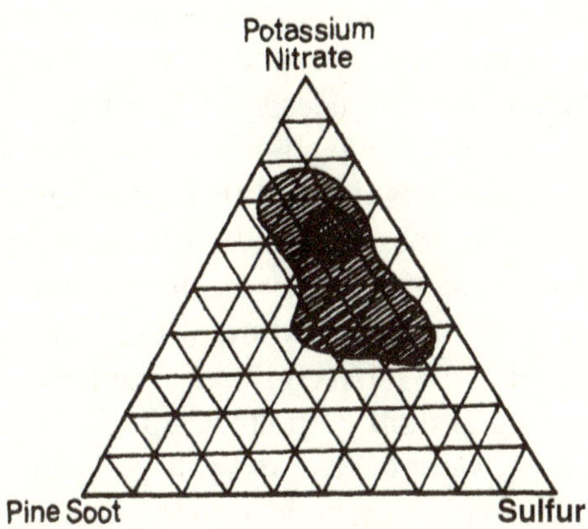

Figure 23.1 Japanese sparkler compositions

lutinated on one end of a rush halm. When it is ignited, it burns violently with a flame at first; then the remaining ash keeping its red-hot state, contracts itself to a small red-hot ball, which is the so-called "fireball." After a few seconds, the temperature of the fireball gradually rises and fine particles begin to fly out of the ball. The particles become more brilliant at a short distance from the ball and explosively branch into pine needle–like sparks. The fireball is red-hot and slightly transparent. The material in the ball moves around, showing that it is reacting. As the reaction of the fireball gradually becomes weaker, the sparks become like willow without branching and at last cease. The temperatures of the fireball measured by Dr. Nakaya and his colleague Sekiguchi are:

Before projecting sparks	850°C
When the sparks are projected most actively	930°C
At the end of active spark projection	830°C

These data derive from the average of the values of 10 measurements, respectively. The sparks projected from the fireball consist of melted and reacting matter. The reaction of the fireball is very active in oxygen and ceases in nitrogen. When the fireball is blown in a stream of air, the sparks become very active at once, and we know from these facts that a proper amount of oxygen is necessary to produce the sparks.

When we analyze the matter that makes a fireball, we cannot find potassium nitrate. The main components are potassium sulfide, potassium sulfate, and carbon. When we heat the potassium sulfide, which is separated from the fireball matter (ash), to redness, it melts, becomes a red-hot ball, and actively bubbles. In this case, the red-hot ball is covered with a thin smoke layer, which is thought from its smell to be sulfur dioxide. When the cooled and solidified fireball matter is heated to red heat with the flame of an alcohol lamp, it becomes active again and produces sparks, but when it is inserted in the reducing part of the flame, the reaction, and accordingly the sparks, suddenly cease. It is thought that the chief components that cause the sparks may be potassium sulfide and carbon in the fireball. If this is true, we also can artificially create the same phenomenon as senko hanabi when we make a mixture of these two components, of an adequate ratio, and heat it. Accordingly, the following experiment was tried: 66% of potassium carbonate, 30% of sulfur, and 4% of lamp soot (pine soot) were heated at a relatively low temperature in a porcelain crucible, being melted to make a mixture of potassium sulfide and carbon. A small amount of the product was fixed at an end of a nichrome wire and heated by the oxidizing flame of an alcohol lamp. As expected, the matter became a red-hot fireball in an active state and produced sparks that were the same as those of senko hanabi.

The mechanism of producing sparks from the fireball is not yet known, and when we burn the senko hanabi, the fireball is apt to fall down from the support, and the viscosity of the matter in the hot state also must be studied further.

Spark composition using metal: The fire branching of sparks is interesting where iron is concerned. Dr. Nakaya studied the iron sparks that are created by a grinder. In this case, pure iron does not branch any sparks, but when the carbon content increases, the sparks branch two or three times; and when the carbon content increases to more than 7%, the shape of the sparks does not change any more. Thus we can tell the amount of carbon in the iron by observing its sparks, when the carbon content is less than 7%. The iron sparks are very similar to those of senko hanabi.

When we use iron or magnesium powder as the ingredient for sparks, it must be coated with some resistant material to prevent it from rusting. Paraffin wax or a benzene solution of Japanese varnish or the like can be used for this purpose. The examples of spark formulas are:

Volcano	Weight (%)
Potassium nitrate	55
Sulfur	9
Hemp coal	13
Coated iron powder	23

This composition is pressed in a rather thick tube or cone with a fuse attached. It is ignited by the fuse and produces sparks through the hole where the fuse is attached.

Fountain	Weight (%)
Ammonium perchlorate	70
Coated magnalium powder	30

It is used in the same way as the volcano. The coating of magnalium must be especially good; otherwise the life of the composition is very short.

ILLUMINANT COMPOSITION

A mixture of magnesium powder and barium nitrate usually is used because it produces the most intensive light, and an aluminum and barium nitrate composition also is used for fireworks. The color of the light generally is white, but in the case of the former, it appears to be whitish green. Magnalium also is sometimes used, but the composition sometimes burns so vibrationally that it is not particularly recommended, although stars that burn vibrationally have been used recently to exhibit special beauty. When we use oxidizers other than barium nitrate or barium salts, the intensity of the light diminishes. Barium oxide in the flame has a strong molecular spectra, and the condensed particles in the flame also emit a strong continuous spectrum. This is why the light intensity of barium composition is high.

Magnesium Illuminant	Weight (%)
Magnesium powder coated with 3% of paraffin	50
Barium nitrate	47
Linseed oil	1.5
Caster oil	1.5

This composition, which is pressed in a paper tube of about 10 cm in diameter by a hydraulic press to a density of 2.00 g/cm^3, burns at a rate of 1.9 mm/sec, and the candlepower of the light is 5,000 to 7,000/cm^2 of the burning surface. To obtain a green light, the ingredients that produce chlorine or hydrogen chloride gas in the flame (e.g., benzenehexachloride or vinyl chloride) are added to the preceding formula. The amount of this ingredient should be 10% to 15%. This type of ingredient also is effective for the following blue and red colored flame compositions. To obtain other colored lights of high intensity, barium nitrate is replaced by potassium perchlorate, and a suitable color-creating ingredient, such as strontium carbonate for the red flame, sodium oxalate for the yellow flame, or Paris green for the blue flame, is used. In the case of blue flame, the amount of magnesium in the composition should be less than 20% to obtain good color. When we use ammonium perchlorate as the oxidizer, the flame color tone is excellent, but the smallest amount of moisture will cause a reaction between magnesium and ammonium perchlorate, thus creating magnesium perchlorate, which is very hygroscopic and damages the composition. Accordingly, the composition that contains magnesium and ammonium perchlorate always must be kept in a perfectly dry state. Strontium nitrate behaves as an oxidizer as well as a color-creating ingredient. It is not easy to store a composition that contains magnesium for long periods, and the pressed composition effloresces gradually on the surface that is in contact with air. To avoid this possibility, the composition is pressed in a thin metal case or is packed with metal foil to protect the composition from moisture in the air. When we make stars with this kind of composition, water must not be used; only water-free organic solvents are permissible.

Aluminum Illuminant for Cores of Stars	Weight (%)
Barium nitrate	67
Aluminum	27
Soluble glutinous rice starch	6

This composition is effectively used as cores of stars. When it is used as an ordinary illuminant pressed in a container, the starch is unnecessary. The stars of this composition must be covered with another composition of high ignitability. It is difficult to obtain a good colored flame with aluminum composition.

SOUND COMPOSITION

There are two kinds of sound composition: report composition and whistle composition. The composition called "thunder" belongs to the former. The explosive sound seems to be sharpened with the rate of explosion. The relationship between the nature of the explosive sound and the rate of explosion is an interesting problem, which has not yet been made clear. The description of sound as "round" or "sharp" may have some relationship to the rate of explosion that is peculiar to the composition. The report composition not only must have a rapid rate of explosion, but also must be easily ignited so that the explosion can be produced easily. Generally, this type of composition has a high sensitivity, and accidents occur frequently with sound composition. Lately, safer compositions without chlorate as the oxidizer have become increasingly popular.

POTASSIUM PERCHLORATE REPORT COMPOSITION

It is said that perchlorates are relatively safe to handle, but ammonium perchlorate is not suitable for report composition because of its hygroscopic nature, and only the potassium salt is used. Only aluminum is used for the heat-creating ingredient. If ferrosilicon is used in place of aluminum, the composition does not easily explode on ignition, and even if it explodes, the sound is very low. It is necessary for the aluminum to be as pure as possible and in very fine powder form. Sulfur or antimony trisulfide is used as the fuel, but there is almost no difference between them. The latter would be recommended only when the material must be kept for a long time. It would be marginally safer to use antimony trisulfide rather than sulfur if potassium chlorate is the oxidizer.

The intensity of sound and the sound-producing zone have been studied in a composition that consists of three components—potassium perchlorate, aluminum, and sulfur—and the results are shown in the triangular diagram in Figure 23.2. The shadowed part with single inclined parallel lines shows the zone in which the sound is produced, and the zone marked with crossed parallel lines shows the composition of the greatest sound. The sound zone is fairly wide. The amount of aluminum generally is increased as much as possible to obtain a strong flash. An example of the composition used is:

Thunder No. 1	Weight (%)
Potassium perchlorate	50
Aluminum	23
Antimony trisulfide	27

When sulfur is used in place of antimony trisulfide, the sound changes very little. In effect, it is necessary to select the most adaptable

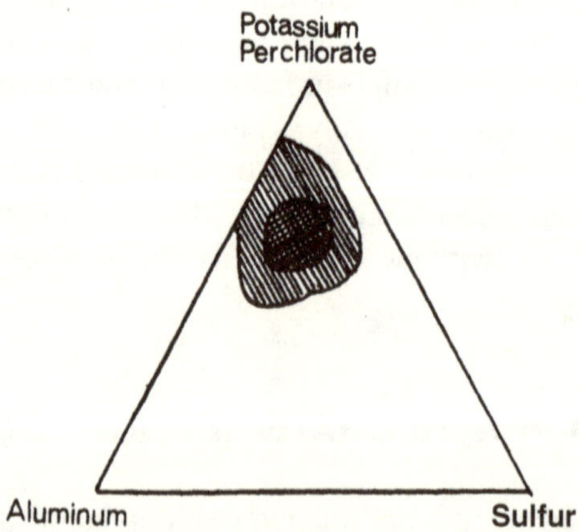

Figure 23.2 Flash composition

and economical composition in the sound-creating zone for our purpose. This kind of composition does not explode but only burns when a small amount is ignited in a free state in the air. When the amount is large, the burning changes instantaneously to detonation. The limit of the amount is likely to be about 50 to 100 g. Accordingly, care must be taken not to accumulate large amounts of composition in the workroom because the destructive power of the detonation caused by this composition is enormous. This kind of composition is not easily exploded by friction or shock.

When this composition is exploded by ignition with a fuse, the resistance provided by the container or the packing has a great influence. With increasing pressure, the detonation limit of the amount of powder becomes smaller than would be the case in a free state (i.e., an unpacked state), and the time interval from the ignition to the detonation is very much shortened. This is a very important characteristic of a powder, especially in the case of aluminum composition. A composition that contains ferrosilicon in place of aluminum is not as strong as aluminum composition. We must make the packing strong and ignition surface as wide as possible by attaching a small amount of blackpowder paste to the end of the fuse in order to obtain a good report.

The influence of loading density on sound is not as clear. From a test, changing the density from 0.5 to 1.2, it seemed that an increase in density gave an increase in sound. A loading density higher than this value will not always give a good result. Aluminum in fine powder form gives a louder sound than when it is of large particle size, but in practice the influence of particle size is very small compared with that of the strength of the container.

POTASSIUM CHLORATE REPORT COMPOSITION

The potassium perchlorate report composition described previously is limited because a minimum quantity is required for detonation, and it cannot move to detonation from ignition with less than this amount. It is difficult to produce a report with a very small amount. Moreover, its characteristically low sensitivity against shock and friction sometimes is a drawback for firework purposes. Potassium chlorate composition remedies these defects, and although it is dangerous, there is still occasion to use it.

Red Explosive Composition	Weight (%)
Potassium chlorate	63
Realgar	37

This is called "red explosive composition" because of its red color, and the preceding formula is an example of this kind. It is very sensitive to shock and friction and is very dangerous to handle. In fact, it is recommended that the wet process be used to handle it safely. The sound produced by this composition gives a finer sound wave than potassium perchlorate composition and is sharper than the latter. On the other hand, the strength of the sound produced by this composition seems to be weaker than that of the potassium perchlorate composition. A very small amount of the red explosive composition can produce detonation. Actually, it detonates perfectly, even when the strength of the container is small and even when the amount of the composition is very small. This composition is used for toy "cracker balls" and the like.

Toy Pistol Cap	Weight (%)
Potassium chlorate	60
Red phosphorus	8
Sulfur (or antimony trisulfide)	32

One example of this is shown above. This composition is more dangerous than the red explosive composition, and it is so sensitive to shock and friction that it must be produced by the wet process. Red phosphorus must be free from white phosphorus and phosphoric acid. In place of sulfur, antimony trisulphide also is used without much change in the character of the composition. The composition of this formula is used for toy pistol caps, but the composition in which sulfur is replaced by antimony trisulfide is used for igniters because of its large flame.

Thunder No. 2	Weight (%)
Potassium chlorate	43
Antimony trisulfide	26
Aluminum	31

This formula looks like the potassium perchlorate composition, but this composition is more sensitive to shock and friction than the perchlorate composition. It is used in Japan because of its low price, its low limit of the minimum amount for detonation, and its ease of detonation even in a rather weak container. It is sometimes called "flash thunder," the name that comes from the dazzling flash of light in contrast to the realgar composition, which produces only a weak light. It is the same with perchlorate composition, and to obtain a strong flash, the amount of aluminum is increased at the expense of the oxidizer, provided that it does not disturb the production of the report. It also is possible to use sulfur in place of antimony trisulfide in the preceding formula.

POTASSIUM PICRATE WHISTLE COMPOSITION

The sound of the explosion is used as a report in the preceding instances. But the following composition is totally different because it gives a sharp musical sound under special burning conditions.

Whistle No. 1	Weight (%)
Potassium picrate	63
Potassium nitrate	37

The Manufacture of Potassium Picrate: A wooden tub (30 L) is prepared, and 20 L of hot water is poured into it. Steam is used to bring the water to the boiling point, and the steam pipe should be made of aluminum. While injecting the steam, 2 kg of picric acid is added and completely dissolved. Then 600 g of potassium carbonate is charged, a little at a time. It reacts with effervescence, and part of thc potassium picrate begins to form because of its low solubility. The end point of the reaction is determined by the sudden cessation of the bubbling. At this time, the charging of potassium carbonate is stopped, and then picric acid is charged to the tub. The overcharged potassium carbonate reacts with effervescence, and the process is finished when the bubbling ceases. The liquid is slightly acidic, and the purpose of ending in the acid state is to be able to filter the liquid easily afterward and to avoid giving a hygroscopic character to the product. The contents of the tub are removed to another aluminum tub, which is cooled from the outside with cold water, the contents being stirred with a wooden stick. When it is cooled too slowly, it produces crystals that are too large, and it is inconveniently dangerous to pulverize them. Thus it must be cooled as quickly as possible to obtain very fine crystals. When the cooling process is finished, the contents are removed to a porcelain funnel that contains a filter paper. The product is spread on a paper sheet and dried in the sun. The yield is about 104 parts of potassium picrate per 100 parts of picric acid by weight. Potassium picrate consists of yellow needle-shaped crystals that are rather difficult to dissolve in water. Its solubility in water is about 6 g per liter at normal temperature. Potassium picrate is rather sensitive to shock and friction, but not as high as that of red explosive composition. Even when it contains a small amount of moisture, it explodes. The burning rate in the air in the free state is very fast, and great care must be taken when handling it. The picrate should be prepared only in small quantities when necessary. The manufacturing process is very simple, but some skill is required. Care must be taken that the abandoned mother liquor of the reaction does not pollute the effluent system.

The standard formula is given as Whistle No. 1. Less potassium nitrate causes an explosion and more produces no whistle. Potassium picrate and potassium nitrate are mixed by a hair sieve in accordance with the mixing process for dangerous compositions described previously. This composition seems safer than potassium picrate alone, according to the results of the drop hammer test.

Gallic Acid Whistle Composition and Other Whistle Composition

The standard formula of the composition that contains gallic acid is:

Whistle No. 2	Weight (%)
Gallic acid	25
Potassium chlorate	75

This composition is very sensitive to shock and friction. The degree of danger in handling it is almost the same as in that of red explosive composition. The mixing process is the same as that of Whistle No. 1. The composition becomes less sensitive when more than 15% of water is added to it.

A mixture of salicylate of sodium and potassium perchlorate is recommended as a safer composition. The whistle is somewhat smaller than that of the preceding composition, and this composition must be protected from moisture because it is very hygroscopic.

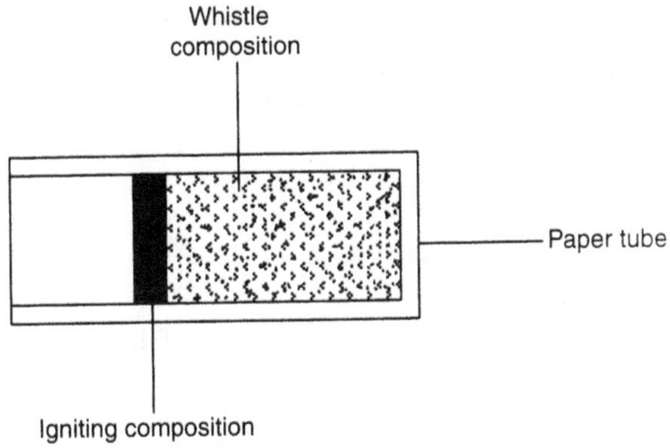

Figure 23.3 Firework whistle

Whistle composition generally is pressed into a paper tube, leaving half of its length unloaded. It is very important to press the composition firmly to avoid an explosion when it is burning. The paper tube therefore should be inserted in a mold and the composition pressed in it several times with a hand press. This process should take place behind a defense plate. The composition is carried in a safe container in small amounts from another room to the workroom to avoid a large unexpected explosion. An example of the product is shown in Figure 23.3. The paper tube should be made of dense material. The ratio of the diameter of the tube to the length should be less than 1:3, but when the tube is too long, it cannot produce a good effect (i.e., the larger the diameter, the louder the whistle, and the tone decreases accordingly).

In the case of a very small tube of 3 mm in diameter, the potassium picrate composition will not whistle, and only the gallic acid composition is usable. The frequency is about 2,600 cycles per second.

SMOKE COMPOSITION

Pyrotechnic smoke for practical use can easily be obtained from a simple smoke device without any special technique and must be safe to handle. Only the smokes that are used for fireworks are discussed here.

The processes are classified as:

1. The burning product grows to form a mist, which absorbs the moisture from the atmosphere. For example, the hexachloroethane

smoke composition creates a vapor of zinc chloride or aluminum chloride, which becomes white smoke when it combines with moisture in the air. Carbon tetrachloride smoke composition is seldom used at present.

2. The material is first vaporized by the heat of combustion and then condenses again to fine solid particles that create smoke. The white smoke caused by sulfur and the yellow smoke caused by realgar belong to this type.
3. Smoke caused by incomplete combustion of a composition. This type of smoke is widely used as black smoke in fireworks. It uses the carbon particles created by the incomplete combustion of naphthalene or anthracene.
4. Colored smoke produced by vaporizing dye. A volatile dye is first vaporized and then it condenses the fine solid particles, which look like colored smoke. Generally, this kind of smoke shows its own beautiful color by reflected light, but it looks dirty when the light is weak. It does not have good color in a cloudy sky. Generally, the color of the aqueous solution of dye from permeated light is the same as the color of cloudy solid particles in reflected light. Physically, it is an interesting problem.

Hexachloroethane Smoke Composition

This composition consists of three components: hexachloroethane, zinc, and zinc oxide. The burning product of this composition is mainly the vapor of zinc chloride, and the temperature of the burning reaction may be about 900°C. The composition generally is contained in a completely moisture-proof can. The burning time of this composition is shown in Figure 23.4; the dimensions of the smoke container were 80 mm in diameter and 114 mm in height.

The influence of the particle size of zinc oxide on the burning rate of the composition is remarkable. Three kinds of zinc oxide were used for a composition of 50% hexachloroethane, 28% zinc dust, and 22% zinc oxide, and were tested to give the following results:

Class of Zinc Oxide	A	B	C
Duration (sec)	227	267	347
Color	grayish	grayish	pure white

The quality of zinc oxide of class C is better than that of class A, and the quality of class B is better than that of class C. In short, the better

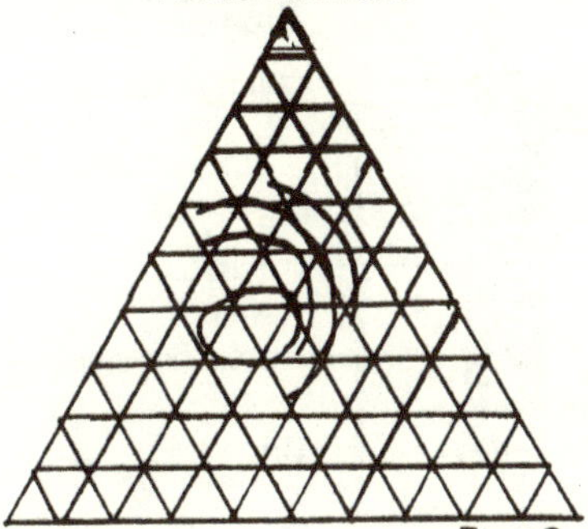

Figure 23.4 Smoke composition

the quality, the longer the duration. Zinc of good quality makes the color of the smoke pure white.

The quality of zinc dust and the particle size of hexachloroethane did not show such a marked influence on the burning time. A particle size of hexachloroethane of less than 2 mm can be used successfully.

A mixture of zinc dust and hexachloroethane reacts violently when water is added to it, drop by drop. It should be noted that this kind of smoke sometimes causes fire in the presence of moisture. This composition never explodes with shock or friction like other explosives. The so-called "Explosion of Smoke" seems to be the explosion of an aluminum composition, which is contained in the igniter for the smoke, or the explosive bursting of the tin by interior expansion, caused by the heat evolved when the water entered the tin.

Smoke composition is so completely sealed in a tin that it must be specially designed to ignite without fail. Generally, a brass cap, the thickness of which is less than 0.2 mm, is soldered into the tin and a quantity of thermite is charged into it. An igniter star is inserted on the top, as shown in Figure 23.5. Ignition takes place when the head of the igniter star is rubbed with a wooden striker that has been painted with a composition of red phosphorus and antimony trisulfide. Next the thermite is ignited, thus generating a large amount of heat, which melts the brass cap, and the hot melted metal drips directly on the smoke composition, infallibly igniting it.

The manufacturing process for smoke composition is not dangerous, but care must be taken to reduce the moisture in the ingredients as much as possible, and frequently the temperature of the composition during mixing should be tested by hand or by some other method.

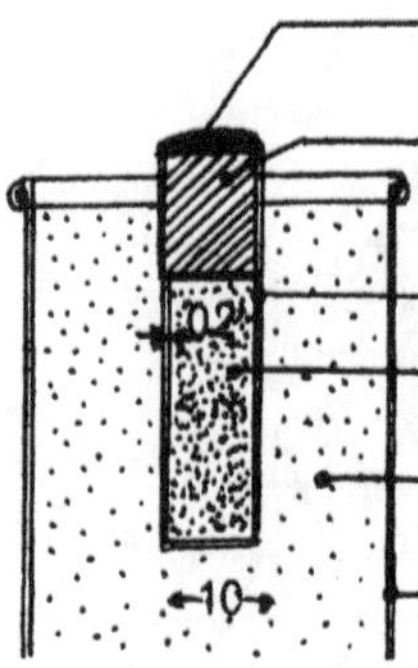

Figure 23.5 Smoke cannister

During the mixing process the light zinc oxide is first charged into the mixer and the heavy zinc dust is charged last. When a rotary mixer is used, the number of revolutions per minute should be fewer than 20. It takes about 20 minutes to mix 100 kg of the composition. The volume of the mixer is designed so as to have two-thirds of the total volume empty with the volume of the charge occupying one-third, the apparent specific gravity of the smoke composition being 1.1 to 1.5.

The composition is sealed in the tin with a canning machine, but it is rather different from the canning of food and is more difficult, especially with a large smoke. The mouth of a can must be an exact circle and must not be warped by the contents. The grove of the cap is filled with gum solution before canning to keep it airtight.

There is another composition in which the zinc powder is replaced by aluminum powder. This composition creates grayish smoke and is not often used in Japan.

Sulfur Smoke Composition

The smoke is white and is easy to generate, and this composition neither reacts with water nor is hygroscopic. Therefore it is used widely in fireworks. The smoke consists of solid particles and is not affected by moisture in the air.

An amount of potassium nitrate and the same amount of sulfur are mixed well together, but a small amount of realgar is added so that it will ignite easily and smoke smoothly.

Sulfur White Smoke	Weight (%)
Potassium nitrate	48.5

Sulfur	48.5
Realgar	3.0

This composition is pressed firmly into a paper tube, both ends of which are sealed with gypsum, and a hole is bored into the composition on the side near to one end. A piece of quickmatch is inserted into the hole. When it is ignited, white smoke issues through the hole. The hole must not be too large or else the smoke will burst into flame. This composition also can be used for white smoke stars when it is mixed with soluble glutinous starch and shaped into cylindrical stars, each of which is covered with gypsum, and then a hole is bored into this gypsum to allow the smoke to come out. Care must be taken not to cause ignition during the drilling operation. A small amount of hemp coal sometimes is added to the composition for good ignition.

Realgar Smoke Composition

A composition that consists of potassium nitrate, sulfur, and realgar can produce smoke of white, light yellow, deep yellow, or purplish gray according to its formula. Yellow smoke is the most popular, and an example is:

Realgar Yellow Smoke	Weight (%)
Potassium nitrate	25
Sulfur	16
Realgar	59

In this formula the percentage of sulur is rather small. The method of generating smoke is the same as that described in (2). The moisture in the composition seems to have a little influence on the burning rate of this composition. In the case of smoke stars for shells, which are used for willow or chrysanthemum, the sulfur is increased much more to make the burning time shorter. For this purpose, an amount of sand sometimes is added to the composition or the stars are covered with gypsum so that they burn quickly under pressure as described in (2). An example of the formula for stars used in a willow shell is:

Yellow Willow	Weight (%)
Potassium nitrate	43

Sulfur	10
Realgar	37
Hemp coal	4
Soluble glutinous rice starch	6

The following composition is used for daylight shells. The smoke looks white, but this may be only a smoke from a burning process, and it may differ from the smoke generators mentioned previously.

White Chrysanthemum	Weight (%)
Potassium nitrate	66
Realgar	13
Lampblack	5
Hemp coal	5
Soluble glutinous rice starch	11

The volume of lampblack is so large that a small amount of starch is used as a binder. The distinguishing feature of this formula is that it contains no sulfur.

Anthracene or Naphthalene Smoke Composition

When a carbon-rich compound such as anthracene or naphthalene mixed with some oxidizer is burned, it creates carbon particles as a result of incomplete burning. (If the amount of oxidizer is too large, a good smoke is not produced.)

For example:

Anthracene Black Smoke I	Weight (%)
Potassium perchlorate	56
Sulfur	11
Anthracene	33

This composition is filled, pressed into a tin, and ignited through a small hole. When a flame blows out of it, a thick black smoke is generated from the flame. This composition is different from the other because the flame appears first and then generates smoke from it. (In many other types, the production of flame generally disturbs the smoke.) The temperature of the flame may be about 800°C. This smoke is adaptable for a black "Dragon" with a parachute, but this composition is a little difficult to ignite and burning is only stabilized under pressure, which is maintained by the smoke case. It is not easy to ignite this composition, even when it is in a powered state, and a special method must be applied (e.g., using thermite). The composition is safer to handle than the following composition, Naphthalene Black Smoke I, and care must be taken because of its explosive nature.

Naphthalene Black Smoke I	Weight (%)
Potassium chlorate	44
Antimony trisulfide	24
Naphthalene	26
Soluble glutinous rice starch	6

This composition is used to make stars that produce a beautiful black smoke, but it also is more explosive than Anthracene Black Smoke I. Naphthalene sublimes out of the star when it stands for a long time, leaving behind potassium chlorate and antimony trisulfide, which increases its explosive nature. Despite this disadvantage, it still is used widely in Japan because it is the most beautiful of all black smokes, although it is better not to use this composition if possible.

Naphthalene (Anthracene) Black Smoke II	Weight (%)
Hexachloroethane	62
Magnesium	15
Naphthalene (or Anthracene)	23

This composition has no oxidizer. The hexachloroethane and magnesium react with each other, producing a smoke of magnesium chloride, and the heat developed causes the isolation of carbon particles from the naphthalene, thus making the color of the smoke black. This composition is loaded into a tin and used as a signal.

Naphthalene (Anthracene) Black Smoke II	Weight (%)
Potassium perchlorate	57
Anthracene	40
Hemp coal	3
Soluble glutinous rice starch	7 (additional %)

This composition is a modification of the composition Anthracene Black Smoke I for stars, but it is more ignitable.

Colored Smoke Composition with Dyestuffs

Dye smokes show their natural color by reflected light. The adjustment of color is achieved by the same principle as that of mixing dyestuffs for printing. A colored smoke composition consists of dyes, a small amount of oxidizer, and some kind of carbohydrate, which adjusts the burning temperature. The composition is pressed into a case or is made into smoke stars, adding a binder (e.g., glutinous rice starch). Potassium chlorate is used as the oxidizer. Starch or wheat flour is used to adjust the temperature and provide a fuel. Sugar, or milk sugar, creates very beautiful colored smoke, but it is not widely used in Japan because of its rather high cost.

Dextrin also is a good ingredient for this purpose. The creation of flame disturbs the generation of smoke, so the smoke must issue out of a small hole in the container after being rapidly cooled, or, in the case of stars, they must be projected rapidly through the air to extinguish the flame. The burning temperature of this type of smoke composition is about 400°C to 600°C, and some part of the dye may be damaged at this temperature. If the temperature rises to a higher level than this during the burning process, the smoke becomes colorless. A few examples of formulas for colored smoke compositions are shown:

Red Smoke	Weight (%)
Potassium chlorate	25
Rhodamine B	24
Para red	36
Wheat flour	15
Blue Smoke:	
Potassium chlorate	28
Methylene blue	17
Indigo pure	40
Wheat flour	15
Green Smoke:	
Potassium chlorate	28
Auramine	10
Methylene blue	17
Indigo pure	30

Wheat flour	15
Violet Smoke:	
Potassium chlorate	26
Indigo pure	22
Rhodamine B	16
Para red	21
Wheat flour	15

In the case of red smoke, if the amount of rhodamine B is increased, the smoke becomes reddish violet; when decreased, it becomes orange red. In the case of green smoke, auramine is somewhat hygroscopic, so it is advisable to protect the composition from moisture.

Some people often add sulfur to the composition to increase the burning of smokes, but it is not recommended in view of the increased sensitivity. To increase the burning rate, it is better to widen the burning surface by granulating the composition, with the smallest sized grains determining the burning time.

Potassium nitrate and potassium perchlorate are not as good oxidizers as potassium chlorate because they cannot create colored smoke as smoothly as potassium chlorate and the percentage in the composition must be greater than that of potassium chlorate. Other heat-producing materials, such as celluloid and nitrocellulose, also are used in place of oxidizer and fuel, but the ratio of the amount of this material to the amount of dye must be increased more than the ratio in the preceding formulas, and the volume of the smoke decreases accordingly. Guanidine nitrate also is used as a low-temperature burning material and is combined with other heat-producing materials (e.g., celluloid, nitrocellulose) to protect the dye from the heat and to keep the ash porous. For example, a mixture of 40% guanidine nitrate, 35% celluloid powder, and 25% dye makes as good colored smoke except in the case of a dye with a high sublimation point, such as indigo. This principle also is applied to insecticidal smoke. Nevertheless, the method of using heat-producing materials in place of oxidizer and fuel is not as adaptable for fireworks because of the rather thin smoke, but it is recommended for some large signals of long duration.

POWDER PASTED PAPER

Powder pasted paper is made of Japanese paper with some kind of composition pasted on it and is widely used as fuse and igniting material.

The paper, on which colored flame composition is pasted, is used especially for the shell "Falling Leaves."

The kind of the composition for powder pasted paper depends on its use, and it must be selected according to the purpose.

Blackpowder Pasted Paper: Four parts of blackpowder is mixed with three parts of water in weight ratio. In winter, potassium nitrate easily oozes out of the pasted blackpowder, and we must reduce this escape by heating it with a safe form of heat. Sometimes soluble glutinous rice starch is added to the paste to give it adhesive power, but when we use grains of gunpowder that are manufactured by milling, the adhesive power is so great that there is no need to add starch to the paste. Newspapers are fixed on a worktable as a protective cover, and Japanese "Kozo" papers are spread out on them. Then the blackpowder paste is uniformly pasted on the papers with a high-quality brush. The pasted papers should be dried on a drying frame in the sun, so a bright, clear, and calm day must be chosen for manufacture. Generally, infrared driers or other heat sources do not give as good a result as the sun because of the lack of uniformity in drying. When pasting, the powder paste must be continuously stirred well to prevent the components from separating from each other. The quantity of the composition pasted on one side of the paper is 0.01 to 0.02 g/cm^2 for fuse, 0.01 g/cm^2 for ordinary use, and 0.02 g/cm^2 for special ignition material. The thickness of the powder pasted with the quantity 0.02 g/cm^2 is rather large, and it must be pasted on one side two or three times to obtain such a thickness. For use as a fuse, paper pasted on both sides generally is used, but paper pasted on one side is used for other purposes. Test pieces (10 mm in width) taken from a thick paper, pasted on both sides with gunpowder, showed the burning rate of 1.7 cm/sec in the atmosphere.

Red Thermite Paper: A composition of red lead oxide (Pb_3O_4) and ferrosilicon, (e.g., in a ratio 4:1) called "red thermite," is mixed to a paste with 5% celluloid solution in amyl acetate. It is pasted on a Japanese "Kozo" paper like the blackpowder pasted paper. It is preferable to dry it in the sun because the use a drier sometimes is dangerous because the vaporized gas of amyl acetate may come into contact with an overheated part of the drier and cause an explosion. Thermite paper is easily ignited by friction, and care must be taken when handling it. This paper can be effectively used for igniting smoke compositions that are difficult to ignite. In this case, the thermite paper is used together with a blackpowder paper, one on top of the other, or sometimes by pasting red thermite on a blackpowder pasted paper. The ratio of the amount of minium to ferrosilicon can be changed to adjust the burning rate of the paper. When the value of the ratio is decreased, the burning rate is delayed.

Colored Fire Paper: This is manufactured the same way as blackpowder pasted paper using a colored flame composition that is used for stars. Sometimes, immediately after pasting the composition on the paper, the pasted composition is covered with another paper to hold the composition well between the two papers. When necessary, it is manufactured in several layers like a sandwich and is used as a part of the shell "Falling Leaves."

Potassium Nitrate Paper: A solution of potassium nitrate in water is spread on a Japanese paper and dried. This paper is used for igniting or improving the ignition of some fireworks. If letters or pictures are drawn with the solution on paper, they disappear when it is dried, but when it is ignited at one point of the painted part, the fire runs only on the drawn line, and the original letters or pictures appear again. This is used as a toy. Hidden pictures also can be made with other oxidizers.

Powder pasted papers often produce fire when they are cut and are very dangerous; the cutting of a number of multifold powder pasted papers at one time should be forbidden. Be sure to cut them in small quantities with a sharp knife, and do not leave the cut pieces near the cutter during the operation. The pieces should be quickly removed from the cutter and the cutting place, and they must be kept in an incombustible container, which must be covered with a lid. The process is successfully carried out by two people working together. There are instances of accidents caused by cutting blackpowder pasted paper, and the author has experienced a fire when cutting red thermite papers.

Fuses: The purpose of the delay fuse for fireworks is to transfer fire and ensure a time delay. It differs somewhat from an industrial fuse.

Handmade Fuse: Powder pasted paper, which is pasted with blackpowder on both sides, is cut to size (250 × 75 mm). A quantity of gunpowder is prepared, crushing the grains, which are moistened first. The paper is then uniformly sprinkled with the powder. A long and slender bamboo stick (3 mm in diameter) is placed on one edge of the longer side of the paper, and the paper is rolled on it by hand. This rolled powder pasted paper becomes the core of the fuse. A Japanese "Kozo" paper (360 × 270 mm) is attached with its one shorter side edge on the rolled core paper and rolled on it so that the paper winds the core as a cover. It is rolled on a wooden work plate with both hands until it is firmly wound, and then the bamboo stick is pulled out and removed. The first securing operation is applied to this prewound paper as follows: After being put on a thick, hard, wooden workplate, it is pressed and rolled repeatedly with another wooden plate (450 × 90 × 32 mm), which is held in both hands. In this case, care must be taken because heat sometimes is generated. The paper is then firmly wound like a stick or a pencil with thick lead. With the hand pressure of an average man, it is necessary to repeat the rolling approximately 100 times with the rolling plate. Another Japanese "Kozo" paper of the same size is wound again on the rolled stick as before to reinforce the cover of the core. The starting end of the paper is fixed on the stick with paste. The same rolling process is performed as a second securing operation. The third securing operation is then applied in the same manner as the second operation. Finally, the end of the paper is fixed on the stick with paste, and the product is dried in the sun. A fuse made by this process is 6.5 mm in diameter, and its core of blackpowder is 3.5 mm in diameter. The burning rate of this fuse is about 1.08 to 1.25 mm/sec, and the deviation is rather wide. When the firework shell explodes in the neighborhood of the highest point of the trajectory, a deviation of such a degree may be allowed because the shell travels slowly when it is higher. When a hand-rolled fuse is compared with an industrial fuse, the ignitability and ignition power of the former are better than that of the latter: The hand-rolled fuse can produce a longer flame than the latter, and on chopping or peeling, the former can hold its powder core intact, and this characteristic enables us to enlarge the ignition power and ignitability, which we cannot do with an industrial fuse. Nevertheless, it is handmade, is expensive, and has rather large time deviations. It therefore is preferable to use a machine-made fuse, but it must be specially made. For example, the cover of the core powder must be free from tarry matter because the core powder sometimes absorbs the tar from the tarred covering material and it makes the fuse unignitable.

A fuse made by the process described here is a normal one, and for 250- and 300-mm shells or even larger shells, the diameter of the core and the thickness of the cover are further enlarged.

The process described here is for the main fuses of firework shells, but it may be simplified for other uses. As a powder core, only the powder of gunpowder is used instead of powder pasted paper. The manufacturing process process in this case may be very simple.

A hand-rolled fuse made in the way described here has rather wide variations, and care must be taken in the way the fuse is selected for timed fireworks. For example, the shell "Five Reports" requires five accurate time intervals, so all five pieces of fuse must be cut from the same length, and care must be taken to separate different batches of fuses.

"Dark Fuse," which creates no visible sparks in the distance during burning, also is made by the same process except that other core compositions are used. For example:

Dark Fuse	A Weight (%)	B Weight (%)
Potassium nitrate	36	56
Realgar	45	34
Paulownia charcoal	10	10
Sulfur	9	—

Fuses that contains realgar must not make direct contact with chlorate composition.

Fuses for Toy Foreworks: The fuse must be easy to ignite with a match; it must be easily ignited by a flame and not have a thick cover. In Japan a twist of paper that contains blackpowder is widely used because of its low cost. Potassium nitrate paper is effectively used in this case.

STARS

Stars may be divided into two broad classes: one group consists of light stars and the other of smoke stars. Stars for night use belong to the former class, and stars for daylight use belong to the latter category. But there are exceptions: A special intensive-light star sometimes is used in daylight, and a smoke star sometimes is used at night, being illuminated by other light stars to give a special mysterious effect. Furthermore, stars are divided into other classes by form (i.e., cubic stars, round stars, and cylindrical stars), the manufacturing processes being different from each other.

Binder for Star Compositions: In general, stars are formed from a colored fire or smoke composition, in which a quantity of binder is mixed. In this case, water or other suitable solvent is added to it. It also is possible to form the composition in a dry state by pressing without the binder, but pressing influences the property of the stars, and the pressing process is laborious and troublesome in practice. The weakness of the star-making process using a binder is that after the stars are formed, they must be dried, warmed, or given an aging time. However, this method can be applied to various kinds of compositions, requires no special forming machine, and is not as laborious. There are two kinds of binders, water-soluble binders and those that are insoluble in water. In Japan water-soluble binders generally are used, and water-insoluble binders are used only in special stars in which water is not permissible. A water-insoluble binder with volatile solvent sometimes is used to dry stars in a short time.

Soluble glutinous rice starch is used mostly as a water-soluble binder in Japan, and it can be mixed in a composition as a powder. Composition that contains this starch becomes a paste when we add water only to the composition and knead it by hand. The process of pasting is simple, and the starch contains no acidic matter, so it can be safely used. The amount of starch in a composition should be small—just enough to solidify the stars. When there is too much starch, it makes stars unignitable or lowers their burning temperature. Generally, the starch unfavorably influences the burning characteristics of the stars. The amount of starch required for solidifying the stars is not fixed definitely but varies according to the kind of composition. For a black-powder-type spark composition or colored flame composition, the amount is 2% to 6%, and for aluminum composition it is necessary to increase the amount to 7% to 10%. Round stars containing aluminum are especially difficult to form when the amount of starch is too small.

Gum arabic and dextrin also are soluble in water, but the viscosity of the solutions is so low that it cannot be compared with that of glutinous rice starch. For example, when 20% solutions (20 g of the matter in 100 cc of water) are compared, gum arabic or dextrin makes a solution of comparatively low viscosity, but glutinous rice starch is gelled without any fluidity. The nature of glutinous rice starch enables us to solidify star compositions with the smallest amount of extra matter, and in Japan the starch is used mostly for forming stars under the name "Mizinko," with little use being made of gum arabic or dextrin.

By an old process, ordinary rice starch is used in place of glutinous rice starch as follows: The rice is first boiled to a paste and then mixed well with the composition. Then the pasted composition is dried and ground to powder in a mill. This powder can be made into stars by kneading it with a quantity of water. It is said that ordinary rice starch gives a smaller amount of ash than glutinous rice starch, but the author has not been able to find such a difference. The residual ash after a star is burned can be produced not only by the binder, but also by the other components.

A solution of shellac in alcohol is recommended as a water-insoluble binder. Shellac as a binder is used for relatively large stars (e.g., of 20–30 mm or more in diameter), and it would take about four days to dry stars of about 40 mm in diameter. When the stars contain hygroscopic materials, absolute alcohol is used. The amount of shellac necessary as a binder is about 5% or more of the composition.

Other kinds of binders, such as a solution of celluloid in amyl acetate or collodion, can be used, and the drying time is very short in this case. Linseed oil also may be used, and in this case the composition may be solidified without drying, but care must be taken that the oil does not damage the ignitability of stars. This method is not used in Japan except in special cases. Plastic binders are not widely used in Japan because of manufacturing problems. The poisonous vapor, the heat of polymerization that sometimes causes fire, and the adhesion of the material to the container prevent its application in our firework processes.

Light Stars: The manufacturing process, using soluble glutinous rice starch as the binder, takes place in the following way:

Cubic Stars: An amount of composition with the binder "Mizinko" is placed in a bowl, and a measured amount of water is poured into it. When the quantity of water is too great, it is difficult to cut the kneaded mass because the composition sticks to the tools; when the amount is too small, the composition cannot be solidified because there is insufficient adhesion. It is better, in fact, to work out before manufacture the amount of water required for each composition or to use the following method: A small part of the composition is taken out, and an amount of water is added to the remainder of the composition, which is then kneaded into a paste. The former part of the composition is then added in small quantities, kneading and adjusting the viscosity at the same time. When the kneading is insufficient, the composition is not uniform and creates a hard surface coating so that it does not dry out. The tools for cutting the kneaded composition are prepared as in Figure 23.6.

First, a thick wooden chopping board (30 cm × 60 cm × 3 cm) is prepared, and around its edges are four thin, long, wooden edgeplates, the thickness of which is the same as that of the cubic stars; the edge-

Figure 23.6 Star making tools

plates are fixed with small nails or brass screws. The kneaded composition is then placed on the chopping board between the edgeplates, pressing with the fingers so that the surface of the pressed composition is little higher than the edgeplate. The surface of the composition is patted using a wooden hammer until it becomes as high as the edgeplates, and then a thin, long knife is pushed on and along the edgeplates from one side of the chopping board to the other, removing the excess composition. The four edgeplates are then removed so that a sheet of composition of uniform thickness remains on the board. The sheet is cut checkerwise into squares as follows: First the sheet of the composition is sprinkled with powder and cut the same thickness as the sheet of composition, thus producing a row of several exactly square sticks of the composition. Each stick is rolled 90° so that the new surface produced by cutting is uppermost, and the old surface on which the composition powder was sprinkled is at the side, to prevent the sticks from adhering to each other. The cut pieces are removed to the other side by sliding them along the board as they are cut to produce a row of parallel strips. The strips are brought together and turned 90° so that they are in close formation. These pieces are cut with the knife at right angles to the strips, the same width as the thickness of each strip. Thus we have regular cubes of composition. The cubes are removed from the chopping-board into a sieve so that small particles of composition can be sifted out and the cubes that stick to each other can be separated. The size of the cube varies according to use. For example, a cube with a 3 mm or 6 mm side is used for the core of round stars, while a 12 mm cube is used for smoke stars. The stars are dried in the sun, but the stars that are relatively large are best dried in the shade because stars that are dried in the sun immediately after the cutting are likely to dry only on the surface, with the result that the inner part of the cube is not easily dried and finally cracks on the surface. A worker should be able to make 30,000 stars of 3 mm or 15,000 stars of 6 mm in a working day. When the cubes are covered with an igniting composition in the way described later, they can be used as complete stars.

Round Stars (Color-Changing Stars): Round stars are made by adding composition layer by layer so that they are enlarged into balls of the proper size, and the cores generally are small cubic stars. The core has a 3-mm side for a completed ball less than 15 mm in diameter, 4.5 mm for a ball of 15 mm to 18 mm in diameter, and 6 mm for a ball of more than 18 mm in diameter. In place of cubic stars, some manufacturers use rapeseed or cereal seed and sometimes shot for the cores. The rapeseeds are so light and so small that much time and labor is required to enlarge them without them sticking to each other at the beginning of the manufacturing process. Therefore they are not very often used at present, although it is said that using such unignitable matter for the cores makes the end of the star-light quite clear, and the artistic view is excellent in chrysanthemum shells.

The manufacturing process of round stars from cubic cores is summarized as follows: The bowl for forming the stars has a spherical bottom and is made of aluminum or other light metal; in fact, any bowl of 30 cm in diameter sold on the market for cooking is suitable for manufacturing stars by hand. For manufacturing a larger quantity of stars at one time, it is better to use a larger bowl with a wider and flatter base. In this case, the bowl is suspended from the ceiling with ropes and operated by hand. Compositions are prepared according to the color-changing plan. A mushy paste of a composition, which is the same as the composition for sprinkling, is prepared. One part of the composition is mixed with water in another small bowl and stirred to a paste. Fireworkers in Japan call this mushy paste "Toro," and the viscosity of the paste is very important. When the viscosity is too high, the cores stick to each other in groups of two or three, especially in the case of light cores; when the viscosity is too low, the cores grow so slowly that the water in the Toro soaks into the cores. The viscosity of the Toro varies greatly according to the type of composition, so the amount of water in the Toro must be adaptable for each composition, especially in the case of a composition that contains aluminum powder, which has a very low viscosity. In this case, it is necessary to add more glutinous rice starch than one would in ordinary compositions. In a word, the vis-

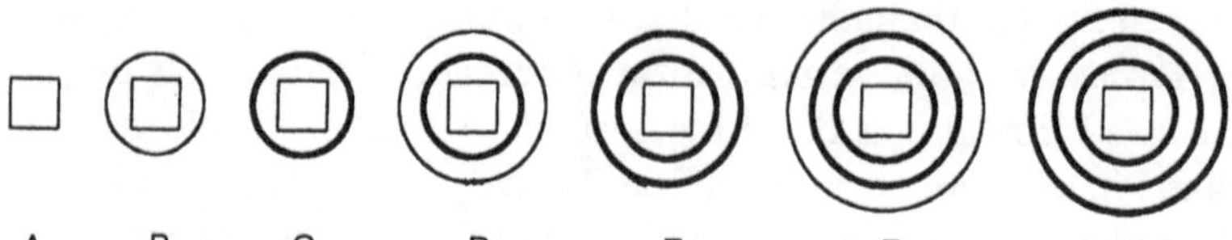

Figure 23.7 Stages in the manufacture of round stars

cosity of the Toro should be as high as possible, provided that the manufacturing process is possible. The development of the core into a star is shown in Figure 23.7.

A correct number of the cores are put into the pasting bowl, and a small amount of the Toro paste is poured over them; then the bowl is shaken around by hand or by other means until the paste completely and uniformly covers the core grains.

A small amount of powder, which consists of the same composition as the paste, is sprinkled on the pasted grains, and the bowl is shaken around again. When the amount of the powder is correct, it is all fixed on the grains, and the bottom of the bowl is so clean that it glistens. When the quantity of paste is too large, the paste adheres to the inside of the bowl and prevents manufacture; when it is too small, the powder remains in the bowl without fixing to the grains. The process of covering the grains with the paste, sprinkling the powder on the grains, and shaking is repeated two or three times, and the grains gradually change shape from cubic to round. Once they have been dried completely in the sun and the operation has been repeated, the grains become an almost perfectly round shape, namely, A becomes B in Figure 23.7. When the composition of the core is "flash," it is difficult to ignite, and in this case the composition of the powder and the paste must be different from that of the cores for ease of ignition. This kind of composition is called "cooperative powder of cores." Next a composition of very weak light intensity is layered on the grains. This layered composition is called "changing relay." This process is done once or twice so that grains become like C in Figure 23.7. The purpose of the changing relay is to make the stars seem as if they change color simultaneously, and clearly, despite their irregularity, although it only appears so because the light intensity of the stars is weakened between the two layers. The grains are then coated with a layer of the next composition by repeating the same process. After coating two or three times, the grains grow like D, and they are well dried in the sun. The grains are coated with the changing relay again like E, and the next composition is layered on the grain like F. A three-color-changing star is shown in Figure 23.7.

Changing Relay	Weight (%)
Potassium nitrate	75
Potassium perchlorate	7
Paulownia coal	8
Antimony trisulfide	3
Accaroid resin	2
Soluble glutinous rice starch	5

In the case of color-changing stars, what kind of composition should be layered inside and what kind of composition should be layered outside a star construction? As a general rule, a dangerous composition is layered inside and safer composition is layered outside, or an expensive composition is layered inside and an inexpensive one is layered outside. In the cores it is preferable to use a composition that produces light that ends clearly in a flash, so, taking into consideration the nature of the trajectory of the stars, a composition of high specific gravity is layered inside and a composition of low specific gravity is layered outside.

Finally, blackpowder-type compositions are layered on the grains as at G, and the stars are complete. Chrysanthemum 8, which has a relatively slow burning rate, is used first, and Chrysanthemum 6, which has a faster burning rate than that of Chrysanthemum 8, is used next. Thus the slower-burning composition is layered first and the faster one next, while the fastest, (i.e., blackpowder) is sprinkled on the grains in the finishing process. The method outlined here is the usual process, but in the case of stars that are difficult to ignite, such as those of aluminum, a specially prepared composition is first layered on the grain before the blackpowder-type compositions mentioned previously, in the finishing process. The problem of ignition is very important, not only with round stars, but also with all other stars, and a new device for igniting may create new kinds of stars.

The speed of the manufacture of stars is closely related to the drying time and the state of the grains to be dried. When the dried grains are again pasted and sprinkled with the next composition, the moisture soaks into the inner layer of the grains that have already dried. It soaks gradually deeper and deeper toward the core of the grain as time passes, so it is necessary to operate quickly and at once put the grains to dry. The rate of soaking of the moisture varies according to the kind of composition, and it especially depends on the amount of glutinous rice starch in the composition and the temperature of the stars. When the amount of starch is too great, there is not very much effect in a short time, but if the grains are left without drying for a long time, the moisture soaks into the center of the grains. Thus the moistened grains make a hard, dry layer on each surface during drying, and the center of each grain cannot be dried any more. The moisture is "driven in." The composition that contains a relatively small amount of glutinous rice starch does not cause such a phenomenon because in this case, although the moisture easily soaks into the center of the grains during every pasting process, they can easily be dried out during every drying process (i.e., the moisture is easily driven out through the layers of compositions). The limit of the amount of starch (e.g., binder) that does not cause the "drive in" phenomenon has not yet been systematically studied, but in the case of ordinary potassium perchlorate colored flame composition or spark composition containing aluminum, it certainly may be about 6%. In the case of aluminum composition, the viscosity of the paste is too low to produce the grains. When the temperature of the grains is low, the rate at which the moisture soaks into the grains can be made slower, so it is better to leave the grains in a cool place after drying by heat and cool the grains sufficiently before the next coating operation.

The thickness of the layer that is added between the two drying steps depends on the weather. On a clear hot summer day it should be rather thick, but on a cloudy winter day it should be rather thin. However, in the case of a composition that contains a relatively large amount of glutinous rice starch and has a high viscosity, success cannot be achieved by this procedure, and the relationship between the thickness of the layer and the weather will be the opposite of the above. In this instance, when the grains are dried in the strong sunshine and the new layer, coated on the grains, is too thick, only the surface of the layer is dried and the inside cannot be dried easily. Even when they are dried on a cloudy day with a rather thick layer, however, the inner and outside parts of the grains are dried successfully and uniformly because the moisture soaks into the grains very slowly in this composition. Thus the thickness of each layer on the grains depends on the weather and the nature of the composition and usually is about 0.5 to 1.0 mm on average, although in blackpowder-type composition the thickness is about half of this. On a day with a clear blue sky in Japan, the coating process is done three times a day in May, four or five times a day in March, and seven or eight times a day in June and July. The coating process must be finished by two or three o'clock in the afternoon in order to dry the grains during daytime.

The preceding process depends very much on the weather, and it is rather difficult to work out a strict program. Accordingly, it is useful to use a drier, (e.g., of infrared rays) to dry the grains as an auxiliary operation. A drier in which the heater is placed under the drying frames cannot be used to dry star grains because of the danger involved.

The complete round star grains, even when they are made very carefully, tend to have some small deviation in their dimensions. For example, a sample of the variations in the diameters of the grains is shown as follows: The values show the measured diameters of the grains. The letters S, K, A, B, and C show the kinds of stars, and the additional figures to the letters are the named diameters of stars.

S15 (mm)	K12 (mm)	A14 (mm)	A13 (mm)	B12 (mm)	C12 (mm)
16.50	13.10	13.40	12.60	12.28	12.70
16.00	12.92	13.20	12.90	12.68	12.39
15.40	12.70	14.80	13.35	12.69	12.37
15.75	12.70	13.75	12.60	12.35	11.73
15.80	12.80	13.70	13.00	12.18	11.92
15.40	12.80	14.30	13.00	12.15	12.00

As the star grains become increasingly larger, the coating cannot be operated in one bowl, and the grains are divided into a few groups. At last they become a number of lots, each lot of stars being classified, packed, and stored, and when the stars are loaded into a shell, it is wise to fill the shell with stars taken from the same batch. When stars of a definite diameter are required, we must first select cores of the right size, and it is necessary to make frequent selections between the coating operations, although in practice the star grains are not selected in this way except in special circumstances.

Stars are manufactured mostly by hand in Japan. Some fireworkers use a coating machine, but there is a need in Japan to develop the demand for fireworks. For small quantities and large numbers of different types of stars, the coating machine is not as adaptable. The machine must be so well cleaned when one composition changes to the next that in the case of a star with layers of different colors or when manufacturing various kinds of stars, the machine operation is not as simple as the hand operation. The principle of manufacturing stars by machine is the same as that by hand, but machine-made stars sometimes are less easy to ignite because they tend to gain a higher specific gravity than those made by hand.

The manufacturing process with water-insoluble binders is almost the same as the preceding process. In this case, the use of large amounts of linseed oil make the stars unignitable. When a volatile solvent is used in place of water for the manufacture of stars, care must be taken to keep the material away from flames, fire, electric heaters, hot electric lamps, and electric switches. Moreover, it must be remembered that the vapors of organic solvents often are toxic.

FIREWORKS SHELLS

Firework shells are classified into two fundamental types, the so-called "Warimono" and "Poka." Warimono is the type that draws a chrysanthemum-like flower in the sky by driving light stars or smoke stars in all directions; it also is named "Chrysanthemum." Chrysanthemum shells have a large amount of bursting charge and a strong outside shell, which helps the action of the bursting charge. The name "Poka" came from the weak sound of the explosion of the shells and is difficult to translate into English. Poka is not a Chrysanthemum type, for its purpose is only to throw out and spread the contents in the sky; it has a small amount of bursting charge, enough to break the shell into two parts along its joining line. It has a weak shell, which is strong enough not to be broken when it is fired. In practice, however, there are shells that lie between the two types. Round shells of European type used to be the same as Poka in their effect.

Warimono (Chrysanthemum)

The most representative construction of Warimono is shown in Figure 23.8. It consists of four main parts (i.e., shell A, star B, bursting charge C, and fuse D). Shell (A) plays a role not only as the container for the stars and the bursting charge, but also to help adjust the ejecting force of the bursting charge. The stars (B) are ignited by the explosion of the bursting charge (C) and fly like the petals of a chrysanthemum flower, burning on the surface. The burning charge is ignited by the fuse (D) and breaks the shell into small pieces by the explosion, which also ignites the stars and ejects them outward in all directions. The fuse (D) is ignited on its outside end by the flame of the lifting charge in the mortar on shooting, and it keeps a delay time until the shell reaches a

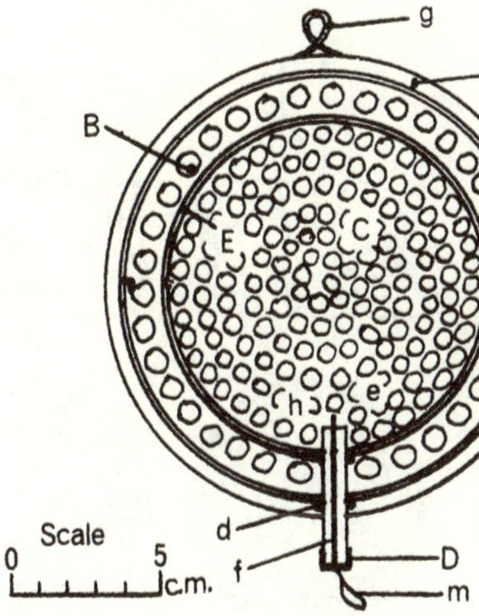

Figure 23.8 Warimono shell

desired height in the sky and ignites the bursting charge just after the delay time has passed. The principle points of these constructions are described as follows:

Shell (A) consists of the outer shell (a) and the inner shell (b). Sometimes only the inner shell is called the "shell," and in this case the outer shell is called the "pasted shell." The inner shell (b) is only the container, and its strength is not very important. As the raw material of shell (b), newspaper or cardboard paper is used, and in Japan the newspaper shells are the most admired because of their uniform brittleness. The outer shell (a) is made of many layers of Japanese paper (Kozo) or kraft paper, which is pasted on inner shell (b) to a thickness that is enough to give a suitable velocity to the stars on the explosion of the bursting charge (C). The structures of the stars, bursting charge, and fuse have been described previously. The process of assembly of a standard type of shell as in Figure 23.9 is shown in order in Figure 23.9. This process is applied to the shells of middle size (e.g., 125 mm, 150 mm, or 175 mm in diameter).

First the inner shell is prepared, a small hole being made in the center of the hemisphere of the shell, and a piece of fuse (f) is inserted through it, being fixed to the hemisphere with hemp (d). The hemp is well pasted and well fixed on the shell by winding it around the fuse. This is a very important process to prevent fire from entering the shell on shooting. The outside end of the fuse is protected by covering it with a paper cap. Generally, the name of the shell or the kind of shell is reg-

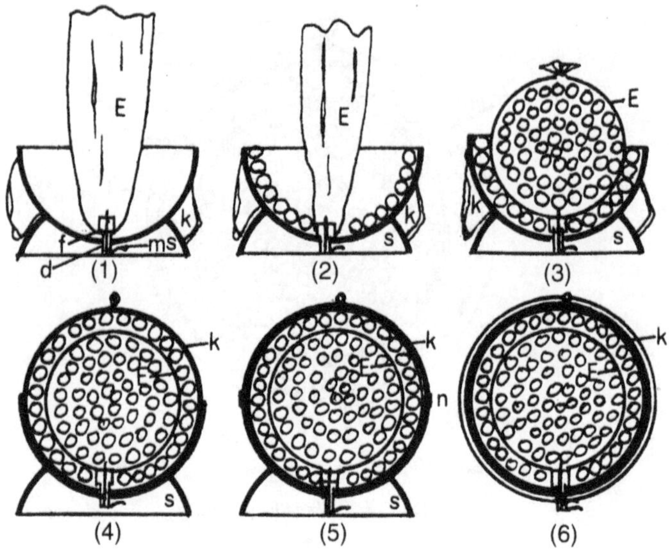

Figure 23.9 Stages in shell manufacture

istered on the cap (m). Before this fuse fixing operation, a bag (E) made of Japanese paper, for the bursting charge, is attached to the inner end of the fuse, and the fuse is inserted into the hole of the hemisphere from the inside. Then an outer bag (K) made of Japanese paper is pasted to the edge of the hemisphere, and it is turned over as in Figure 23.9 (1). The bag (K) is used for the arrangement of the stars of the upper hemisphere. Two bags, (E) and (K), generally are made of strong Japanese paper (Kozo) as described, but recently we have found that they often are made of a rather weak copy paper. When we use a bursting charge of chlorate powder, the inner bag (E) must always be made of strong paper so that it cannot be broken during the arrangement of the stars. This prevents the bursting charge from coming into contact with the stars and causing an accident, since the chlorate in the bursting charge together with the sulfur in the stars increases the sensitivity. The inner shell that is thus prepared is put on a rest (S) as in Figure 23.9 (1). Figure 23.9 (2) shows the arrangement of stars in the lower hemisphere. Then, as shown in (3), an amount of the bursting charge is charged into the bag (E) and the hemisphere is tapped on the outside with a small wooden stick to avoid any looseness between the stars and the bursting charge. The bag is then sealed at the top, being bound with a piece of hemp; the outer bag (k) is restored upward, and the stars of the upper hemisphere are arranged with the help of bag (k) as in (4). This arrangement is covered with another hemisphere on the top, and it is pressed downward by hand, being tapped with the wooden stick on the outside so that the two hemispheres meet each other at their edges. Sometimes the shell hemispheres have tally marks at the joining point of the two to make the operation easier. When the shell is correctly assembled, the hemispherical shells meet each other with only a little pressure from the hand, thus avoiding any looseness between the stars and the bursting charge. If the hemispherical shells meet each other loosely, it shows that the amount of the bursting charge is too small, and it must be increased. The shell thus assembled has a ribbon tape pasted over the junction, as in (5) shown by (n), or it is fixed with another hemispherical shell, which is attached on the side of the assembled shell to prevent separation while the shell is sent to the next process, pasting. Figure 23.9 (6) shows the complete product.

The process described here is most suitable for shells of middle size, but for smaller or larger shells the process must be managed conveniently according to the situation. The process shown in Figure 23.10 is applied to a small shell of less than 4 inches in diameter. In this case, the two hemispheres are prepared individually and then combined into one sphere.

Figure 23.11 shows the process for larger shells of about 250 mm to 310 mm in diameter, and in this case (1), first the lower hemisphere is prepared, and (2) then the upper hemisphere, in which a loading hole is

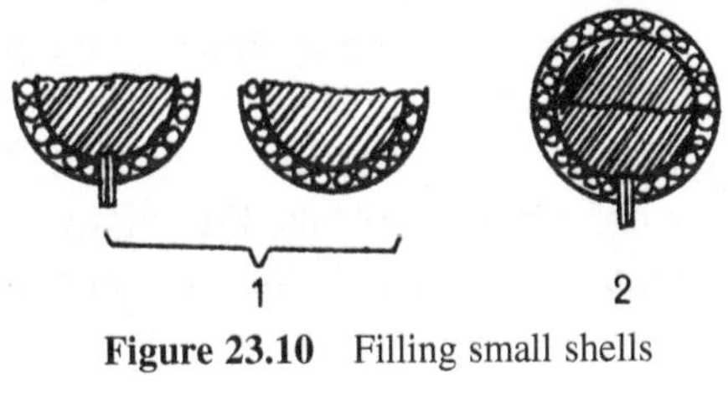

Figure 23.10 Filling small shells

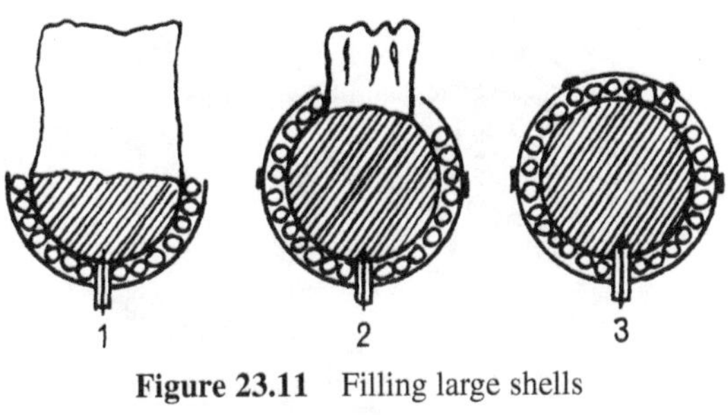

Figure 23.11 Filling large shells

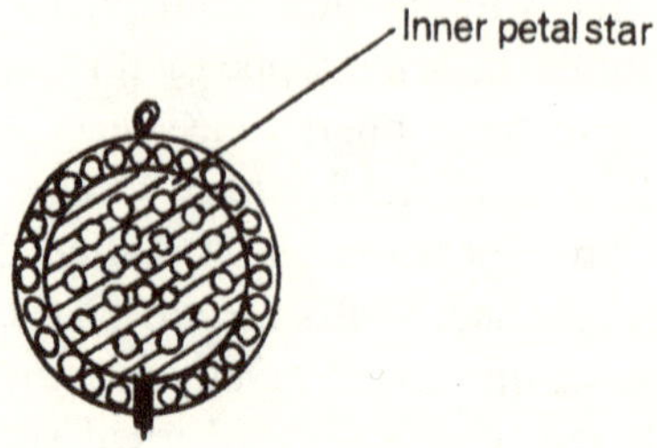

Figure 23.12 Shells with inner petal stars

made, is fixed to the lower hemisphere. The bursting charge and the stars are loaded alternately through this hole, keeping the boundary between the bursting charge and the stars by the paper bag. (3) The hole is then sealed with the piece of shell, which was previously removed from the upper hemisphere when the loading hole was made. During the process the sphere is tapped on the outside with a small wooden stick to assemble the contents tightly.

The processes described here are for the Chrysanthemum with single-petaled flowers. The processes for double-petaled flowers or special multipetaled flowers are described as follows: Figure 23.12 shows the simplest form, and in this case the inner petal stars are mixed in the bursting charge. Figure 23.13 shows the bursting charge composition pasted on the inner petal stars. Figure 23.14 shows a more elaborate shell in which a part of the bursting charge first is shaped into a sphere by being wrapped with thin paper and wound on the outside with string.

The inner petal stars are then arranged outside the bursting charge sphere and wrapped again on the outside by the same process. This is the inner ball, which creates inner petals. Putting this ball at the center of the shell, we assemble and complete the shell in much the same way as described previously.

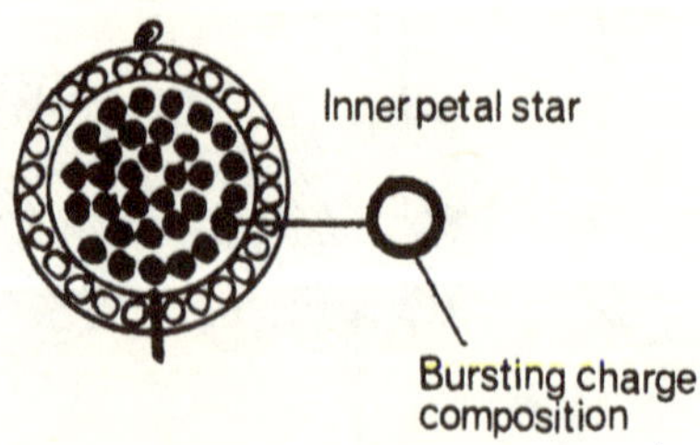

Figure 23.13 Shells with inner petal stars

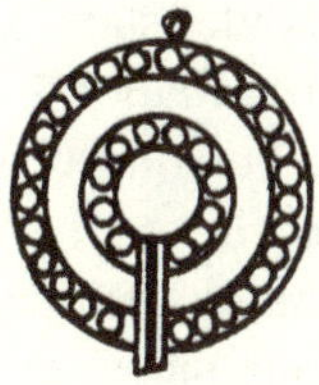

Figure 23.14 More complex inner petal star shell

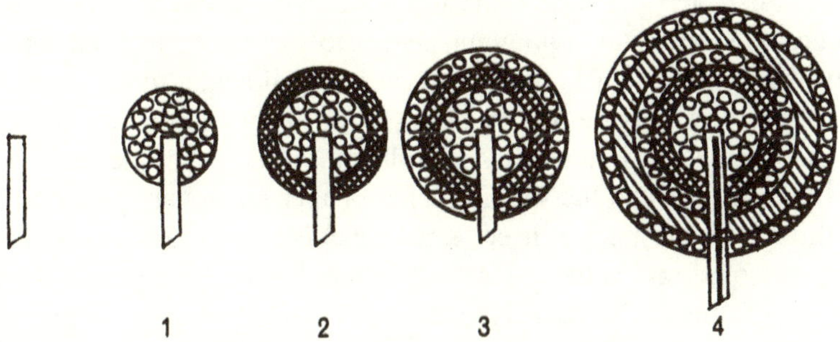

Figure 23.15 Assembly of a multipetaled shell

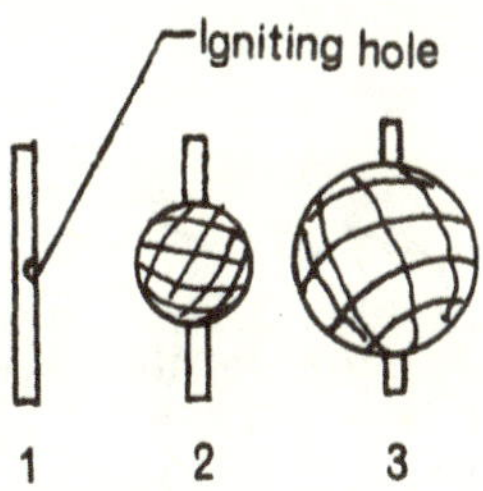

Figure 23.16 Shell with an axis

Figure 23.15 shows the process for a multipetaled Chrysanthemum, and it is assembled in the way described previously, the only difference being to first make the inner double-petaled ball. Figure 23.16 shows the process for another elaborate inner multipetaled ball, that is, a paper tube, which becomes the axis of the ball, and perforated at the center. The bursting charge and petal stars are arranged around the tube alternately so that the hole is at the center of the ball. One end of the tube is connected to the fuse on assembly. This kind of ball with an axis is called a "Nukishin."

The stars that are placed at the center do not fly as far when the exploding charge bursts, and the duration of the inner stars must be made shorter than that of the outer stars. Moreover, the grain size of the inner bursting charge must be smaller than that of the outside stars to strengthen the action of the bursting charge. If it is not planned this way, we do not create a good flower.

The pasting process of making the outer shell is as follows:

As Figure 23.17 shows, long pieces of paper are pasted on the inner shell like the stripes on a watermelon. In this case, A–A' is taken as an axis, and the points A and A' become the poles. The procedure is as follows: First, a paper tape is pasted around the assembled inner shell along the seam of the two hemispheres to prevent them from separating. The axis is taken so that it becomes perpendicular to the seam. The paper pieces are made by cutting a sheet of paper so that the direction of the fibers is at right angles to the length of the paper pieces in order to fit closely around the shell. Japanese paper previously pasted in two or three folds normally is used. Kraft paper is so inexpensive and strong that it also is used as the outer shell material. The paper pieces are prepared as follows: The paste is made of one part of wheat flour and three parts of water, but some prefer to use a paste made of nonglutinous rice. The paste is spread on one side of the paper with a wooden spatula, and the pasted surface is folded in half, with the pasted sides together. The

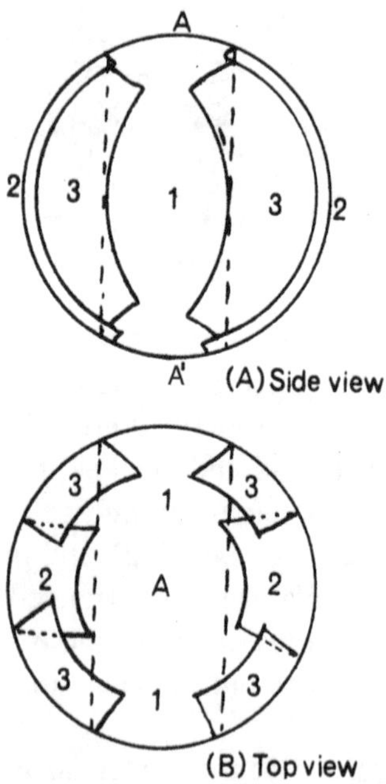

Figure 23.17 Making outer shell covers

purpose of this is to protect the paste from drying during the pasting operation of the outer shell. The pasted paper is cut into long pieces, keeping them folded. The pasted shell with one layer is shown in Figure 23.17. The width of the paper piece with another piece on the shell should be avoided if possible, as shown in (B), so that the shell may be uniform. In Figure 23.17, the figure on each piece of paper pasted on the shell shows the order of pasting. Then two other poles are taken so that the new axis is placed at right angles to the axis A–A' and the pasting process is repeated the same way as before. When two or three layers of paper have been added, a process called "gorokake" is applied; the pasted shell is placed on a thick wooden plate and another small wooden plate is placed on the shell, which is then pressed and rolled between the two plates. The edges of the pieces of paper on the shell are then perfectly fixed on the shell, which becomes a perfectly round sphere. After this process is finished, the shell is dried in the sun. When it is well dried, the same process (i.e., pasting, gorokake, and drying) is repeated until the paper folds (i.e., the thickness of a layer of pasted paper) reach the required degree. The necessary number of layers of pasted paper is proportional to the diameter of the shell, according to the experience of fireworkers, and in ordinary cases, with Japanese Kozo paper five or seven layers per 24 mm in diameter is recommended. For example, in the case of a 100-mm shell (in diameter), the number of all layers is $4 \times 5 = 20$, and in the case of a 150-mm shell (in diameter) it is $6 \times 5 = 30$, when we take the five layers per 24 mm.

These two shells have the same breaking strength against the bursting pressure according to the experience of fireworkers, and this is proved by the calculation based on the stress strain theory. Specifically, the layer of the pasted paper must be increased proportionally with the shell size to keep the same breaking strength. Figure 23.18 shows a section of a shell that is pasted in this way. In the case of kraft paper, the strength is twice that of Japanese "Kozo" paper, and the necessary layers become half of the above values.

Poka

The construction of Poka. The construction of the shell "Poka" is quite different from that of the shell "Warimono" (i.e., chrysanthemum). The shell has chambers in which various aerial floating objects are packed. The principle of the arrangement of the chambers in the shell is shown in Figure 23.19. A is the simplest one and consists of one fire chamber, but B has two chambers—a fire chamber and a fire prevention cham-

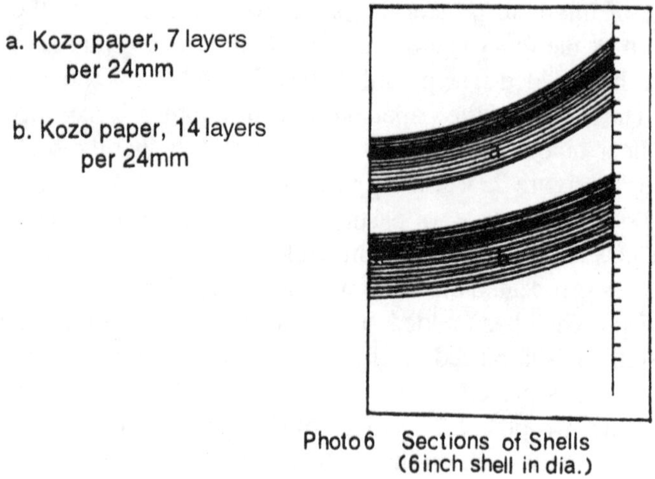

Figure 23.18 Sections of shell

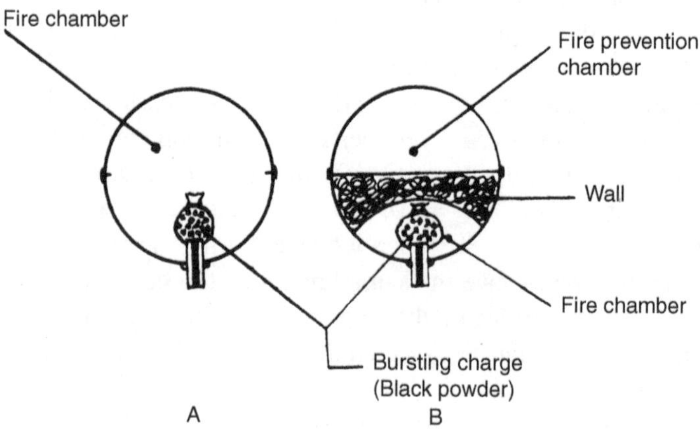

Figure 23.19 Poka shell construction

ber. The fire chamber is constructed so that the flame of the bursting charge easily touches the contents. The fire chamber contains stars, reports, small flowers, falling leaves, comets, whistles, and so forth, which must be ignited without fail. The fire prevention chamber is constructed in such a way that a wall is placed between the two chambers to protect the contents from the fire of the bursting charge when the shell explodes. This “wall” consists of cardboard or cottonseeds; the wall also acts as stuffing to prevent the contents from moving. The method of fixing the fuse to the shell is the same as in the Warimono shells.

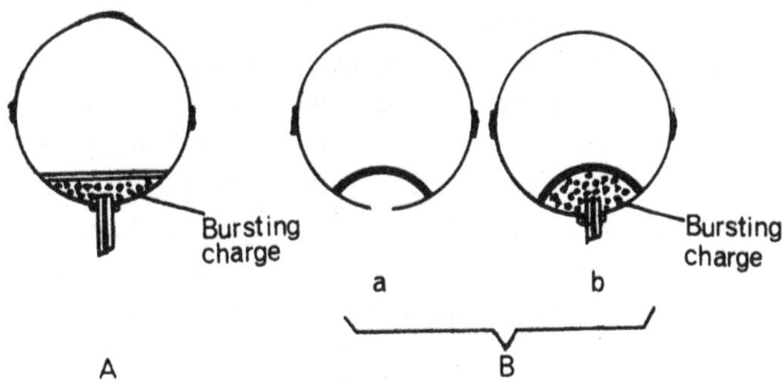

Figure 23.20 Poka shell construction

The amount of the bursting charge for Poka should be enough to break the shell into two pieces (i.e., the hemispheres), and it is fixed in the shell as shown in Figures 23.19 and 23.20. This method in Figure 23.19 is to fix the charge on the inside end of the fuse, and is rather troublesome, but the bursting charge is well ignited by the fuse without failure.

Figure 23.20 (A) shows a more simple method. In this case, a few pieces of "powder-pasted" paper are inserted in the charge for perfect ignition. Figure 23.20 (B) shows the shell with no bursting charge as in (a), which is afterward charged and fixed with a fuse as in (b), when it is used. It can be handled as "paper" without the bursting charge during storage and transportation, and it is very convenient when it is used for paper balloons called "Fukuromono" and flags.

Figure 23.21 shows the constructions of representative Poka shells. A is a report shell, "thunder," which makes three or five reports with the same time intervals. B is a flag or balloon, and C is a flare, in which the fire prevention wall is made of cottonseeds to protect the parachute from the fire. The construction of chains are the same as that of flares.

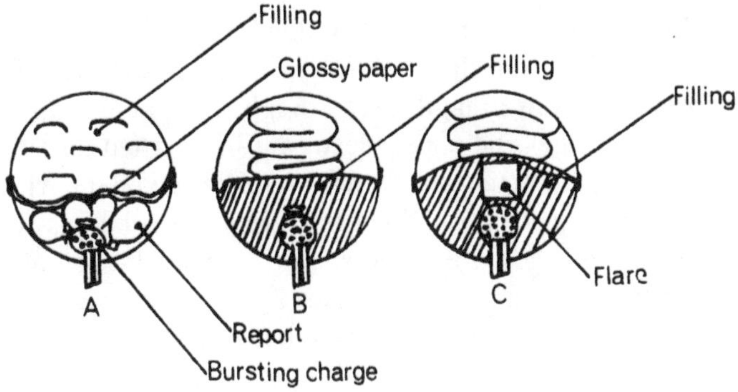

Figure 23.21 Poka shells with flare reports or paper figures

Planning of Poka. Here the relationship between the strength of the shell and the amount of the bursting charge must be considered. The shell must be strong enough not to be broken when it is fired, and if the weight of the contents is large, we also must think of the possibility of "head breaking," which means that the shell is broken by the shock of the firing and the contents are immediately projected out of the shell. It is necessary to maintain the strength of the shell during firing, but it is wise to have the minimum amount of the bursting charge, enough to break the shell into two hemispheres. The layers of paper pasted on the shell are just adequate for these conditions. A few examples of the amount of the bursting charge are:

The Bursting Charge of Poka (in the case of blackpowder)

Diameter of Shell (mm)	Amount of Charge (g)	Diameter of Shell (mm)	Amount of Charge (g)
75	2.0	175	6.0
100	3.0	200	7.0
125	4.0	250	8.0
150	5.0	300	10.0

The process for pasting paper on the assembled shells is the same as that of Chrysanthemum. The folds of the paper layer of the outer shell is 1.7 to 2.2 per 24 mm in diameter, on the seam of the two hemispheres, which gives an average value of about 2.1 on the same part. The weakest position of the completed shell is the junction of the two hemispheres, and the shell is broken along this line when it explodes. The inner shell is made of newspaper and, as described previously, its strength is unimportant in the case of the Chrysanthemum, but is important in the case of Poka because the ratio of the strength of the inner shell to the whole is higher than that of the Chrysanthemum. A newspaper shell 3 mm in thickness corresponds to the eight folds of Japanese Kozo paper in strength. Accordingly (e.g., in the case of 100-mm shell in diameter), the paper pieces are first pasted on and along the seam two or three times, and then they are pasted on the whole surface of the sphere just once. When the strength of the pasted paper along the junction is too great, the shell cannot be broken along the line, but is perforated in some other part by the explosion gas, which escapes from the shell through this hole but does not project the contents out of the shell.

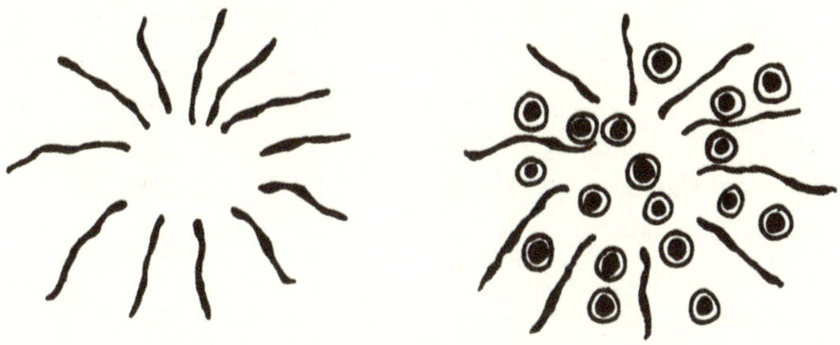

Figure 23.22 Development of daylight chrysanthemum

DAYLIGHT CHRYSANTHEMUM

The Daylight Chrysanthemum is used as an example of a manufacturing technique. This most beautiful and elegant effect uses smoke and light in contrast to each other.

The effect of this shell is developed in two stages: In the first stage, only thick smoke lines appear in red, yellow, green, and blue. Dazzling red star light pistils then suddenly appear on the petals as the second stage. The smoke lines must be short and thick, but the star lights must be very intensive.

Important points for the shell construction:

1. The burning time for the smoke stars is about 4 seconds and must not exceed 5 seconds.
2. The light stars must be an ammonium perchlorate and magnesium composition. The stars are covered with sulfurless blackpowder as a delay. The delay should be about 3 seconds. The light from the delays must be invisible in daylight.
3. The number of paper layers on the outside of the shell are half of an ordinary chrysanthemum, Warimono. The diameter of the flower must not be too large (i.e., about a half the normal size).

Manufacture

Smoke Stars: The compositions must be granulated in order to obtain the broad and short smoke traces. The granules are loaded into a paper tube that is ignited with a sulfurless match.

The following are examples of smoke composition:

	Red Weight (%)	Yellow Weight (%)	Green Blue Weight (%)	Blue Weight (%)
Potassium chlorate	28	32	33	37
Milk sugar (lactose)	20	28	27	27
Rhodamin B conc	30	—	—	—
Oil orange	22	—	—	—
Oil yellow	—	40	20	—
Phthalocyanine blue	—	—	20	40
Dextrine (additional %)	3	3	3	3
For 500 grams add				
Water	270 mL	180 mL	180 mL	260 mL
Ethanol				
to knead the mixtures	—	—	—	78 mL

The materials are mixed by hand and then kneaded after the addition of water. The blue does not mix well with water, but the addition of alcohol makes mixing easier. When the mass is well mixed, the dough is pressed onto a sieve of about 2 mm aperture with a wooden spatula so that the mass comes through the mesh like vermicelli. The sieved material is spread on paper and dried in the open air.

These smoke compositions are made with potassium chlorate, and although they are not very sensitive, they should not be handled roughly. The sieve mesh tends to become loose with use and it is best self-made. Making smoke composition is very dirty work, but much trauma can be avoided with skill. This technique is an application of a smokeless powder, where the burning time of the powder is adjusted by the thickness of the granules.

SMOKE STARS

The smoke granules are loaded into a kraft tube of about 30 mm inside diameter, 30 mm long, and with a wall thickness of 2 mm. A small spherical shell case also can be used.

The star construction is illustrated as follows:

The black match is only lightly fixed to the smoke outlet with a stopper that consists of a small amount of nitrocellulose solution in acetone. It is there only to stop the fuse from falling out.

The fuse composition ideally is made with sulfurless blackpowder to avoid the contact of potassium chlorate with sulfur. A suitable mixture

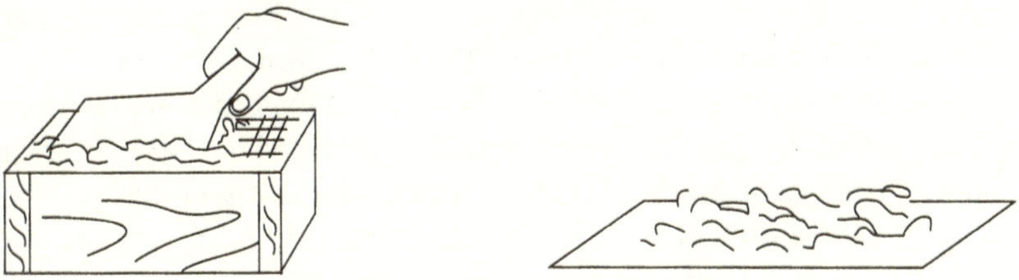

Figure 23.23 Making smoke grains

would be: potassium nitrate 80% and pine charcoal 20%. This is wet with 3 parts of gum arabic in 40 parts by weight of hot water.

LIGHT STARS FOR PISTILS

The most beautiful effects in this type of shell are somewhat limited. The use of ammonium perchlorate composition is recommended, although it is somewhat troublesome to manufacture. The magnesium must be coated with potassium dichromate to protect it from the ammonium perchlorate. This is done as follows:

50 g of potassium dichromate dissolved in 300 mL of hot water; 1,000 g of magnesium preheated to a temperature of 100°C.

The hot dichromate solution is quickly poured into the magnesium in its aluminium bowl and mixed with a stick; rubber gloves should be

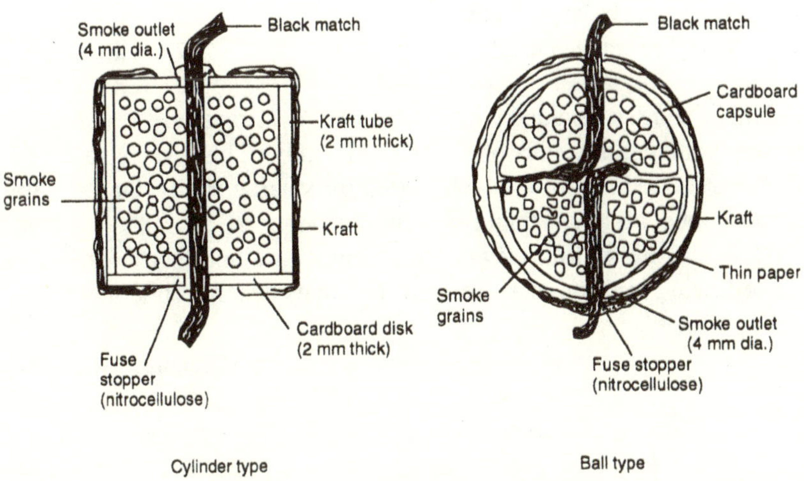

Figure 23.24 Construction of a smoke star

worn. When it is uniformly brown, it is dried after spreading on sheets of paper. When it is dry, it is passed through a 30 mesh sieve.

Potassium dichromate is on the list of toxic materials and is known to badly affect the mucous membrane. People in the industry come into contact with this substance for short periods and not everyday, and it is unlikely to be a problem provided that gloves and dust masks are used. By comparison, smoking should be more harmful.

The waste fluid should be treated with iron (11) sulfate.

$$H_2Cr_2O_7 + 6FeSO_4 + 6H_2SO_4 \rightarrow Cr_2(SO_4)_3 + 3Fe_2(SO_4)_3 + 7H_2O$$

In this way the most poisonous hexivalent chrome is reduced to the tetravalent form, which is less poisonous. Further treatment is to treat the liquor with slaked lime:

$$Cr_2(SO_4)_3 + 3Ca(OH)_2 \rightarrow 2Cr(OH)_3 + 3CaSO_4$$

The value for the solubility of chrome hydroxide in 100 g of water is 9.1 × 10–7 g.

The Composition of the Light Star	Weight (%)
Ammonium perchlorate	43
Magnesium passing 80 mesh coated with potassium dichromate	35
Accaroid resin	10
Strontium carbonate	10
Potassium chromate (stabilizer)	2
Guanidine nitrate (stabilizer)	2 (additional)

The mixture is kneaded with a binder (10% nitrocellulose solution in acetone) and cut into stars of about 15 mm cube. This is the size for 12" shells; some reduction will be required for smaller shells. Stars for this purpose must be larger than for night shells.

The stars are well dried at room temperature and then coated with the following ignition mixture until the cubes become spherical.

	Weight (%)
First ignition coat:	
Potassium perchlorate	71
Accaroid resin	12

Pine charcoal	5
Potassium dichromate	3
Strontium carbonate	9
The last coating for the 3-second delay:	
Potassium nitrate	65
Pine charcoal	35

A 10% solution of nitrocellulose in acetone is used for the coatings. These coatings are complete when the last coating has a delay time of 3 seconds. The purpose of this delay is to make the flying stars invisible at the first stage of the flower display.

BURSTING CHARGE

Smoke stars and light stars are very voluminous and thus rather different from ordinary daylight shells. The bursting charge therefore is granulated so that the granules can fill up the space between the stars. The granules should be sieved to remove any dust.

Bursting Charge for the Daylight Shell	Weight (%)
Potassium perchlorate	68
Potassium nitrate	2
Hemp charcoal	30
Glutinous rice starch (additional)	3

The completed daylight chrysanthemum shell is illustrated in the following figure:

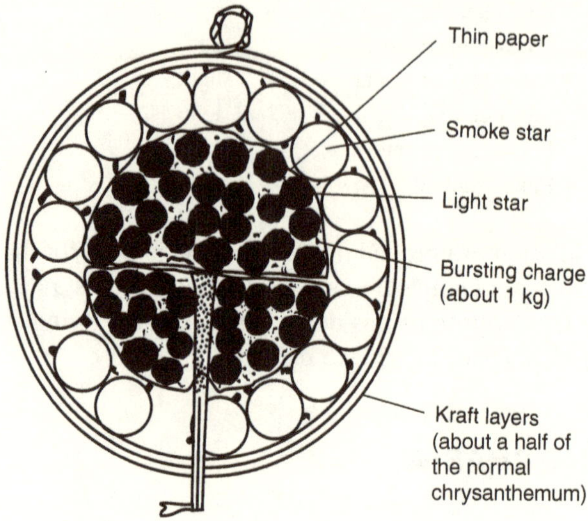

Figure 23.25 Six-inch daylight chrysanthemum

GLOSSARY

Term	Description
1.1G (0333)	The UN classification of fireworks packaged for transport that pose a mass-explosion (detonation) hazard (See also: UN number)
1.2G (0334)	The UN classification of fireworks packaged for transport that pose a projectile hazard (See also: UN number)
1.3G (0335)	The UN classification of fireworks packaged for transport that pose a fiery projectile or thermal radiation hazard (See also: UN number)
1.4G (0336)	The UN classification of fireworks packaged for transport that pose a limited hazard (See also: UN number)
1.4S (0337)	The UN classification of fireworks packaged for transport that pose a very limited hazard, the effects of accidental ignition are confined within the package itself (See also: UN number)
Aerial bomb	Preferred term: Aerial shell (See also: Aerial shell)
Aerial firework	In general a firework that functions above the immediate area of the ground—i.e., rockets, shells, Roman candles, and mines (See also: Ground firework)
Aerial shell	A shell designed to function at high altitude. cf. water shell (See also: Water shell, Shell)

Alloy	A combination, usually of two metals, that takes on some of the characteristics of its components. Alloys cannot be separated into their constituent parts by normal physical methods. (See also: Magnalium, Ferro-Titanium)
Amorce	A type of cap used in toy pistols
Aqueous	Pertaining to water. In fireworks, aqueous usually refers to solutions used for damping stars in manufacture. cf. Organic (See also: Organic)
Atomic pattern	In a shell burst, usually taken to be three contiguous circles representing the orbits off electrons around a central nucleus (rather than the atomic "hazard" symbol) (See also: Pattern shell)
Bag mine	A mine without a rigid case that is fired from a mortar. The advantage of bag mines is their very low debris pattern. (See also: Mortar mine)
Banger	Usually a complete firework, designed to produce a loud bang rather than a component of a larger firework (e.g. a mine)—which are better referred to as crackers. (See also: Cracker)
Bare match	Black match without a sleeve, preferred term: Black match (See also: Black match, Raw match)
Barrage	A combination of several fireworks, most usually Roman candles and/or mines, designed to be fired with a single ignition (See also: Battery)
Battery	In fireworks a combination of, say, Roman candles, fused together for increased effect and/or duration. (See also: Barrage)
Battle in the Clouds	A shell producing a series of salutes after bursting

Bengal	A pyrotechnic colored flare
Bickford Fuse	A slow-burning fuse used either for preparation of internal shell delays or for timing sequential firings (See also: Safety fuse, Delay)
Black match	Usually a cotton thread coated with blackpowder, in its raw state. Black match contained within a paper tube is usually referred to as piped match. (See also: Raw match, Piped match)
Black shell	Preferred term: Blind shell (See also: Blind shell)
Blackpowder	A composition, consisting of potassium nitrate, sulfur, and charcoal in the ratio 75:15:10 widely used in fireworks manufacture as a propellant and as the basis for compositions containing metal powders. It is considered by most people that blackpowder does not detonate on ignition, but merely burns extremely fast! (See also: Composition, Propellant)
Blasting powder	Blasting powder may be made with either potassium nitrate (type A) or sodium nitrate (type B) as the oxidant. (See also: Gunpowder, Smokeless powder)
Blind shell	A shell that fails to burst, having been successfully launched from its mortar. Potentially very dangerous
Blinker	An effect of periodic burning giving the effect of a flashing composition or strobe
Bomb	Inappropriate term for shell (See also: Aerial shell)
Bombardo racks	Usually a bottom fused multiple firing assembly (See also: Top fused, Finale barrage)
Bombette	In essence, a mini shell, usually found as a component of a Roman candle, and less

often as a component of a mine or even as a subcomponent of a shell. (Pupadelle) (See also: Roman candle, Mine)

Bottom fused — The normal method of fusing of a shell, in which the shell delay is ignited by the lifting charge of the shell. Also, for cakes where fusing is at the base of each tube (See also: Top fused)

Bottom shot — Typically a maroon as the last shot of a multibreak shell (See also: Multibreak shell)

Bounce — A charge of blackpowder at the base of a gerb—used to give an audible "crack" at the end of the burning of the gerb and to enhance the effect (See also: Gerb)

Boxed finale — A rapid firing array, usually of shells, with a single point of ignition. Physically they consist of a number of preloaded mortars, very often with titanium salute shells (See also: Finale barrage)

BPA — British Pyrotechnics Association—a trade association concerned with all aspects of fireworks safety and use in the United Kingdom. Recently split into Retail and Display sections.

Break — A normal shell is referred to as "single break." In a multibreak shell there are many sequential bursts, each a separate entity (cf. shell of shells, for instance) (See also: Multibreak shell, Repeater shell)

British standard — Prepared in the late 1980s for consumer fireworks. The standard sets performance, labeling, and constructional requirements for a variety of consumer fireworks available to the public in the United Kingdom and prescribes test regimes and methods for compliance. (See also: DOT classification, European standard)

Brocade	Long-burning star similar to but brighter and shorter-burning than a kamuro star (See also: Kamuro)
Burning	Typically an exothermic oxidation/reduction reaction. For fireworks the oxidant is usually a solid oxygen-rich ionic salt such as potassium nitrate (See also: Detonation)
Bursting charge	The internal charge in a shell designed to break the shell at the predetermined time, spreading and igniting the contents of the shell. Bursting charges typically are made of blackpowder (for effects shells) or flash powder (for color shells) (See also: Lifting charge)
Butterfly burst	A burst of a cylindrical tube from a central point, thus producing an effect akin to the wings of a butterfly. The term is also used for the more complicated burst pattern of a "butterfly" shell, although in many ways the theory of action is similar
Cake	Colloquial term for a multishot battery, arising from the outward appearance of many of the smaller items (e.g., 90 shot cakes) (See also: Multishot battery)
Caliber	In firework terms, usually the inside diameter of the firing tube, although strictly the diameter of the projectile
Candle	Abbreviated term for Roman candle (See also: Roman candle)
Cannonade	Usually an aerial shell containing several shots fused to explode at the same time or sequentially after the shell bursts. Also a popular generic name for a multishot battery from China.
Capping	Usually a rolled kraft paper tube used to connect several fuses together in a spark-proof join. (See also: Kraft paper)

Case	Typically the tube containing the pyrotechnic composition of the firework.
Category 1 firework	Indoor firework as defined by British standard 7114; part 2 (See also: British standard, European standard)
Category 2 firework	Garden firework as defined by British standard 7114; part 2 (See also: British standard, European standard)
Category 3 firework	Display firework as defined by British standard 7114; part 2 (See also: British standard, European standard)
Category 4 firework	Fireworks defined in the British Standard as being not suitable for sale to the general public. Generally, but erroneously, taken to mean larger display fireworks. (See also: British standard, European standard)
Catherine wheel	The traditional name for the generic wheel. The name derives from St. Catherine (See also: Wheel)
Celebration cracker	Usually a roll of many hundreds or thousands of individual cracker units designed to be unrolled and hung from a solid object before lighting. These items, traditionally part of Chinese New Year celebrations, are now widespread. However, recent legislation has limited their sale in the United Kingdom
Chain fused	A method of fusing several elements, particularly in a finale box or shell sequence
Changing relay	A low light intensity composition used, particularly in Japanese shell manufacture, to accentuate and regularize the color change in stars of a chrysanthemum or peony shell
Charging	Usually the process of filling a tube with composition or units (e.g., a gerb or Roman candle)

Cherry bomb	A small powerful banger containing flash powder now banned in the United States. The item was usually round and red—hence "cherry" (See also: M-80, Banger)
Chinese cracker	syn. Celebration cracker (See also: Celebration cracker)
Choke	The narrowing of a tube, most often associated with fountains and rockets. Chokes may be made by physically distorting the tube, by means of an end piece, or by clay or other material. (See also: Gerb, Fountain)
Chrysanthemum	A spherical burst, typically Japanese, in which the stars leave a visible trail. cf. peony shell. (See also: Peony shell, Dahlia shell)
Chuffing	The sound produced by the unstable burning of some rocket motors—usually a sign of instability.
Class B firework	The U.S. categorization for display fireworks (See also: DOT classification)
Class C firework	The US categorization for consumer fireworks (See also: DOT classification)
Closed circuit	A complete electric circuit, usually in the context of a circuit ready to fire. (See also: Open circuit, Short circuit)
Coconut shell	Usually a shell containing large comets (gold, silver, or crackle) that produce a typical coconut palm–type effect on bursting. Typically the shell will also be fitted with a complementary rising tail (See also: Palm burst)
Color-enhancing agent	Usually a chlorine donor such as PVC or Saran added to a color composition to enhance the intensity of the color. The chlorine forms metal-Cl species in the flame that emit strongly in the visible part of the

	spectrum. It is thought that potassium chlorate/perchlorate may play a similar, though diminished, role (See also: PVC)
Comet	Usually a solid cylinder of composition, manufactured in a mold by hand or by machine. The effect is that of a large star rising (from, say, a Roman candle). The comet is completely self-consuming and thus particularly suitable for sites where debris is a problem (See also: Roman candle)
Comet pump	A larger version of a star pump typically used to make Roman candle stars and splitting comets (See also: Comet, Splitting comet)
Composition	The generic and widely used term for all pyrotechnic mixtures. More specifically, composition is taken to mean the list of ingredients in a particular pyrotechnic mixture. All compositions contain at least an oxidant and a fuel, together with additional ingredients for color/effect production, etc. (See also: Oxidant, Fuel)
Cone	A specialized type of fountain in the shape of a cone. The advantages of a cone are predominantly ease of filling and the fact that the burning area increases as the fireworks proceeds, thus compensating for the increase in diameter of the choke. (See also: Fountain)
Confinement	The process by which some explosives, e.g., blackpowder, can change from extremely rapid burning to something approaching detonation. For instance, blackpowder confined in a tube will produce a loud report when lit, while blackpowder burning loose does not. (See also: Explosive)
Continuity	An electric circuit is said to be continuous when it is complete—thus a continuity

	check of a circuit is carried out to ensure that the circuit is not open (See also: Ohm meter, Closed circuit)
Convolute wound tube	A tube wound from a piece of paper the same width as the tube is long. Convolute tubes tend to be stronger than spiral wound tubes, although they are also more expensive to produce (See also: Spiral wound tube)
Covalent bond	A type of chemical bond in which electrons are shared by the participating atoms. This type of bond typically occurs between non-metallic elements. In fireworks, the important occurrence is in high energy species in the flame producing colors (See also: Ionic bond)
Cracker	A better term, and less emotive, than banger. Also an assembly of many crackers often referred to as a "Chinese cracker." A novelty cracker, commonly used at Christmas in the United Kingdom, is another use of the term
Crackle	A relatively recent effect comprising many small sharp bangs, thrown from a relatively low intensity comet. Chemically, most crackle compositions contain either lead, copper, or bismuth oxides
Croaker	syn. Screecher (See also: Screecher)
Cross match	Typically a piece of thin raw match used to facilitate ignition of a shell's internal time fuse. Generally made by either splitting or punching the time fuse
Crossette	The American term for a splitting comet. (See also: Splitting comet, Mosaic)
Crossing stars	Typically a pyrotechnic effect formed by fitting two stars together in a tube with a central bursting charge. Also known as

	French splits (See also: Splitting comet, Butterfly burst)
Crown	As in "Crown Chrysanthemum" shell—syn. Kamuro (See also: Kamuro, Brocade)
Crown chrysanthemum	syn. Diadem chrysanthemum. Typically a chrysanthemum-like shell bursts with longer burning stars that continue to fall to the ground after the normal maximum burst diameter. Very often the stars have a color change at the end of their flight. (See also: Kamuro, Brocade)
Crown wheel	syn. Flying saucer (See also: Flying saucer, Tourbillion, Girandola)
Cut star	A star, usually cuboid in form, prepared from a rolled sheet of composition (See also: Round star, Pillbox star)
Cylinder shell	An aerial shell, typically of European manufacture, that is cylindrical in form. Very often a "stack" of cylinder shells is combined, with suitable modification, to produce a typical multibreak shell. Cylinder shells are usually "spiked" to produce a harder burst. (See also: Aerial shell, Spiking)
Dahlia shell	A spherical shell burst, similar to a peony, but usually with fewer, brighter, stars (See also: Chrysanthemum shell, Peony shell)
Dark fire	In Roman candle terminology, the low light-emitting composition surroundng Roman candle stars for priming and rapid ejection. The term has also been applied to the composition applied between colors in color-changing stars (See also: Changing relay, Prime)
Daylight shell	A shell designed to be fired in daylight, and thus incorporating one or more of the following effects: noise units (crackers, whis-

	tles etc.), smoke, magnesium stars (See also: Aerial shell)
Deflagration	A particular type of explosive propagation in which propagation is faster than mere burning, but is not detonation. (See also: Detonation, Burning)
Delay	Usually a pyrotechnic composition that burns at a predetermined rate and used for timing either within a firework assembly (e.g., a Roman candle) or between firework elements (e.g., in a shell sequence). (See also: Sequence)
Delay fuse	A pyrotechnic composition designed to give a delay before functioning the next device in the explosive train. The most common use for a delay fuse is to provide a number of seconds for the operator to retire from the device before it functions. Also the internal delay within a shell used to ignite the bursting charge (See also: Fuse, Shell delay)
Detonating cord	A high-powered explosive material encased in a plastic or cloth sleeve that burns by propagation of a detonating shock wave (typically 5,000–7,000 meters/sec) (See also: Safety fuse)
Detonation	An exothermic chemical reaction in which the propagating front travels at supersonic speeds and thus an explosion always results (See also: Deflagration)
Detonator	Not to be confused with a firework igniter, or squib, a detonator is used to initiate high explosives. As such, detonators are security attractive items and their possession is controlled in many countries (See also: Igniter, Squib)
Display area	Usually the area in which the rigging of the display takes place (syn. firing area), but

	more generally the entire area encompassing spectator area, firing area, safety area, and fallout area (See also: Fallout area, Firing area)
Display firework	Usually a large firework intended for use at large public/private displays. In the United States it is erroneously synonymous with UN 0335 (1.3G) fireworks (See also: British standard, Transportation)
DOT	Abbreviation for the U.S. Department of Transportation. In the United Kingdom the similar department is now called the Department of Environment and the Regions (Abbr. DETR)
DOT classification	The assigning of fireworks by the U.S. DOT to one of three classes (See also: British standard, European standard)
Double base	Homogeneous propellants that usually contain nitrocellulose in propellant nitroglycerine and typically used in small arms ammunition and military rockets but rarely in fireworks
Draw-out shell	A two-break shell in which the first burst is usually color, the second color and report
Driver	A specialized gerb, usually more powerful than a gerb used on a static set piece, whose primary purpose is in turning a wheel or similar item. In the past, turning cases were invariably gold, usually made with neat blackpowder with the addition of charcoal, and produced very few sparks. Modern drivers often include titanium for additional visual effect. (See also: Gerb, Turning case)
DTI	In the United Kingdom the Department of Trade and Industry, responsible for aspects of the sale of fireworks to the general public

EIG	The Explosive Industry Group of the British Confederation of British Industry. The EIG is not a trade organization and as such does not actively promote the firework industry. Its primary purpose is liaison with government on safety and legislative matters (See also: BPA)
Electric firing	The process of firing a display electrically. Many varied systems have been developed ranging from simple "nail boards" to automatic, computer-controlled systems.
Electric igniter	The preferred term for the device used to ignite pyrotechnics electrically (See also: Squib)
Electric match	syn. Electric igniter
Electrostatic-sensitivity	The tendency of a composition to ignite (usually accidentally) from the sensitivity energy supplied by an electric spark. (See also: Sensitivity)
European standard	A proposed standard (CEN 212) for consumer fireworks in the EU. The standard is due to come into force in 1999. (See also: British standard)
Explosive	Technically any material that is capable of undergoing a self-contained and self-sustained exothermic chemical reaction at a rate that is sufficient to produce substantial pressures on their surroundings, thus causing physical damage. ALL fireworks are classified as explosives (See also: High explosive, Firework)
Explosive train	The progress of fire from one explosive element to another. For instance, within a hand-lit shell, the train is Delay Fuse → shell leader → lifting charge → shell delay → bursting charge → star prime → star
Fallout area	The area designated for debris to fall at a firework display. Obviously, the position

and size of the fallout area are critically dependent on the wind direction and strength at the time of the display. Careful planning at the design stage must allow for variations in the fallout area and position. (See also: Safety area, Firing area)

Ferro-Titanium — An alloy of iron and titanium that is finding increasing use in firework manufacture. Different ratios of Fe:Ti are available, although generally all burn with a much more silver flame than Fe alone. (See also: Alloy)

Finale barrage — A rapid firing, prefused sequence (usually of aerial fireworks) that is typically fired at the end of a display.

Firecracker — syn. Cracker (See also: Cracker)

Firework — Technically an explosive assigned one of five UN numbers (0333–0337). For our purposes, a device that is designed for entertainment and that comprises pyrotechnic composition (See also: UN classification)

Firing area — The best term for the actual area of firing (rather than display area)

Firing current — The current that is applied to an electric igniter that causes it to function (See also: Safe current)

First fire — A composition used, particularly in gerbs, to initiate the explosive train. It is not synonymous with prime (See also: Prime)

Fix — Old English term for a gerb that is not a turning case. Very often these gerbs had a "bounce" (See also: Turning case, Bounce)

Flanked — Usually applied to racks or mortars or Roman candles on a frame in which three or more tubes are angled to produce a dispersed effect.

Flare — A pyrotechnic device used to produce colored light when ignited. In the United States this typically is a tube, similar to a large lance. In the United Kingdom the term is often applied to distress signals (See also: Torch, Fusee)

Flash paper — A form of nitrocellulose, easily ignited and used to produce a puff of flame

Flash powder — An extremely powerful pyrotechnic composition, typically made from potassium perchlorate (or rarely potassium chlorate) and powdered aluminum (or magnesium). In fireworks, flash powder is often used for powerful maroon shells and for bursting color shells

Flash rocket — A rocket that usually contains only flash powder as its payload and thus functions with a loud report and a flash. Flash rockets should never be fired in multiples from cones for risk of detonation. Flash rockets find much use for bird scaring (See also: Flight rocket)

Flight rocket — Usually a small caliber (approx. 14 mm) rocket fired in a large number simultaneously from a rocket cone or rocket frame to produce a characteristic fanlike effect. (See also: Volley, Rocket cone)

Flitter — A spark effect (usually silver/gold) produced by the incorporation of relatively coarse metal powders (usually aluminum). The glitter effect is similar but distinct (See also: Glitter, Strobe)

Flower pot — A shell misfunction in which the shell bursts within the mortar propelling the shell contents upward as if from a mine. cf Muzzle break (See also: Mine, Muzzle break)

Flying saucer	An unusual firework device, usually constructed from a ring of plastic or wood, with turning cases and lifting cases. The functioning of the device usually involves rotation around a vertical axis, followed by ascent into the air. "Double acting" saucers fall and then reascend to the crowd's delight! Also called Girandola (See also: Crown wheel, Tourbillion)
Flying squib	A toy firework of erratic flight now banned in many countries. Not to be confused with the electrical squib
Fountain	A device comprising pyrotechnic composition charged into a tube that may or may not be choked. The composition may be hand charged or, more commonly nowadays, machine charged (See also: Cone, Gerb)
Friction sensitivity	The tendency for a composition to ignite as the result of frictional energy (i.e., rubbing). (See also: Sensitivity)
Front	Usually an arrangement of fountains, mines, set pieces, or Roman candles along a line parallel to the spectators and fired simultaneously (See also: Line)
Fuel	In a pyrotechnic composition, that which the oxidant oxidizes. Common fuels include charcoal, sulfur, aluminum, and magnalium. All common pyrotechnic compositions contain at least an oxidant and a fuel (See also: Reducing agent)
Funnel and wire	One method of charging tubes with firework composition (See also: Charging)
Fuse	The generic term for the means of transferring fire to a firework, or from one part of a firework to another (See also: Delay fuse, Shell delay)

Fuse cover	The protective cover for the initial fuse of a firework. Often colored to aid identification in the dark
Fusee	A long-duration flare, usually red, that may be used as a warning flare on the highway or railway. Fusees may also be used to light fireworks. cf. Portfire (See also: Flare, Portfire)
Garden firework	A firework, usually of limited power and composition weight, intended to be used in restricted areas outdoors (See also: European standard, British standard)
Garniture	Aerial effects carried in the head of rockets
Gerb	Usually a relatively thick-walled tube filled with composition and having a choke. A gerb functions by throwing out a shower of sparks. From French—gerb—sheaf of corn (See also: Fountain)
Girandole	syn. Flying saucer (See also: Flying saucer)
Glitter	An effect that produces drossy droplets of molten composition that reach with the air to produce a sparkling or glittering effect that is not as distinct as a strobe effect. Similar but distinct from flitter (See also: Flitter, Strobe)
Glutinous rice starch	A binding agent much favored by Japanese star makers (See also: Paste)
Greek fire	Used in combat, Greek fire was an early use of pyrotechny, although some descriptions of compositions do not contain an oxidizer. It consisted of sticky long-burning composition usually fired from catapults
Green man	The symbol of the Pyrotechnics Guild International. The "Green men" were performers (c. 1600) who carried flaming clubs and scattered fireworks (probably

	sparks) as they headed processions (See also: PGI)
Ground burst	A low-level burst of a shell and potentially very dangerous
Ground firework	A firework designed to function at ground level
Ground maroon	A single powerful cracker designed to produce a loud report and a flash.
Ground salute	syn. Ground maroon (See also: Ground maroon)
GRP mortar	Glass Reinforced Plastic—a relatively recent addition to the design of mortars. GRP mortars, usually spirally wound are light, inexpensive, and extremely strong. However, there is some doubt as to their suitability for cylinder shells, especially in larger calibers (See also: HDPE mortar, Steel mortar)
Gums	Usually applied to binding agents soluble in water (See also: Resins)
Gun	A poor term for mortar (See also: Mortar)
Gun cotton	High-grade nitrocellulose containing 13% or more of nitrogen. Firework powders may contain 12.2%.
Gunpowder	Fireworkers prefer the term, although chemically and physically the two are the same (See also: Blackpowder)
Hammer shell	A shell, typically multibreak, consisting of color breaks and reports timed to break in alternation
Hanabi	Japanese word for fireworks, roughly translated as "flowers of fire"
Hanging fire	A fuse or pyrotechnic composition that continues to burn very slowly, often almost invisibly, rather than at its design speed. As

	such, a hangfire presents a serious danger to firers (See also: Misfire)
HDPE mortar	High Density Polyethylene—an extremely useful material for mortars. Belling rather than fragmentation of HDPE mortars tends to occur with failure of normal (not salute) shells
High explosive	An explosive that is capable of detonating when unconfined (See also: Low explosive, Propellant)
HSE	The British Health and Safety Executive—the legislative and enforcement body in the United Kingdom
Hummer	A device that produces a humming sound, usually made from a thick-walled tube filled with composition, sealed at both ends, and pierced tangentially to the inner diameter. The sound is made as the device spins rapidly in flight
Hygroscopic	The property of a material that causes it to absorb and retain moisture from the air. As such, hygroscopic compounds find only limited use in firework manufacture
Igniter	Shortened term for electric igniter (See also: Electric igniter)
Igniter cord	Some types are more properly called plastic igniter cords. Generally made for the blasting industries in several speeds. The slow cord finds use in fireworks manufacture, particularly for fitting of delay fuses
Ignition	The initiation of burning of a pyrotechnic
Indoor firework	In terms of the British and European standards, devices of very limited power suitable for use indoors. Types include sparklers, snakes, and other novelty items. (See also: British standard, European standard)

Ionic bond	A type of chemical bond characterized by transfer of electrons from one atom to another. Thus common salt is written Na+Cl–. Most oxidants and coloring agents for firework compositions are ionic compounds. (See also: Covalent)
Japanese style shell	The ultimate spherical burst shell. The Japanese strive to produce perfect symmetry and patterns in their shells. Japanese shells are also noted for the contrasting colored pistils that form part of the burst of many effects (See also: Peony shell, Pattern shell)
Kamuro	A long-burning star, usually silver or gold, that falls a substantial distance from the shell burst before, perhaps, changing color at the end of its flight (See also: Brocade, Crown chrysanthemum)
Kraft paper	A strong paper used for pasting shells and for capping (See also: Capping, Paste)
Lance	Usually a small, thin-walled, tube containing colored composition used to make lancework (See also: Lancework)
Lancework	Usually a message, logo, or design made on a wooden lattice work frame consisting of many lances fused together
Leader	The initial fuse of a shell that transfers fire from the delay fuse (if any) to the lifting charge of the shell. For small caliber shells, the leader may be used to lower the shell to the bottom of the mortar tube, but this is not good practice with larger caliber shells
Lifting charge	The charge beneath an aerial shell (or Roman candle unit) that propels the unit into the air. The listing charge almost universally used in firework manufacture is granulated blackpowder (See also: Bursting charge)

Line	In electrical firing, one "line" is a single circuit, perhaps consisting of many individual ignitions, that are fired simultaneously (See also: Electric firing)
Line rocket	A rocket designed to travel along a wire or rope (See also: Pigeon)
Low explosive	An explosive that burns or deflagrates on ignition rather than detonating. Almost all pyrotechnic compositions are low explosives (See also: High explosive, Propellant)
M-80	A type of small, but powerful, device containing flash powder. M-80s are now banned from sale in the United States (See also: Cracker, Thunderflash)
Machine	A construction, commonly used in the nineteenth and early twentieth centuries, to "enhance" the spectacle of a fireworks display. Great efforts were made to disguise the presence of fireworks within statues and ornaments, which would then be ignited to produce the intended, but concealed, firework effect
Magnalium	The most commonly used alloy in firework making. Magnalium is usually a 1:1 mixture of magnesium and aluminum and is described chemically as a eutectic mixture of Al_2Mg_3 in Mg_2Al_3. (See also: Alloy)
Manufacture	The process of making fireworks from the raw materials. The term is more generally applied to any manipulation of firework components (e.g., fusing shells).
Maroon	An exploding device that produces a loud bang. Aerial maroons are the most common, the composition being with blackpowder or flashpowder. From French—marron—chestnut (from the noise they make in a fire) (See also: Salute, Flash powder)

Match	The generic term for quickmatch, black match, etc. (See also: Quickmatch, Piped match)
Meal powder	Finely divided blackpowder available in several grades. (See also: Blackpowder)
Mesh size	The designation of the number of wires of standard thickness per inch used to make a sieve. For instance, a 60 mesh sieve has an aperture of 250 microns (See also: Sieve size)
Metal salt	The combination of an electropositive metal ion with an electronegative anion. For instance, potassium nitrate (See also: Ionic bond)
MIDI	A method of computer control of firework displays in which cues are programmed like notes on a score. MIDI is an internationally recognized coding standard usually used for composing music (See also: Electric firing)
Mine	Typically a complete article with firing tube, but generally the firework itself (See also: Mortar mine, Bag mine)
Mine bag	syn. mortar mine. (See also: Mortar mine)
Mini mine	A Roman candle in which each shot produces a mine effect with many stars, rather than the more typical single star per shot (See also: Roman candle)
Misfire	In general, any failure of a firework to function as predicted. Modern usage restricts the term to the failure of a firework fuse (See also: Hang fire)
Mixture	Usually synonymous with "composition," but may mean the list of ingredients of a composition (See also: Composition)
Mortar	The tube used to fire an aerial shell or mine. Mortars can be constructed from paper,

	plastic, GRP, or metal (See also: Aerial shell)
Mortar mine	A mine fired from a mortar (See also: Mine)
Mosaic	The French term for splitting comet (See also: Splitting comet, Crossette)
Multibreak	An aerial shell consisting of more than one section producing a separate effect in seshell and ignited by the bursting of the preceding section. The public may incorrectly refer to a "shell of shells" as a multibreak effect (See also: Shell of shells, Repeater shell)
Multishot battery	The generic term for a collection of pyrotechnic pieces lit at a single ignition point, but often used exclusively for items referred to as "cakes" (See also: Cake)
Muzzle break	A malfunction of a shell where the bursting charge operates just as the shell leaves the mortar. This is a common point of shell failure as the pressure changes that act on the shell are great at this point (See also: Flower pot)
Niagara Falls	Brocks often fitted Niagara Falls with a loud whistle accompanying the visual effect (See also: Waterfall)
No-fire current	The upper limit for a current that will not fire an igniter, and thus the upper limit for a test current for electrical circuits (See also: Safe current)
Noise mine	A mine in which the principal effect is ejection of pyrotechnic noise units (e.g., crackers or whistles) (See also: Star mine)
Nomatch	A specialized system for igniting fireworks using a shock tube. The advantage of nomatch is the extremely high speed of propagation leading to almost simultaneous ignition of several pieces at great distances.

Ohm meter	A device for measuring the resistance of a circuit, and typically built into electrical firing panels. The current applied by the Ohm meter must be less that the no-fire current! (See also: No-fire current)
Open circuit	An electric circuit that is not complete—i.e., will not fire when a current is applied (See also: Closed circuit)
Orange book	The United Nations book on the Classification and Testing of Dangerous Goods (See also: UN classification)
Organic	In our terms, a solvent that is not based on water (e.g., acetone or cyclohexanone) (See also: Aqueous)
Oxidant	The component of a firework composition that supplies the oxygen to the reaction (e.g., potassium nitrate) (See also: Fuel)
Oxidizing agent	In firework compositions syn. Oxidant
Palm burst	The central burst, similar to a coconut shell, of a color shell. For instance, a "red peony with palm core"
Parallel circuit	An electrical circuit in which the current is divided to pass through several igniters. Parallel circuits are less easy to test for line breaks and short circuits than series circuits. (See also: Series circuit)
Paste	The most common usage is that referring to the pasting of aerial shells to enhance the burst of the shells (See also: Spiking)
Pattern shell	A shell, usually with many fewer stars than a chrysanthemum shell of the same caliber, whose burst pattern is such that a pattern rather than a sphere of stars is produced. Pattern shells come in many levels of complexity, but perhaps the most pleasing is the simple single circle (See also: Atomic pattern, Saturn pattern)

Pellet	An alternative term for a star, usually restricted to pumped, cylindrical form, stars (See also: Star)
Peony shell	A typical Japanese style of shell in which the stars do not leave a trail of sparks (See also: Chrysanthemum shell, Dahlia shell)
PGI	The American Pyrotechnics Guild International
Phlegmatize	To slow down or to make less active.
PIC	Plastic igniter cord (See also: Igniter cord)
Pigeon	A specialized type of novelty firework in which a rocket motor is forced to run horizontally along a wire or rope, usually accompanied by a whistling effect. Often, the pigeon will make the journey several times, first in one direction, then the other (See also: Line rocket)
Pillbox star	A star made from pressing (usually by hand) composition into a small thin-walled cardboard tube. Pill box stars are rarely made nowadays, but their effect can be dramatically different from that of round or pumped stars. Pill box stars usually have a longer burning duration than pumped or round stars (See also: Star)
Piped match	Raw match enclosed in, usually, a paper tube used for transferring fire from one firework to another. Piped match also forms the leader of a shell (See also: Raw match, Black match)
Pistil	In typical Japanese shells, a central core to the burst of a contrasting or complementary color to the main burst (See also: Chrysanthemum shell, Peony shell)
Plug	Typically the closure of a mortar tube, but more generally the closure of any tube (e.g., a Roman candle tube)

Poka shell	A weak busting shell of Japanese design commonly used for deploying parachutes or tissue-paper flags (See also: Warimono shell, Japanese style shell)
Polverone	syn. Pulverone
Portfire	Usually a thin-walled tube filled with slow-burning composition used to ignite other fireworks. It is similar to a fusee, but its flame is usually less fierce and varies in color. A test for a good portfire is that it should continue to burn after being dropped vertically onto its lit end at arm's length!
Post	A geographic position on a firing site used to identify the layout of the site. For instance, there may be three posts of Roman candles spread along the front of a site (See also: Line)
Press	A machine used to fill composition into tubes (e.g., gerb press) or for making fireworks (e.g., Roman candle press)
Prime	Often a slurry of blackpowder, a binder, and water, occasionally with added ingredients (e.g., silicon) to increase the burning temperature used for ensuring ignition of reluctant compositions (See also: First fire)
Priming	A process carried out to ensure ignition of a pyrotechnic composition when the composition itself is difficult to ignite. For instance, round stars are often primed for use in shells where the ignition time is short, whereas the same stars may be used without priming in a mine where the ignition time is longer
Propellant	A composition used, typically, in a rocket motor to provide force. In more general terms, any composition used to propel a firework into the air (See also: Explosive, Rocket motor)

Pulverone	Granulated rough powder mixture (usually of the same composition as blackpowder) used as the bursting charge of a shell
Pumped star	A star produced by compressing composition in a mold. Pumped stars are usually cylindrical in form (See also: Round star)
Punk	A wick for lighting small fireworks (See also: Portfire)
PVC	Poly vinyl chloride—one of many chlorine donors used as color-enhancing agents in firework compositions (See also: Color-enhancing agent)
Pyrotechnic	The generic term for any item (or composition) that reacts in a self-sustaining chemical reaction and generally produces an effect of light, smoke, noise, or heat. Pyrotechnic articles are classified differently from fireworks and the term usually is restricted to theatrical effects and specialized items such as mole smokes or thermite charges (See also: Firework)
Quickmatch	syn. Raw match (See also: Raw match, Piped match)
Rack	An apparatus, usually for firing rockets. The term also may be applied to "racks" of mortars (See also: Rocket rack)
Rain	Usually silver rain or gold rain, in modern fireworks the long-lasting stars from a shell or rocket that fall all the way to the ground. Care must be taken in the use of rain shells. In older terminology a "golden rain" was a particularly attractive type of hand-held fountain (See also: Glitter)
Ram	The rod that is used to compress powder within a tube. The ram usually is quite a tight fit to the tube (cf. funnel and wire)

Ramming	The process of filling a firework case with composition. Ramming usually is applied to a mechanical process rather than to a manual process.
Raw match	Blackpowder-coated thread used for linking fireworks (See also: Quickmatch)
Reducing agent	The chemical role of a fuel in a firework composition. As the oxidizing agent oxidizes the fuel, the fuel can be said to reduce the oxidant (See also: Oxidizing agent, Fuel)
Repeater shell	Usually a cylinder shells with several timed color bursts at regular intervals. Repeater shells often are fired in sequence—1 break, 2 break, 3 break, 4 break, etc. cf. Multibreak shell (See also: Multibreak shell)
Report	Loud sharp explosion, specifically at the end of a burning device
Resins	Usually applied to binding agents soluble in organic solvents, e.g., accaroid resin (See also: Gums)
Resistance	The property of a material that acts to impede the flow of electrical current. In electrical firing of fireworks, the resistance of a line is measured to prevent accidental "open" or "short" circuits. (See also: Ohm meter)
Ring shell	An aerial shell that produces a symmetric ring of stars on bursting. Ring shells can be stabilized in flight with a rope "tail" to control the orientation of burst (See also: Pattern shell)
Rising effect	Often synonymous with "tail effect," but also may be applied to shells in which, for instance, whistles or small shells (rising flowers) have been attached and which

	function on the shell's ascent (See also: Tail effect)
Rocket	An aerial device propelled into the air by a motor (cf. shell); many of the public will describe any aerial firework as a "rocket"
Rocket cone	A device for firing flight rockets usually made from sheet steel curved into the characteristic cone shape (See also: Flight rocket)
Rocket motor	The power unit of a rocket, typically manufactured nowadays by pressing blackpowder into a choked tube without a spindle. Rocket motors occasionally find other uses in pyrotechnics—as wheel drivers and as short duration fountains (See also: Rocket)
Rocket rack	A rack, usually made of wood or metal, for mounting many rockets before firing
Rocket spindle	The spike (usually metal) used to form the older type of rocket motor with a central cavity for increased burning pressure (See also: Rocket, Choke)
Roman candle	A tube, usually cardboard, in which several charges are loaded, each with its own delay fuse and lifting charge, which function in a sequential manner
Round shell	An aerial shell in the form of a sphere. Round shells usually contain colored stars (See also: Cylinder shell)
Round star	A star prepared by rolling, thus applying layer upon layer of composition onto a central core
Roundel shell	An aerial shell consisting of several maroons that burst in a ring pattern one after another (See also: Siatene shell)
Safe current	The current level that it is safe to test an electric igniter without ignition (See also: Firing current)

Safety area	The area around a display site, usually not including the fallout area, which is considered separately (See also: Fallout area)
Safety cap	syn. Fuse cover
Safety fuse	A specialized fuse, designed for commercial blasting of construction similar to Bickford fuse but with a heavy waterproof coating (See also: Bickford fuse)
Salute	American term for maroon. (See also: Maroon)
Saturn pattern	Usually refers to a "Chrysanthemum in Circle" type shell rather than an "Atomic" pattern shell (See also: Pattern shell, Atomic pattern)
Scratch mix	A finely sieved mixture of potassium nitrate, charcoal, and sulfur primarily used as a prime for stars (See also: Prime)
Screecher	Physically a whistle with a hole through it, producing a much more "rasping" sound. In a screecher the instability arising from the oscillations of burning interfere with each other almost to the point of causing the resulting firework to detonate
Senko hanabi	A delicate pyrotechnic sparking effect, commonly produced in Japan, produced from the burning of a sulfur-rich blackpowder composition. When burned, the droplets of molten composition that form react further with air to produce attractive branching sparks
Sensitivity	The ease of ignition of a firework composition. Highly sensitive compositions (e.g., flash powder) require extremely careful handling. (See also: Electrostatic sensitivity, Thermal stability)
Sequence	Usually refers to the pattern of firing of a section of a display. For instance, a

	sequence could comprise 10 × 3" gold shells followed by 10 × 4" gold shells followed by 5 × 5" gold shells
Sequencer	An electrical firing system used to send regular electric pulses to fire a number of fireworks in a very accurately controlled manner
Series circuit	The preferred method of linking multiple electric igniters. Series circuits are arranged so that the current runs through each igniter in a sequential way. Series circuits are much easier to test for continuity and correct wiring than parallel circuits (See also: Parallel circuit)
Serpent	Usually a small tube filled with composition and possible a report charge, that is fired en masse from shells, mines, or rarely Roman candles. The serpents fly about in a random fashion before bursting with a report or stars
Set piece	A generic term for a ground firework but usually distinguished from lancework. The set piece may be static or revolving and is made up from gerbs and/or noise and color units. (See also: Lancework)
Shell	The most spectacular of fireworks consisting of a lifting charge (to propel the shell into the air) and a bursting charge to eject stars or subassemblies in the air after a predetermined delay. Shells are fired from mortars (See also: Bursting charge, Lifting charge)
Shell delay	A more precise term than delay fuse, this refers to the internal delay within a shell to permit it to ascend to its desired height before igniting the bursting charge. Shell delays are commonly made from composition pressed into a card tube (for cylinder

	shells, especially those with plastic molded cases) and variations of Bickford fuse (See also: Delay fuse, Bursting charge)
Shell of shells	An aerial display shell that contains internal shells that are ignited when the main shell bursts, and subsequently produce secondary bursts
Short circuit	Usually the accidental completion of an electrical circuit that causes the current not to flow through the electric igniters and thus leads to line failure. Short circuits usually can be discovered readily in series circuits by electrical testing of the circuit with an ohmmeter
Shot	Usually refers to the single functioning of, say, a Roman candle. Thus typically Roman candles are referred to as "8 shots"
Siatene shell	An aerial shell consisting of several maroons that burst in a ring pattern at the same time (See also: Roundel shell)
Sieve size	The size of the hole in a sieve. (See also: Mesh size)
Smoke	An air suspension of particles usually from incomplete combustion of a composition
Smokeless	A pyrotechnic mixture containing nitrocellulose and nitroglycerine, so called powder because, unlike blackpowder, it does not produce much smoke on burning. In this way, it found favor as a propellant in small arms devices, although its use in fireworks is rare
Spark	The typical effect caused by incandescent particles ejected from the burning surface of a composition
Sparkler	Usually a wire coated with pyrotechnic composition that gives off small sparks when burned. Sparklers, although consid-

	ered safe, are the cause of the greatest number of hospitalized accidents in the United Kingdom each season
Spider shell	An aerial shell having a small number of relatively large stars producing a web effect. Spider shells having 24 large comets are sometimes called octopus shells. (See also: Palm burst)
Spiking	syn. Stringing. (See also: Cylinder shell)
Spiking horse	The device used to facilitate the spiking, or stringing, of shells
Spiral wound tube	A paper tube wound from several narrow paper strips at an angle. Roman candles made with spiral tubes are prone to failure if fire can be transferred by loose composition trapped in the spiral winding (See also: Convolute wound tube)
Splitting comet	A comet in which there is an internal charge (usually of flash powder) which when ignited splits the comet into several pieces. The effect is of a comet that travels for some period and then fragments. Splitting comet stars typically are found in shells, mines, and especially Roman candles. syn. Crossette
Spolette	A shell or Roman candle delay fuse usually made from pressing blackpowder into a small bore tube (See also: Roman candle, Shell delay)
Squib	syn. Electric igniter
Star	Pellets of composition (usually cylinders, cubes, or spheres) used in mines, shells, Roman candles, rockets, and occasionally gerbs (See also: Round star, Cut star)
Star mine	A mine in which the projection of colored stars is the principal effect (See also: Mini mine, Noise mine)

Steel mortar — A mortar made from steel tube, usually with a welded steel base. Steel mortars are increasingly rarely used due to worries about their fragmentation should a powerful shell burst within the tube. However, for some shells (particularly cylinder shells), they are still the material of choice for most people (See also: HDPE mortar, GRP mortar)

Storage — The holding of fireworks before their use. In most countries storage of fireworks above a certain quantity requires a license (See also: Transportation)

Streamer — A star that leaves a trail of burning material.

Stringing — syn. Spiking. The process of winding a strong string around the outer surface of a shell to produce a more regular bursting pattern (See also: Spiking, Cylinder shell)

Strobe — The effect of a strobe is the regular pulsing "on-off-on-off" of light as a firework composition burns. There are several proposed explanations of this effect. Strobe effects are most often seen in ground fireworks (strobe pots) or as stars in an aerial shell or rocket (See also: Blinker)

Tail effect — Usually a term applied to a shell in which a star (comet) has been attached to the outside and which produces a rising column of sparks on the shell's ascent. "Tail" may also be applied to rockets, Roman candle stars or even whistle units where a persistent (usually silver) spark follows the flight of the device (See also: Weeping willow shell, Trunk)

Temple — syn. Machine

Thermal stability — The tendency for a composition to ignite from the energy applied by heat. Thermal stability testing is routinely carried out as part of the authorization procedure for fire-

	works in many countries (See also: Sensitivity)
Thermite	A mixture of aluminum and iron oxide (Fe_30_4) still used for in situ welding of railway tracks
Thunderflash	A generic term for a report with flash (See also: M-80, Cracker)
Tiger tail shell	Usually a solid sphere of composition fired in exactly the same manner as a shell. The effect produced is of an extremely thick rising comet. There is a small shell burst at the apex of its flight (See also: Tail effect)
Titanium	A silver metal much used for producing brilliant white sparks (e.g., in a maroon or gerb). Titanium does not corrode (cf. aluminum), but is extremely hard and may increase the friction sensitivity of a firework composition (See also: Maroon)
TNT equivalent	A measure of explosive strength used as a comparison to TNT, usually for determining safe loading of buildings
Top fused	Usually an aerial shell in which the time fuse (shell delay) for the functioning of the bursting charge is physically at the top of the shell and lit independently to the lifting charge (See also: Bottom fused)
Torbillion	Also Tourbillion. Either very similar to a serpent unit, or a larger aerial firework comprised of a saxon and wing, designed to rise into the air on ignition (See also: Crown wheel, Serpent)
Torch	syn. Flare
Torpedo	A flying squib or throwdown usually containing silver fulminate. The term also has been applied to railway signals (See also: Flying squib)

Transportation	The process of consigning a load of fireworks, usually taken to apply once the consignment has left the factory gates. Transportation of fireworks is subject to heavy legislative control
Tremalon	From Italian—*tremolante*—flickering, twinkling. (See also: Glitter, Flitter)
Trunk	The rising effect seen on willow shells and increasingly on many other shells (See also: Weeping willow shell, Coconut shell)
Turning case	A specialized type of gerb used for driving wheels. Typically turning cases are made from composition containing a larger proportion of blackpowder than the equivalent gerb (See also: Gerb, Fix)
UN classification	The assignment of a packaged firework into one of the UN's five classes for fireworks (See also: European standard, UN mark)
UN compatibility group	The "G" of 1.3G. The compatibility group, largely irrelevant for most firework usage, prescribes which explosives may be transported with which others. For instance, detonators should not be transported with primary explosives, and explosives containing toxic agents should not be transported AT ALL!
UN hazard code	syn. UN number
UN mark	A complicated index assigned to the PACKAGING of dangerous goods. (cf. UN number) (See also: UN classification)
UN number	A four digit number assigned to any hazardous goods after classification in its TRANSPORT PACKAGING according the methods prescribed in the "Orange Book." For fireworks the relevant numbers are 0333 (1.1G), 0334 (1.2G), 0335 (1.3G), 0336 (1.4G), and 0337 (1.4G). The UN

	number should always be quoted as it uniquely identifies an item AND its hazard. (See also: UN classification, Orange book)
Very	A type of pistol flare named after Edward W. Very (died 1910).
Visco fuse	A fuse, commonly used on consumer fireworks as the delay fuse, that usually is made by wrapping a core of blackpowder with thread and lacquer
Volley	A term usually applied to a mass firing or rockets (See also: Flight rocket, Rocket cone)
Warimono shell	A Japanese term for the type of shell that produces a spherical burst of stars. Most shells are of this type. cf. Poka shell. (See also: Poka shell, Japanese style shell), Warimono means "chrysanthemum"
Water firework	The generic term for any firework fired on the surface of water to maximize the visual effect of its reflections (See also: Water shell, Water gerb)
Water gerb	Usually a gerb or fountain weighted at one end and attached to a piece of cork designed to function on the surface of water. A water gerb may be lit by hand and thrown onto the water's surface or fired like a shell from a mortar (in each case with a suitable delay fuse).
Water shell	A shell designed to function on the surface of water (e.g., a lake) producing a hemisphere of stars. Water shells may be fired from mortars angled at a low angle or may be set up on the water's surface before the start of the display (See also: Shell)
Waterfall	Usually an extended curtain of silver sparks from vertical or horizontally burning tubes filled with a composition containing alu-

	minum. Waterfall shells produce the same effect and are best fired en masse to produce a spectacle
Weeping willow shell	syn. Willow shell
Wheel	A rotating set piece, usually powered by gerbs or turning cases, and most often rotating in a vertical plane (See also: Set piece, Turning case)
Whistle	Usually a tube containing a composition made using potassium benzoate, sodium salicylate, or, rarely nowadays, potassium picrate. On burning, the composition burns in a rapidly oscillating manner, and the resulting pressure waves are amplified by the tube in a manner similar to an organ pipe (See also: Screecher)
Whizzer	American alternative name for hummer
Willow shell	An extremely attractive shell consisting of stars made with a high percentage of charcoal. The effect is of long-burning golden stars that often (but undesirably) fall all the way to the ground. The shell may optionally be fitted with a "trunk" (syn. Weeping willow shell)

REFERENCES

1. A. St.H.Brock, *Pyrotechnics*, O'Connor (1922).
2. A. St.H.Brock, *A History of Fireworks*, Harrap (1949).
3. J. R. Partington, *A History of Greek Fire and Gunpowder*, Heffer (1960).
4. A. A. Shidlovsky, *Osnovy Pirotekhniki (Fundamentals of Pyrotechnics)*, 1st ed., Moscow (1943).
5. H. E. Ellern, *Modern Pyrotechnics*, Chemical Publishing (1961). Military and Civilian Pyrotechnics (1968).
6. G. W. Weingart, *Pyrotechnics*, 1st ed., Chemical Publishing (1947).
7. T. L. Davis, *Chemistry of Powder and Explosives*, Wiley (1941).
8. A. Izzo, *Pirotecnia e Fuochi Artificiali*, Hoepli, Milan (1950).
9. H. B. Faber, *Military Pyrotechnics*, Government Printing Office, Washington, D.C. (1919).
10. F. Ullmann, *Encyklopadie der Technischen Chemie*, München.
11. Kirk-Othmer, *Encyclopedia of Chemical Technology*, Interscience Publishers, New York.
12. British Intelligence Objectives Sub-Committee Reports, Number 461, 477, 1233, 1313. Reports of the Combined Intelligence Objectives Sub-Committee.
13. S. Fordham, *High Explosives and Propellants*, Pergamon (1966).
14. D. B. Chidsey, *Goodbye to Gunpowder,* Alvin Redman (1964).
15. Edwards & Wray, *Aluminum Paint and Powder*, Aluminum Co. of America.
16. *Shellac*, Angelo Rhodes Ltd.
17. *Guide to the Explosives Acts*, Her Majesty's Stationary Office.
18. N. Heaton, *Outlines of Paint Technology*, Griffin.
19. T. Kentish, *The Pyrotechnists Treasury*, Chatto & Windus (1887).
20. W. H. Browne, *The Art of Pyrotechny*, The Bazaar (1880).
21. H. G. Tanner, Instability of Sulfur-Potassium Chlorate Mixture. *Journal of Chemical Education*, Vol. 63, No.2, Feb. 1959.
22. Watkins, Cackett & Hall, *Chemical Warfare, Pyrotechnics and the Firework Industry*, Pergamon (1968).
23. Shimizu, Dr. T. *Hanabi (Fireworks) in Japanese Tokyo* (1957).
24. British Explosive Industry, International Congress of Applied Chemistry (1909).
25. B. J. Buchanan, *Gunpowder*. The History of an International Technology, ISBN 0-861971248, Bath University Press (1996).
26. Bryan Earl, *Cornish Explosives* Trevithick Society.
27. R. M. Winokur, *The Pyrotechnic Phenomenon of Glitter Pyrotechnica* No. 2 (1978).
28. Shimizu, Dr. T, *Studies in Firefly Compositions*, Pyrotechnica XII (1988).
29. Cardwell and Shimizu, *Studies in Strobe Compositions* Pyrotechnica, V (1979) and VII (1982).

   Anon., *Traditional Cylinder Shell Construction*, Pyrotechnica IX (1984) and XI

(1987).

Shimizu, Dr. T., *Studies in Microstars*, Pyrotechnica X (1985).

Selcuk Oztap, *Pyrotechnic Whistles and Applications*, Pyrotechnica XI (1987) and XIII (1990).

Shimizu, Dr. T., *Investigating the Glitter Phenomenon*, Pyrotechnica XIV (1992).

O'Neill, R, *Performance of Charcoal in Pyrotechnic Composition*, Pyrotechnica XVII (1997).

Pyrotechnica: Occasional Papers in Pyrotechnics, 1977 to present. Seventeen numbers published so far. Pyrotechnica Publications, PO Box 51162, Midland, Texas 79705, pyropubs@aol.com

30. Shimizu, Dr. T., *Fireworks from a Physical Standpoint*, Hower Verlag Hamburg. English translation Pyrotechnica.
31. Arthur Lotz, *Das Feuerwerk* (1941). Reprint Olms. Zürich.
32. J. H. McLain, *Pyrotechnics from the Viewpoint of Solid State Chemistry*, Franklin Institute Press (1980).
33. H. J. Yallop, *Explosive Investigation*, The Forensic Society, Harrogate, U.K. (1980).
34. Shimizu, Dr. T., *Fireworks: The Art, Science and Technique*. Maruzen, Tokyo (1981). Third ed (1996). ISBN: 0-929388-05-4, Pyrotechnica Publications.
35. Hartig, Dr. H., *Zündwaren, Fachbuchverlag* Leipzig (1971).
36. Finch & Ramachandran, *Matchmaking: Science, Technology and Manufacture*, Ellis Horwood, Chichester, U.K. (1983).
37. George A. Plimpton, *Fireworks: A History and Celebration*. Doubleday, New York (1984).
38. John A. Conkling, *Chemistry of Pyrotechnics*, Dekker, New York (1985).
39. M. C. Philip, *A Biography of Firework Books*, St Paul's Biographies, Winchester, U.K. (1985).
40. Ed Sievernich, Sybille & Budde, *Das Buch der Feuerwerkskunst*, Greno Verlag, Nördlingen, Germany.
41. Buchwald, *Feuerzauber*, Nico Pyrotechnik.
42. Yang Shuo & Ding Jing, *The Origin and Development of Ancient Gunpowder*, Dept. of Mechanical Engineering, Beijing Inst. of Technology, P.O. Box 327, Beijing, China.
43. Ed Dolan & Langer, *Explosives in the Service of Man*, Royal Institute of Chemistry, ISBN 0-85404-732-8.
44. Shimizu, Dr. T., *A New Method of Coating Magnesium*, Proceedings 19th International Pyrotechnics Seminar.
45. Krone, Dr. U., *Strahlungsemission in Intervallen-oscillierende Verbrennung pyrotechnisce sätze*, Institut für Chemie der treibe und Explosivstoffe. Karlsruhe (1975).
46. Simienowicz, *The Great Art of Artillery*, SR Publishers, U.K., ISBN 0-85409-663-9
47. Gray, Marsh, & McLaren, *A Short History of Gunpowder and the Role of Charcoal in Its Manufacture*, Journal of Material Science, Chapman & Hall, U.K.
48. Wharton, Dr. R., R. J. Rapley, and J. A. Harding, *The Mechanical Sensitiveness of Titanium/Blackpowder Pyrotechnic Compositions, Propellants, Explosives, Pyrotechnices*, Vol. 18, pp. 25–28 (1993)
49. Maxwell, W. R. *Pyrotechnic Whistles*, Fourth International Symposium on Combustion (1952).

50. Hernandez, Dr. P., *New Technology to Improve Lance and Quickmatch Reliability to Setwork in Extreme Environments*, International Symposium on Fireworks, Disney World (1996).
51. J. F. Bennett, *Fireworks*. A twice-yearly periodical for pyrotechnophiles, 47 issues at Winter 2005. Editor and publisher J. F. Bennett, P.O. Box 40, Bexhill-on-Sea, East Sussex. TN40 1GX. U.K. Email: editor@fireworks-mag.org; Web: http://www.fireworks-mag.org.
52. United Nations Recommendations for the Transport of Dangerous Goods, ISBN 92113903 4 and ISBN 921139033 8.
53. Default classification—letter from HSE ref: XI/9111/1/1B (March 1992).
54. *The Collins Concise Dictionary*, ISBN 000433160-5 (1988).
55. *Fireworks and the Law. A Guide for Users of Fireworks*. Available from Explosive Industry Group, CBI, Centrepoint, New Oxford Street, London.
56. See http://homepages.enterprise.net/saxtonsmith/robens/robens.htm.
57. *Guide to the Explosives Act 1875*, ISBN011880796X.
58. Health and Safety at Work Act (1974).
59. *List of Classified and Authorise Explosives*. Available from HSE.
60. Faversham Papers, No. 4, by Arthur Percival, Faversham Society.
61. *A Guide to the Classification and Labelling of Explosives Regulations*, ISBN 0118837066 (1983).
62. *A Guide to the Packaging of Explosives for Carriage Regulations*, ISBN 011885282 (1991).
63. The Control of Explosives Regulations (1991).
64. The Road Traffic (Carriage of Explosives) Regulations, see ISBN 011885479 (1989). This regulation is modified on a biannual basis.
65. European Agreement Concerning the International Carriage of Dangerous Goods by Road (ADR), ISBN 011550901.
66. Consumer Protection Act, ISBN 0105443875 (1987).
67. General Product Safety Regulations. Available from DTI.
68. Firework Safety Regulations (1987).
69. Manufacture and Storage of Explosives Regulations. Approved Code of Practice L139. ISBN 0-7176-2816-7.
70. The Manufacture and Storage of Explosives Regulations, www.hse.gov.uk/explosives/.
71. Enacted in the U.K. as the Control of Major Accident Hazards Regulations (COMAH).
72. BS 7114. Available from BS1, Milton Keynes, U.K.
73. CEN/TC 212, Draft European Standard for Fireworks.
74. Many thanks to Dr. John Conkling for supplying information for this section. See proceedings from the 1996 fireworks symposium at Orlando for further information.
75. Administered by the Explosives Division, Natural Resources, Canada.
76. "Experiences of Fireworks in Europe," commissioned by the U.K. Department of Trade and Industry, Consumer Safety Unit, 1994, and reproduced with their permission.
77. *Journal of Pyrotechnics*, Issue 15 (2002). 1775 Blair Road, Whitewater, CO 81527, U.S.A.
78. Hardt, Alexander P., *Pyrotechnics*, Pyrotechnica Publications, P.O. Box 51162, Midland, TX 79705 U.S.A., pyropubs@aol.com

79. Ian von Matitz. *Blackpowder Manufacturing, Testing & Optimizing*, American Firework News, 203, ISBN 0-929931-21-1, HC67-Box 30, Dingmans Ferry, PA, ISBN 18328-95-06, www.fireworksnews.com
80. Enschede Accident Report PML 2002-C70, TNO Prins Maurits Laboratory, The Netherlands.
81. Debusk, Keaffaber, Schwegler, & Repoff, *Environmental Effects on Fireworks on Bodies of Water*, CANMET Conference Montreal 1992. Also in *Journal of Pyrotechnics*, No. 3 (1996).
82. D. Chapman, R. K. Wharton, J. E. Fletcher, Studies of the Thermal Stability and Sensitiveness of Sulfur/Chlorate Mixtures, *Journal of Pyrotechnics*, Issues 11 and 12.
83. R. K. Wharton, Observations on the Heights Attained by Spherical Firework Shells, *Journal of Pyrotechnics*, No. 1 (1995).
84. S. G. Myatt and R. K.Wharton, *A Survey of Factors Telating to the Use of Mortars in Firework Displays in the UK*, Pyrotechnica, XVI (1995).
85. S. G. Myatt and M. R. Edwards, Measurement of the Effectiveness of Various Mitigation Methods of Reducing the Projectile Hazards from Fragmenting Steel Firework Mortar Tubes, *Journal of Pyrotechnics*, No. 10 (1999).
86. Marc Speer, Firework Shell Drift & Fuse Failure, *Journal of Pyrotechnics*, Issue 17 (2003).
87. Didier Brunel, *Le grand livre des feux d'artifice*, CNRS Editions, Paris, ISBN 2-271-06192-X,. www.cnrseditions.fr.
88. *Journal of Pyrotechnics*, No. 8 (1998).
89. *American Firework News*, ISBN 0-929931-20-3 (2002).

INDEX

Note: Entries for figures or tables are indicated with an "f" or a "t," respectively.

C

www.ingramcontent.com/pod-product-compliance
Lightning Source LLC
LaVergne TN
LVHW101452200926
848483LV00001B/112
* 9 7 8 0 8 2 0 6 0 4 0 7 7 *